Lecture Notes in Physics

For information about Vols. 1–151, please contact your bookseller or Springer-Verlag.

Vol. 152: Physics of Narrow Gap Semiconductors. Proceedings, 1981. Edited by E. Gornik, H. Heinrich and L. Palmetshofer. XIII, 485 pages. 1982.

Vol. 153: Mathematical Problems in Theoretical Physics. Proceedings, 1981. Edited by R. Schrader, R. Seiler, and D.A. Uhlenbrock. XII, 429 pages. 1982.

Vol. 154: Macroscopic Properties of Disordered Media. Proceedings, 1981. Edited by R. Burridge, S. Childress, and G. Papanicolaou. VII, 307 pages. 1982.

Vol. 155: Quantum Optics. Proceedings, 1981. Edited by C.A. Engelbrecht. VIII, 329 pages. 1982.

Vol. 156: Resonances in Heavy Ion Reactions. Proceedings, 1981. Edited by K.A. Eberhard. XII, 448 pages. 1982.

Vol. 157: P. Niyogi, Integral Equation Method in Transonic Flow. XI, 189 pages. 1982.

Vol. 158: Dynamics of Nuclear Fission and Related Collective Phenomena. Proceedings, 1981. Edited by P. David, T. Mayer-Kuckuk, and A. van der Woude. X, 462 pages. 1982.

Vol. 159: E. Seiler, Gauge Theories as a Problem of Constructive Quantum Field Theory and Statistical Mechanics. V, 192 pages. 1982.

Vol. 160: Unified Theories of Elementary Particles. Critical Assessment and Prospects. Proceedings, 1981. Edited by P. Breitenlohner and H.P. Dürr. VI, 217 pages. 1982.

Vol. 161: Interacting Bosons in Nuclei. Proceedings, 1981. Edited by J.S. Dehesa, J.M.G. Gomez, and J. Ros. V, 209 pages. 1982.

Vol. 162: Relativistic Action at a Distance: Classical and Quantum Aspects. Proceedings, 1981. Edited by J. Llosa. X, 263 pages. 1982.

Vol. 163: J. S. Darrozes, C. Francois, Mécanique des Fluides Incompressibles. XIX, 459 pages. 1982.

Vol. 164: Stability of Thermodynamic Systems. Proceedings, 1981. Edited by J. Casas-Vázquez and G. Lebon. VII, 321 pages. 1982.

Vol. 165: N. Mukunda, H. van Dam, L.C. Biedenharn, Relativistic Models of Extended Hadrons Obeying a Mass-Spin Trajectory Constraint. Edited by A. Böhm and J.D. Dollard. VI, 163 pages. 1982.

Vol. 166: Computer Simulation of Solids. Edited by C.R.A. Catlow and W.C. Mackrodt. XII, 320 pages. 1982.

Vol. 167: G. Fieck, Symmetry of Polycentric Systems. VI, 137 pages, 1982.

Vol. 168: Heavy-Ion Collisions. Proceedings, 1982. Edited by G. Madurga and M. Lozano. VI, 429 pages. 1982.

Vol. 169: K. Sundermeyer, Constrained Dynamics. IV, 318 pages. 1982.

Vol. 170: Eighth International Conference on Numerical Methods in Fluid Dynamics. Proceedings, 1982. Edited by E. Krause. X, 569 pages. 1982.

Vol. 171: Time-Dependent Hartree-Fock and Beyond. Proceedings, 1982. Edited by K. Goeke and P.-G. Reinhard. VIII, 426 pages. 1982.

Vol. 172: Ionic Liquids, Molten Salts and Polyelectrolytes. Proceedings, 1982. Edited by K.-H. Bennemann, F. Brouers, and D. Quitmann. VII, 253 pages. 1982.

Vol. 173: Stochastic Processes in Quantum Theory and Statistical Physics. Proceedings, 1981. Edited by S. Albeverio, Ph. Combe, and M. Sirugue-Collin. VIII, 337 pages. 1982.

Vol. 174: A. Kadić, D.G.B. Edelen, A Gauge Theory of Dislocations and Disclinations. VII, 290 pages. 1983.

Vol. 175: Defect Complexes in Semiconductor Structures. Proceedings, 1982. Edited by J. Giber, F. Beleznay, J. C. Szép, and J. László. VI, 308 pages. 1983.

Vol. 176: Gauge Theory and Gravitation. Proceedings, 1982. Edited by K. Kikkawa, N. Nakanishi, and H. Nariai. X, 316 pages. 1983.

Vol. 177: Application of High Magnetic Fields in Semiconductor Physics. Proceedings, 1982. Edited by G. Landwehr. XII, 552 pages. 1983.

Vol. 178: Detectors in Heavy-Ion Reactions. Proceedings, 1982. Edited by W. von Oertzen. VIII, 258 pages. 1983.

Vol. 179: Dynamical Systems and Chaos. Proceedings, 1982. Edited by L. Garrido. XIV, 298 pages. 1983.

Vol. 180: Group Theoretical Methods in Physics. Proceedings, 1982. Edited by M. Serdaroğlu and E. İnönü. XI, 569 pages. 1983.

Vol. 181: Gauge Theories of the Eighties. Proceedings, 1982. Edited by R. Raitio and J. Lindfors. V, 644 pages. 1983.

Vol. 182: Laser Physics. Proceedings, 1983. Edited by J. D. Harvey and D. F. Walls. V, 263 pages. 1983.

Vol. 183: J. D. Gunton, M. Droz, Introduction to the Theory of Metastable and Unstable States. VI, 140 pages. 1983.

Vol. 184: Stochastic Processes – Formalism and Applications. Proceedings, 1982. Edited by G. S. Agarwal and S. Dattagupta. VI, 324 pages. 1983.

Vol. 185: H. N. Shirer, R. Wells, Mathematical Structure of the Singularities at the Transitions between Steady States in Hydrodynamic Systems. XI, 276 pages. 1983.

Vol. 186: Critical Phenomena. Proceedings, 1982. Edited by F.J.W. Hahne. VII, 353 pages. 1983.

Vol. 187: Density Functional Theory. Edited by J. Keller and J.L. Gázquez. V, 301 pages. 1983.

Vol. 188: A. P. Balachandran, G. Marmo, B.-S. Skagerstam, A. Stern, Gauge Symmetries and Fibre Bundles. IV, 140 pages. 1983.

Vol. 189: Nonlinear Phenomena. Proceedings, 1982. Edited by K. B. Wolf. XII, 453 pages. 1983.

Vol. 190: K. Kraus, States, Effects, and Operations. Edited by A. Böhm, J. W. Dollard and W. H. Wootters. IX, 151 pages. 1983.

Vol. 191: Photon Photon Collisions. Proceedings, 1983. Edited by Ch. Berger. V, 417 pages. 1983.

Vol. 192: Heidelberg Colloquium on Spin Glasses. Proceedings, 1983. Edited by J. L. van Hemmen and I. Morgenstern. VII, 356 pages. 1983.

Vol. 193: Cool Stars, Stellar Systems, and the Sun. Proceedings, 1983. Edited by S. L. Balliunas and L. Hartmann. VII, 364 pages. 1984.

Vol. 194: P. Pascual, R. Tarrach, QCD: Renormalization for the Practitioner. V, 277 pages. 1984.

Lecture Notes in Physics

Edited by H. Araki, Kyoto, J. Ehlers, München, K. Hepp, Zürich
R. Kippenhahn, München, H. A. Weidenmüller, Heidelberg
and J. Zittartz, Köln
Managing Editor: W. Beiglböck, Heidelberg

226

Non-Linear Equations in Classical and Quantum Field Theory

Proceedings of a Seminar Series Held at DAPHE, Observatoire de Meudon, and LPTHE, Université Pierre et Marie Curie, Paris, Between October 1983 and October 1984

Edited by N. Sanchez

Springer-Verlag
Berlin Heidelberg GmbH

Editor

N. Sanchez
Observatoire de Paris, Section d'Astrophysique
5, place Jules Janssen, F-92195 Meudon Principal Cedex

ISBN 978-3-540-15213-2 ISBN 978-3-540-39352-8 (eBook)
DOI 10.1007/978-3-540-39352-8

PREFACE

The quantum theory of fields has, in recent years, led to important theoretical progress in elementary particle physics and also in statistical mechanics. At the same time, current developments in general relativity and quantum field theory are becoming more and more interconnected. A knowledge of the geometrical structures so essential for a discussion of the classical theory of general relativity is also becoming increasingly useful in the study of other classical and quantum field theories.

Non-linear differential equations play a central role in almost all interesting physical theories. Integrable theories and the methods to solve them are of conceptual and practical importance. Common features to these theories are the existence of an infinite number of conserved quantities, an associated linear problem and Bäcklund transformations. A large class of integrable theories can be solved by the (classical and quantum) inverse method. The underlying (dynamical) symmetries allowing this exact solvability have associated bilinear (Yang-Baxter) and Kac-Moody algebras. In addition, a number of analogies and links between different integrable non-linear field equations have been found, e.g. between self-dual Yang-Mills fields, static monopoles, non-linear sigma models and the gravitational field with two Killing vectors. These links have been very useful in allowing methods developed for one problem to be applied directly to another.

A seminar series "Séminaires sur les équations non-linéaires en théorie des champs" intended to follow current developments in mathematical physics, and particularly in the above-mentioned domains, was started in the Parisian region in October 1983. The seminars take place alternatively at DAPHE - Observatoire de Meudon - and LPTHE - Université Pierre et Marie Curie (Paris VI) - and they encourage regular meetings between theoretical physicists of different disciplines and a number of mathematicians. Participants come from Paris VI and VII, IHP, ENS, Collège de France, CPT-Palaiseau, GAR-Meudon, IHES, LPTHE-Orsay and CPT-Marseille. The present volume "Non-Linear Equations in Classical and Quantum Field Theory" accounts for the first twenty-two lectures delivered up to October 1984 in this series.

It is a pleasure to thank all the speakers for accepting our invitations and for their interesting accounts, whether they be of a review nature or an exposition of recent work. We thank all the participants for their interest and for their stimulating discussions. We are especially indebted to Héctor J. de Vega at LPTHE - Paris VI, jointly responsible with us for these seminars, for his efficient collaboration which has made this series possible. We also thank M. Dubois Violette at Orsay, J.L. Richard at Marseille, and B. Carter and B. Whiting at Meudon for their cooperation and encouragement. We acknowledge Mrs. C. Rosolen and Mrs. D. Lopes for their practical assistance in the organisation and for their typing of part of these proceedings.

These seminars are financially supported by the CNRS. We particularly thank the Scientific Direction "Mathématiques-Physique de Base" which has made this series possible. We extend our appreciation to Springer-Verlag for their cooperation and efficiency in publishing these proceedings and hope that the possibility of making our seminars more widely available in this way will continue in the future.

Meudon, November 1984

Norma SANCHEZ.

Organising Commitee

H.J. de Vega	(LPTHE - Paris VI)
M. Dubois Violette	(LPTHE - Orsay)
J.L. Richard	(CPT - Marseille)
N. Sánchez	(DAPHE - Meudon)

TABLE OF CONTENTS

(*) Lecture given by this author.

A NEW CLASS OF UNITARIZABLE HIGHEST WEIGHT REPRESENTATIONS OF INFINITE DIMENSIONAL LIE ALGEBRAS.

H.P. Jakobsen
Mathematics Institute
Universitetsparken 5
DK - 2100 Copenhagen Ø
Denmark.

V.G. Kac
Department of Mathematics
M.I.T.
Cambridge, Mass 02139
U.S.A.

0. Introduction.

The representation theory of infinite-dimensional Lie algebras has emerged in the past few years as a field that has remarkable applications to many areas of mathematics and mathematical physics. All these applications show that the following two assumptions about the representation in question are fundamental :

1) unitarizability ;
2) existence of a highest weight vector.

In more detail, let $\mathfrak{g}$ be a complex (possibly infinite-dimensional) Lie algebra, let $\mathcal{U}(\mathfrak{g})$ denote its universal enveloping algebra, let $\mathfrak{p}$ be a subalgebra of $\mathfrak{g}$ and let ω be an antilinear anti-involution of $\mathfrak{g}$ (i.e. $\omega.[x,y] = [\omega.y, \omega.x]$ and $\omega.(\lambda x) = \bar{\lambda}(\omega.x)$) such that

$$(0.1) \qquad \mathfrak{p} + \omega.\mathfrak{p} = \mathfrak{g} \quad .$$

Let $\lambda : \mathfrak{p} \to \mathbb{C}$ be a 1-dimensional representation of $\mathfrak{p}$. A representation $\pi : \mathfrak{g} \to \mathfrak{gl}(V)$ is called a *highest weight representation* with highest weight λ if there exists a vector $v_\lambda \in V$ with the following properties :

$$(0.2) \qquad \pi(\mathcal{U}(\mathfrak{g}))\, v_\lambda = V$$

(0.3) $\pi(b)v_\lambda = \lambda(b)v_\lambda$ for any $b \in \mathfrak{p}$

(Of course, (0.2) is satisfied automatically if the representation π is irreducible) A Hermitian form H on V such that

(0.4) $H(v_\lambda, v_\lambda) = 1,$

(0.5) $H(\pi(g)u, v) = H(u, \pi(\omega.g)v)$ for all $g \in \mathfrak{g}$ and $u, v \in V$

is called *contravariant* (it is determined uniquely by (0.4) and (0.5)). It is easy to show that, under some natural assumptions, for any highest weight $\lambda : \mathfrak{p} \to \mathbb{C}$ there exists a unique highest weight representation with a non-degenerate contravariant Hermitian form. The non-trivial problem is whether this form is positive definite; if this is the case, the representation π is called *unitarizable*.

For example, let $\mathfrak{g}$ be the infinite-dimensional Heisenberg algebra, i.e. a Lie algebra with a basis p_i, q_i $(i \in \mathbb{Z})$ and c, with commutation relations : $[p_i, q_i] = c$ and all the other brackets zero. Put $\mathfrak{p} = \mathbb{C}c + \sum \mathbb{C}p_i$ and let $\lambda : \mathfrak{p} \to \mathbb{C}$ be defined by $\lambda(c) = a \in \mathbb{C}^\times$ and $\lambda(p_i) = 0$. Then any representation of $\mathfrak{g}$ with highest weight λ is irreducible and equivalent to the canonical commutation relations representation L(a) (ie. $q_i \to x_i$, $p_i \to a\frac{\partial}{\partial x_i}$, $c \to a$). Let ω be an antilinear anti-involution of $\mathfrak{g}$ defined by $\omega(p_i) = q_i$, $\omega(q_i) = p_i$, $\omega(c) = c$. Then L(a) is unitarizable if and only if a is a positive real number.

The unitarizable highest weight representations of finite-dimensional semisimple Lie algebras have been classified quite recently, and the answer is highly non-trivial [1], [3].

The present paper grew out from an attempt to solve the analogous problem for affine Kac-Moody algebras. Recall that, given a simple finite-dimensional Lie algebra $\dot{\mathfrak{g}}$, the associated affine Kac-Moody algebra is

$$\mathfrak{g} = \mathbb{C}[z, z^{-1}] \otimes_{\mathbb{C}} \dot{\mathfrak{g}} + \mathbb{C}c,$$

with the following commutation relations :

(0.6) $[z^m \otimes a, z^k \otimes b] = z^{m+k} \otimes [a, b] + m\delta_{m,-k}(a,b)c \; ; \; [\mathfrak{g}, c] = 0$

Here $a, b \in \dot{\mathfrak{g}}$, (a,b) is the Killing form on $\dot{\mathfrak{g}}$, and $m, k \in \mathbb{Z}$. Let $\dot{\mathfrak{b}}$ be a Borel subalgebra of $\dot{\mathfrak{g}}$ and $\dot{\omega}$ a compact antilinear anti-involution (i.e. the real subalgebra $\{x \in \dot{\mathfrak{g}} \mid \dot{\omega}.x = -x\}$ is the compact form of $\dot{\mathfrak{g}}$) such that (0.1) holds. The conventional choice of the "Borel subalgebra" $\mathfrak{b}$ of $\mathfrak{g}$ is

(0.7) $\mathfrak{b} = \mathbb{C}c \oplus (1 \otimes \dot{\mathfrak{b}}) \oplus (z \otimes \dot{\mathfrak{g}}) \oplus (z^2 \otimes \dot{\mathfrak{g}}) \oplus \dots$

Let ω be the compact antilinear anti-involution of $\mathfrak{g}$, i.e. $\omega\cdot(z^m\otimes a+\lambda c)=$ $=z^{-m}\otimes\dot{\omega}(a)+\bar{\lambda}c$ and let $\mathfrak{p}=\mathfrak{b}$. Then the affine algebra $\mathfrak{g}$ admits a remarkable family of unitarizable highest weight representations, called integrable highest weight representations. An exposition of the theory of these representations along with some of its beautiful applications may be found in the book [4].

On the other hand, a simple computation shows that for $\mathfrak{p}$=(conventional $\mathfrak{b}$) and any other choice of ω there is no unitarizable highest weight modules except the trivial one, in sharp contrast to the finite-dimensional theory.

However, again in contrast to the finite-dimensional theory, an affine Lie algebra has several conjugacy classes of Borel subalgebras, and the next natural step is to try "non-conventional" Borel subalgebras.

As a result, we found the following unitarizable highest weight representations of the Lie algebra $\mathfrak{g}=s\ell_2(\mathbb{C}[z,z^{-1}])=\mathbb{C}[z,z^{-1}]\otimes s\ell_2(\mathbb{C})$ (the central charge, i.e. the eigenvalue of c, is trivial) :

Let $V=\mathbb{C}[x_k;k\in\mathbb{Z}]$ be the space of polonomials in indeterminates x_k . Put $f_k=\begin{pmatrix}0&0\\z^k&0\end{pmatrix}$, $h_k=\begin{pmatrix}z^k&0\\0&-z^k\end{pmatrix}$, $e_k=\begin{pmatrix}0&z^k\\0&0\end{pmatrix}$

Let $\mathfrak{p}=\left\{\begin{pmatrix}a(z)&b(z)\\0&-a(z)\end{pmatrix}\;\middle|\;a(z),\,b(z)\in\mathbb{C}[z,z^{-1}]\right\}$,

and let ω be an antilinear anti-involution of the Lie algebra $s\ell_2(\mathbb{C}[z,z^{-1}])$ defined by

$$\omega\cdot f_k=-e_{-k}\;,\qquad \omega\cdot e_k=-f_{-k}\;,\qquad \omega\cdot h_k=h_{-k}\,. \tag{0.8}$$

Let m be a finite measure on the circle S^1, not concentrated in a finite number of points ; put $\lambda_k=\int_{S^1}z^k dm$ (e.g. $\lambda_k=\delta_{k,0}$ if m is the Lebesgue measure). Then the map

$$f_k\to x_k\;;\;h_k\to-\Big(\lambda_k+2\sum_{j\in\mathbb{Z}}x_{j+k}\frac{\partial}{\partial x_j}\Big), \tag{0.9}$$
$$e_k\to-\Big(\sum_i\lambda_{i+k}\frac{\partial}{\partial x_i}+\sum_{i,j\in\mathbb{Z}}x_{i+j+k}\frac{\partial}{\partial x_i}\frac{\partial}{\partial x_j}\Big)$$

defines a unitarizable irreducible representation π_m of the Lie algebra $s\ell_2(\mathbb{C}[z,z^{-1}])$ on the space V, the polynomial 1 being an eigenvector for $\mathfrak{b}$, if m is positive. The real form $\{x\in s\ell_2(\mathbb{C}[z,z^{-1}])\,|\,\omega\cdot x=-x\}$ of $\mathfrak{g}$ is the real Lie algebra $su(1,1)^{S^1}$ of polynomial maps of the circle into su(1,1). Thus, we have a unitary representation of $su(1,1)^{S^1}$ on the space V, and this construction can easily be generalized to the case of $su(n,1)^{S^1}$. Moreover, we show that these representations together with integrable highest weight representations and representations "concentrated" in a finite number of points, are the only unitarizable highest weight representations of all affine Kac-Moody algebras.

Finally, in a similar fashion, we can construct unitarizable highest weight representations of the Lie algebra $su(n,1)^X$, where X is a set with a finite measure. The corresponding formula for the Hermitian form is identical to that for the truncated

correlation function in quantum field theory.
We will discuss elsewhere the question of integrability of these representations to the corresponding group $SU(n,1)^X$.

1. Generalities.

Let $\mathfrak{g}$ be a Lie algebra with an antilinear anti-involution ω ; then ω extends uniquely to an antilinear anti-involution of the universal enveloping algebra $\mathcal{U}(\mathfrak{g})$. Let $\mathfrak{p}$ be a subalgebra of $\mathfrak{g}$ satisfying (0.1). Choose a subspace $\mathfrak{n} \subset \mathfrak{g}$ such that $\mathfrak{g} = \mathfrak{p} \oplus \mathfrak{n}$. Then we have the decompostition into a direct sum of vector spaces : $\mathcal{U}(\mathfrak{g}) = \mathfrak{n}\,\mathcal{U}(\mathfrak{g}) \oplus \mathcal{U}(\mathfrak{p})$. Denote by β the projection on the second summand. Let $\lambda : \mathfrak{p} \to \mathbb{C}$ be a 1-dimensional representation of $\mathfrak{b}$ and extend it to the whole $\mathcal{U}(\mathfrak{p})$. Put $\mathcal{U}^\lambda(\mathfrak{p}) = \{ b \in \mathfrak{p} \mid \lambda(b) = 0 \}$. Define a sesquilinear form H on $\mathcal{U}(\mathfrak{g})$ by

$$(1.1) \qquad H(u,v) = \lambda(\beta((\omega.v)u))$$

It is straightforward that

$$(1.2) \qquad H(u'u, v) = H(u, (\omega.u')v) \text{ and,}$$

$$(1.3) \qquad H(\mathcal{U}^\lambda(\mathfrak{p}), \mathcal{U}(\mathfrak{g})) = 0 \quad .$$

In order to have the Hermitian property of H $\left(\text{i.e } H(u,u') = \overline{H(u',u)}\right)$ we need to assume

$$(1.4) \qquad \lambda(\beta(u)) = \overline{\lambda(\beta(\omega.u))} \quad \text{for } u \in \mathcal{U}(\mathfrak{g})$$

We put

$$(1.5) \qquad M(\lambda) = \mathcal{U}(\mathfrak{g}) / \mathcal{U}(\mathfrak{g})\,\mathcal{U}^\lambda(\mathfrak{p})$$

and define a representation $\tilde{\pi}_\lambda$ of $\mathfrak{g}$ on $M(\lambda)$ via left multiplication. The representation $(M(\lambda), \tilde{\pi}_\lambda)$ is called a (generalized) *Verma module*. We denote by $v_\lambda \in M(\lambda)$ the image of 1. Then (0.2) and (0.3) are satisfied, so that $M(\lambda)$ is a highest weight representation with highest weight λ . Furthermore, since, by (1.3), the kernel of the Hermitian form H contains $\mathcal{U}^\lambda(\mathfrak{p})$, we obtain a contravariant Hermitian form on $M(\lambda)$, also denoted by H. By uniqueness of the contravariant form, H is independent of the choice of $\mathfrak{n}$ (satisfying (1.4)).
Let $I(\lambda)$ be the Kernel of H on $M(\lambda)$ and put $L(\lambda) = M(\lambda)/I(\lambda)$. Then $\tilde{\pi}_\lambda$ induces a highest weight representation π_λ of $\mathfrak{g}$ on the space $L(\lambda)$ and H induces a non-

degenerate contravariant Hermitian form on $L(\lambda)$, also denoted by H. It is clear that conversely, if $L(\lambda)$ is a highest weight representation with a non-degenerate Hermitian form, it is obtained from the Verma module $M(\lambda)$ as above.

Thus, we arrive at the following

Lemma 1.1. Let $\mathfrak{g}$ *be a Lie algebra with an antilinear anti-involution* ω, *and let* $\mathfrak{p}$ *be a subalgebra of* $\mathfrak{g}$ *satisfying* (0.1). *Let* λ *be a 1-dimensional representation of* $\mathfrak{p}$. *Suppose that we can choose a subspace* $\mathfrak{n} \subset \mathfrak{g}$, *such that* $\mathfrak{g} = \mathfrak{p} \oplus \mathfrak{n}$, *and such that the corresponding projection* $\beta : \mathcal{U}(\mathfrak{g}) \to \mathcal{U}(\mathfrak{p})$ *satisfies* (1.4). *Then there exists a unique highest weight representation* π_λ *with highest weight* λ *of* $\mathfrak{g}$ *on a vector space* $L(\lambda)$ *with a non-degenerate contravariant Hermitian form.* □

We shall sometimes write $M_{\mathfrak{p},\omega}(\lambda)$, $L_{\mathfrak{p},\omega}(\lambda)$, $\pi_{\lambda;\mathfrak{p}}$ and H_ω instead of $M(\lambda)$, $L(\lambda)$, π_λ, and H, in order to emphasize the dependence on $\mathfrak{p}$ and ω.

Remark : Let $\mathfrak{g}$, $\mathfrak{p}$, ω, λ and $\mathfrak{n}$ be as in Lemma 1.1, and let $\mathfrak{b} \subset \mathfrak{p}$ be a subalgebra such that $\mathfrak{b} + \omega\mathfrak{b} = \mathfrak{g}$ and such that there exists a subspace $\mathfrak{u} \subset \mathfrak{p}$ with $\mathfrak{p} = \mathfrak{b} \oplus \mathfrak{u}$ and

(1.6) $\lambda|_{\mathfrak{u}} = 0$.

Let $\bar{\lambda} = \lambda|_{\mathfrak{b}}$. Then the highest weight representations $\pi_{\lambda;\mathfrak{p}}$ and $\pi_{\bar{\lambda},\mathfrak{b}}$ are equivalent. Indeed, since $\mathcal{U}^{\lambda}(\mathfrak{p}) \supset \mathcal{U}^{\bar{\lambda}}(\mathfrak{b})$, there exists a surjective $\mathfrak{g}$-map $\tilde{\varphi} : M_{\mathfrak{b},\omega}(\bar{\lambda}) \to M_{\mathfrak{p},\omega}(\lambda)$. Using (1.6), one easily checks that Ker $\tilde{\varphi}$ is contained in the Kernel of H_ω . It follows that $\tilde{\varphi}$ induces an equivalence of representations $\varphi : L_{\mathfrak{b},\omega}(\bar{\lambda}) \xrightarrow{\sim} L_{\mathfrak{p},\omega}(\lambda)$. □

Example : The canonical commutation relations representation L(a) of the infinite-dimensional Heisenberg algebra (see the introduction) is a Verma module. Choose $\mathfrak{u} = \sum_i \mathbb{C} q_i$; then condition (1.4) holds whenever $a \in \mathbb{R}$. Hence L(a) carries a Hermitian form whenever $a \in \mathbb{R}$ (it is clear that this condition is also necessary). The representation L(a) is irreducible iff $a \neq 0$. Thus L(a) is an (irreducible) highest weight representation with a contravariant Hermitian form if $a \in \mathbb{R} \setminus \{0\}$. □

2. Involutions and Borel subalgebras.

For basic definitions and facts of the theory of Kac-Moody algebras we refer to the book [4].

Let $\mathfrak{g} = \mathfrak{g}'(A)$ be a Kac-Moody algebra associated to the generalized Cartan matrix A. This is a Lie algebra with *Chevalley generators* $e_i, f_i, \alpha_i^\vee$ $(i = 1, \dots, n)$ satisfying certain well-known relations. Let Δ be the set of roots of $\mathfrak{g}$ and let

$$\mathfrak{g} = \bigoplus_{\alpha \in \Delta \cup \{0\}} \mathfrak{g}_\alpha$$

be the *root space decomposition* of $\mathfrak{g}$.

We call a subset Δ_+ of Δ a *set of positive roots* if the following three properties hold :

(2.1) if $\alpha, \beta \in \Delta_+$ and $\alpha + \beta \in \Delta$, then $\alpha + \beta \in \Delta_+$;

(2.2) if $\alpha \in \Delta$, then either α or $-\alpha$ lie in Δ_+ ;

(2.3) if $\alpha \in \Delta_+$, then $-\alpha \notin \Delta_+$.

Example : Let Π^{st} be the set of roots corresponding to the generators e_i $(i=1,\dots,n)$ of $\mathfrak{g}$; we call Π^{st} the *standard set of simple roots*. Let $\Delta_+^{st} = \{ \sum_i k_i \alpha_i \mid k_i \in \mathbb{Z}_+ = \{0,1,2,\dots\}$ and $\alpha_i \in \Pi^{st} \}$. Then Δ_+^{st} is the conventional set of positive roots (see [4]), which we call the *standard set of positive roots*. □

Given a set of positive roots Δ_+, one associates to it the *Borel subalgebra* $\mathfrak{b} = \oplus_{\alpha \in \Delta_+ \cup \{0\}} \mathfrak{g}_\alpha$. The Borel subalgebra associated to Δ_+^{st} is denoted by $\mathfrak{b}^{st}$.

A subalgebra $\mathfrak{p}$ of $\mathfrak{g}$ containing $\mathfrak{b}$ is called a *parabolic subalgebra*.

An antilinear anti-involution ω of $\mathfrak{g}$ is called *consistent* if $\omega . \mathfrak{g}_\alpha = \mathfrak{g}_{-\alpha}$

It is clear that, replacing e_i by λe_i and f_i by $\lambda^{-1} f_i$, one can bring ω to the following form :

$$\omega . e_i = \pm f_i \qquad (i = 1, \dots, n)$$

An important example is the *compact antilinear anti-involution* ω_c defined by :

$$\omega_c . e_i = f_i \qquad (i = 1, \dots, n)$$

Note that if ω is a consistent antilinear anti-involution and $\mathfrak{p}$ is a parabolic subalgebra of $\mathfrak{g}$, then condition (0.1) holds. One can show that conversely, if (0.1) holds for some $\mathfrak{p}$ and ω, then ω is conjugate to a consistent antilinear anti-involution (cf. [5]).

Let W be the Weyl group of the Kac-Moody algebra $\mathfrak{g}$; for a real root α, let $r_\alpha \in W$ denotes the reflection with respect to α .

We start with the classification of sets of positive roots in the finite-dimensional case.

Lemma 2.1. If Δ is a finite root system, then a set of positive roots Δ_+ is W-conjugate to Δ_+^{st} .

Proof. If $\Pi^{st} \subset \Delta_+$, then $\Delta_+ = \Delta_+^{st}$ and there is nothing to prove. Otherwise, there exists $\alpha \in \Pi^{st} \setminus \Delta_+$, and $|(r_\alpha . \Delta_+) \cap -\Delta_+^{st}| < |\Delta_+ \cap -\Delta_+^{st}|$. After a finite number of such steps we get $\Delta_+ = \Delta_+^{st}$. □

By Lemma 2.1, a Borel subalgebra of a finite-dimensional simple Lie algebra $\mathfrak{g}$ is conjugate to $\mathfrak{b}^{st}$.

The situation is different for infinite-dimensional Kac-Moody algebras. For example, let $\mathfrak{g}$ be an affine Kac-Moody algebra associated to $\dot{\mathfrak{g}}$ ("non-twisted" case) ; then the subalgebra $\mathfrak{b}$ defined by (0.7) is $\mathfrak{b}^{st}$. On the other hand, putting $\dot{\mathfrak{b}} = \dot{\mathfrak{h}} \oplus \dot{\mathfrak{n}}$, where $\dot{\mathfrak{h}}$ is a Cartan subalgebra and $\dot{\mathfrak{n}}$ a maximal nilpotent subalgebra of $\dot{\mathfrak{g}}$, we have another Borel subalgebra of $\mathfrak{g}$, which is not conjugate to $\mathfrak{b}^{st}$, namely the *natural* Borel subalgebra :

$$\mathfrak{b}^{nat} = \mathbb{C}c \oplus (\mathbb{C}[z] \otimes_{\mathbb{C}} \dot{\mathfrak{h}}) \oplus (\mathbb{C}[z, z^{-1}] \otimes_{\mathbb{C}} \dot{\mathfrak{n}}).$$

A "twisted" affine algebra is a fixed point set in $\mathfrak{g}$ of a non-trivial symmetry of Chevalley generators, and we take $\mathfrak{b}^{nat}$ to be the intersection with this set of the natural Borel subalgebra of $\mathfrak{g}$.

To show that $\mathfrak{b}^{st}$ and $\mathfrak{b}^{nat}$ are not conjugate, note that (similar fact holds for any Kac-Moody algebra) :

$$\mathfrak{b}^{st} = (\dot{\mathfrak{h}} + \mathbb{C}c) \oplus [\mathfrak{b}^{st}, \mathfrak{b}^{st}] \quad ;$$

on the other hand one has

$$\mathfrak{b}^{nat} = (\mathbb{C}c + \mathbb{C}[z] \otimes_{\mathbb{C}} \dot{\mathfrak{h}}) \oplus [\mathfrak{b}^{nat}, \mathfrak{b}^{nat}].$$

Now we turn to the classification, up to W-equivalence, of the subsets of positive roots of an affine root system ; this is equivalent to the classification of Borel subalgebras of an affine Kac-Moody algebra $\mathfrak{g}$ up to conjugation.

Let Δ be the root system of an affine Kac-Moody algebra $\mathfrak{g}$ and let $\Pi^{st} = \{\alpha_0, \alpha_1, \dots, \alpha_\ell\}$ be the standard set of simple roots (the ordering of simple roots is that of [4]). Put $\dot{\Pi} = \{\alpha_1, \dots, \alpha_\ell\}$, $\dot{\Delta} = \{\alpha \in \Delta \mid \alpha = \sum_{i=1}^{\ell} k_i \alpha_i\}$,

$$\dot{\Delta}_+ = \{\alpha \in \dot{\Delta} \mid \alpha = \sum_{i=1}^{\ell} k_i \alpha_i \text{ with } k_i \geq 0\}.$$

Then $\dot{\Delta}$ is the root system of the "underlying" finite-dimensional simple Lie algebra $\dot{\mathfrak{g}}$. Let δ be the unique indivisible imaginary root from Δ_+^{st} . Recall that the sets Δ and Δ_+^{st} can be easily reconstructed in terms of the finite root system $\dot{\Delta}$ and the root δ [4, Chapter 6] :

Example. Let $\mathfrak{g}$ be an affine Lie algebra associated to $\dot{\mathfrak{g}}$. Then :

$$\text{(2.4)} \quad \Delta = \{\alpha + n\delta \mid \alpha \in \dot{\Delta} \cup \{0\}, \ n \in \mathbb{Z}\} \setminus \{0\} ,$$

$$\text{(2.5)} \quad \Delta_+^{st} = \{\alpha + n\delta \mid \alpha \in \dot{\Delta} \cup \{0\}, \ n > 0\} \cup \dot{\Delta}_+ . \quad \square$$

Given a root space decomposition $\dot{\mathfrak{g}} = (\bigoplus_{\alpha \in \dot{\Delta}} \dot{\mathfrak{g}}_\alpha) \oplus \dot{\mathfrak{h}}$ we have :
$\mathfrak{g}_{\alpha + k\delta} = z^k \otimes \dot{\mathfrak{g}}_\alpha$ if $\alpha \in \dot{\Delta}$; $\mathfrak{g}_{k\delta} = z^k \otimes \dot{\mathfrak{h}}$.

For a "twisted" affine Lie algebra, the set of roots is a subset of $(\Delta \cup \frac{1}{2}\Delta) \cap Q$, invariant under the shift by $k\delta$ (k = 2 or 3), where Δ is defined by (2.4) (see [4, Chapter 6] for details).

Let Δ_+ be a set of positive roots of an affine root system Δ . Replacing Δ_+ by $-\Delta_+$ if necessary, we can assume that $\delta \in \Delta_+$. A root $\alpha \in \Delta_+$ is called *bad* if all the roots of the form $\alpha+n\delta$ $(n\in\mathbb{Z})$ lie in Δ_+ ; otherwise, a root $\alpha\in\Delta_+$ is called *good*. It is clear that δ is a good root for any Δ_+ and that all the roots from Δ_+^{st} are good.

Lemma 2.2. Let Δ_+ be a set of positive roots of an affine root system Δ, such that $\delta\in\Delta_+$.

(a) If $\alpha\in\Delta_+$ is good, then there exists $s\in\mathbb{Z}_+$ such that a root $\alpha+n\delta$ lies in Δ_+ iff $n\geqslant -s$.

(b) If $\alpha,\beta,\alpha+\beta\in\Delta_+$ and α is bad, then $\alpha+\beta$ is bad.

(c) If $\alpha,\beta,\alpha-\beta\in\Delta_+$ and α and β are good, then $\alpha-\beta$ is good.

(d) If $\alpha,\beta,\alpha+\beta\in\Delta_+$ and α and β are good, then $\alpha+\beta$ is good.

Proof. (a) and (b) are obvious and (c) follows from (b). If α and β are good roots for Δ_+, but all the roots of the form $\alpha+\beta+n\delta$ $(n\in\mathbb{Z})$ lie in Δ_+, then all the roots of the form $-\alpha-\beta+n\delta$ lie in $-\Delta_+$. But $\alpha+s\delta\in-\Delta_+$ for some s since α is good, hence all the roots of the form $-\beta+n\delta$ lie in $-\Delta_+$ and all the roots of the form $\beta+n\delta$ lie in Δ_+. This contradiction proves (d). □

Let X be a subset of $\dot{\Pi}$. We associate to X a subset of positive roots Δ_+^X of Δ as follows. In the non-twisted case we put :

$$\Delta_+^X=\{\alpha+n\delta \mid \alpha\in\dot{\Delta}_+\setminus\mathbb{Z}X,\ n\in\mathbb{Z}\}\ \cup \tag{2.6}$$
$$\{\alpha+n\delta\mid\alpha\in(\dot{\Delta}\cap\mathbb{Z}X)\cup\{0\},\ n>0\}\cup\dot{\Delta}_+$$

In the twisted case, we put $\Delta_+^X=(\Delta_+\cup\frac{1}{2}\Delta_+)\cap\Delta$, where Δ_+ is the set defined as in (2.6).

Proposition 2.1. If Δ is an affine root system, then every set of positive roots is $W\times\{\pm 1\}$-conjugate to one of the sets Δ_+^X.

Proof. Let Δ_+ be a subset of positive roots of Δ ; we can assume that $\delta\in\Delta_+$. Let $X=\{\alpha_i\in\dot{\Pi}\mid$ either α_i or $-\alpha_i$ is a good root of $\Delta_+\}$. Put $\dot{\Delta}_X=\mathbb{Z}X\cap\dot{\Delta}$,

$$\Delta_X=\{\alpha+n\delta\mid\alpha\in\dot{\Delta}_X\cup\tfrac{1}{2}\dot{\Delta}_X\cup\{0\},\ n\in\tfrac{1}{2}\mathbb{Z}\}\cap\Delta,$$

$$\Delta_{X+}=\Delta_X\cap\Delta_+.$$

By lemma 2.2, for each $\alpha\in\Delta_X$, either α or $-\alpha$ is a good root of Δ_+. It follows that $|\Delta_{X+}\cap-\Delta_+^{st}|<\infty$. Applying the same argument as that in the proof of Lemma 2.1, we may assume that $\Delta_{X+}\subset\Delta_+^{st}$. Applying again this argument we may assume that $\dot{\Pi}\subset\Delta_+^{st}$. But then, by Lemma 2.2, Δ_{X+} is the set of good roots of Δ_+. It follows that $\Delta_+=\Delta_+^X$. □

3. A list of unitarizable highest weight representations.

Let $\mathfrak{g}$ be an arbitrary Kac-Moody algebra, let $\mathfrak{b}$ be the standard Borel subalgebra, of $\mathfrak{g}$, let $\omega=\omega_c$ and let $\lambda:\mathfrak{b}\to\mathbb{C}$ be a 1-dimensional representation of $\mathfrak{b}$

defined by

$$\lambda(e_i) = 0 \quad , \quad \lambda(\alpha_i^\vee) = m_i \in \mathbb{Z}_+ \quad (i = 1,\dots,n) \; .$$

Then the representation $\pi_{\lambda;\mathfrak{b},\omega_c}$ is unitarizable ([4, Chapter 11]). These representations are called *integrable* highest weight representations. In particular, if $\mathfrak{g}$ is finite-dimensional, thesse representations are precisely all finite-dimensional irreducible representations of $\mathfrak{g}$.

Furthermore, it is well-known that if $\mathfrak{g}$ is a finite-dimensional simple Lie algebra, then an infinte-dimensional highest weight representation $\pi_{\lambda;\mathfrak{b},\omega}$ is unitarizable only if ω is a consistent antilinear anti-involution which corresponds to a Hermitian symmetric space, and all possibilities for λ are listed in [1],[3].

There is the following "elementary" way to construct a unitarizable highest weight representation π_λ of an affine Lie algebra $\mathfrak{g}$. First, we take $\pi_\lambda(c) = 0$ so that π can be viewed as a representation of the Lie algebra $\mathbb{C}[z,z^{-1}]\otimes_{\mathbb{C}}\dot{\mathfrak{g}}$ in the "non twisted case" or its subalgebra in the "twisted" case. Furthermore, fix N non-zero complex numbers $c_1,\dots,c_N$ and denote by $\varphi_i : \mathbb{C}[z,z^{-1}]\otimes_{\mathbb{C}}\dot{\mathfrak{g}} \to \dot{\mathfrak{g}}$ the evaluation map at c_i i.e. $\varphi_i(z^k\otimes g) = c_i^k g$. Fix a Borel subalgebra $\dot{\mathfrak{b}}$ of $\dot{\mathfrak{g}}$ and a consistent antilinear anti-involution $\dot{\omega}$ of $\dot{\mathfrak{g}}$. Let π_{λ_i} $(i=1,\dots,N)$ be a unitarizable highest weight representation of $\dot{\mathfrak{g}}$ on $L(\lambda_i) = L_{\dot{\mathfrak{b}},\dot{\omega}}(\lambda_i)$. Then $\pi_{\lambda_i}\cdot\varphi_i$ is a unitarizable highest weight representation of $\mathfrak{g}$.

Let $\mathfrak{p} = \mathbb{C}[z,z^{-1}]\otimes\dot{\mathfrak{b}}$, define a representation $\mathfrak{p}\to\mathbb{C}$ by $\lambda(z^k\otimes b) = \sum_i c_i^k\lambda_i(b)$ and an antilinear anti-involution ω of $\mathfrak{g}$ by $\omega.(z^k\otimes g) = z^{-k}\otimes(\dot{\omega}.g)$

Then we have :

$$L_{\mathfrak{p},\omega}(\lambda) = L(\lambda_1)\otimes\dots\dots\otimes L(\lambda_N) \, ,$$

$$\pi_{\lambda;\mathfrak{p}} = (\pi_{\lambda_1}\cdot\varphi_1)\otimes\dots\dots\otimes(\pi_{\lambda_N}\cdot\varphi_N)$$

Thus, the representation $\pi_{\lambda;\mathfrak{p}}$ of $\mathfrak{g}$ on the space $L_{\mathfrak{p},\omega}(\lambda)$ is unitarizable. We call these representations *elementary*.

Finally, let $\mathfrak{g} = sl_{\ell+1}(\mathbb{C}[z,z^{-1}])$ (i.e. we assume that the center $\mathbb{C}c$ acts trivially), let $\mathfrak{p} = \{(a_{ij}(z))\in\mathfrak{g} \mid a_{ij}=0 \text{ for } i>j\} \supset \mathfrak{b}^{nat}$, and let $\omega.(a_{ij}(z)) = (\varepsilon_{ij}\overline{a}_{ji}(z^{-1}))$, where $\varepsilon_{ij}=1$ if $i\neq 1$ or $j\neq 1$ or $i=j=1$, and $\varepsilon_{ij}=-1$ otherwise (for $a(z) = \sum c_i z^i \in \mathbb{C}[z,z^{-1}]$ we write $\overline{a}(z) = \sum \overline{c}_i z^i$).

Let m be a finite positive measure on the unit circle $S^1\subset\mathbb{C}$. Define a linear functional $e_m : \mathbb{C}[z,z^{-1}]\to\mathbb{C}$ by $e_m(a(z)) = \int_{S^1} a(z)\,dm$. Define a representation $\lambda_m : \mathfrak{p}\to\mathbb{C}$ by $\lambda_m((a_{ij}(z)) = -e_m(a_{11}(z)))$.

Then the representation π_{λ_m} of $sl_{\ell+1}(\mathbb{C}[z,z^{-1}])$ on the vector space $L_{\mathfrak{p},\omega}(\lambda_m)$ is called *exceptional*. We will show, as a part of a more general result, that they are unitarizable.

Now we can state our first main result.

Theorem 3.1. *Let $\mathfrak{g}$ be an affine Lie algebra, let ω be a consistent antilinear anti-involution of $\mathfrak{g}$ and let $\mathfrak{b}$ be a Borel subalgebra of $\mathfrak{g}$. Let $\lambda : \mathfrak{b} \to \mathbb{C}$ be a 1-dimensional representation of $\mathfrak{b}$. Then the representation $\pi_{\lambda;\mathfrak{b},\omega}$ of $\mathfrak{g}$ on the space $L_{\mathfrak{b},\omega}(\lambda)$ is unitarizable if and only if it is equivalent to either an integrable representation, or an elementary representation, or an exceptional representation.*

4. Elimination.

We now begin to prove Theorem 3.1. This section is devoted to the negative, i.e. non-unitarizable, aspects.

Let us first look at the affine Lie algebra $\hat{L}(sl_2)$ associated to $sl_2(\mathbb{C})$. We have that $sl_2(\mathbb{C}) = \text{span}\{e, f, h\}$ with commutation relations

$$(4.1)\quad [e,f] = h \quad ; \quad [h,e] = 2e \quad ; \quad [h,f] = -2f$$

We write elements $z^n \otimes X$ of $\mathbb{C}[z,z^{-1}] \otimes sl_2$ as $z^n X$. We have that

$$(4.2)\quad \hat{L}(sl_2) = \mathbb{C}[z,z^{-1}] \otimes_{\mathbb{C}} sl_2 \oplus \mathbb{C}c \quad ,$$

where, in particular,

$$(4.3)\quad [z^n e, z^m f] = n\, \delta_{n,-m}\, c$$

According to Proposition 2.1 there are, up to conjugacy, two distinct sets of positive roots. In either case, a consistent involution ω is of the form

$$(4.4)\quad \omega(e) = \varepsilon_1 f \quad ; \quad \omega(zh) = \varepsilon_2 zh$$

where $\varepsilon_1^2 = \varepsilon_2^2 = 1$. There are essensially four distinct parabolic subalgebras compatible with (some of) the involutions above :

$$(4.5)\quad \mathfrak{b}^{st} = \mathfrak{h} \oplus \text{span}\{z^k e \mid k \geq 0\} \oplus \text{span}\{z^k f, z^k h \mid k > 0\},$$

$$\mathfrak{p}^{nat} = \text{span}\{z^k e, z^k h \mid k \in \mathbb{Z}\},$$

$$\mathfrak{b}^{nat} = \text{span}\{z^k h \mid k \geq 0\} \oplus \text{span}\{z^k e \mid k \in \mathbb{Z}\}, \text{ and}$$

$$\mathfrak{p}^{kt} = \text{span}\{ z^k h, z^k e, z^k f \mid k \geq 0 \}$$

Assuming that $L(\lambda) = L_{\mathfrak{b},\omega}(\lambda)$ is unitarizable, we now analyze the restrictions this requirement imposes on the data. First we observe, as remarked after Lemma 1.1, that $\mathfrak{p}^{kt}$ does not give other unitarizable modules than does $\mathfrak{b}^{st}$. A second simplifying observation is, that due to the symmetry of the Hermitian form, in the case of $\mathfrak{b}^{nat}$, $\lambda(z^n h) = -\delta_{n,0}\, a$ for some $a \in \mathbb{R}$. Hence, the treatment of this case is covered by that of $\mathfrak{p}^{nat}$.

Returning to (4.4), it follows that

$$(4.6) \quad \omega(z^n h) = \varepsilon_2^n z^{-n} h \qquad \omega(z^n f) = \varepsilon_1 \varepsilon_2^n z^{-n} e.$$

We let $(z^n h)\cdot v_\lambda = \lambda(z^n h)\, v_\lambda = \lambda_n v_\lambda$, and $c \cdot v_\lambda = c\, v_\lambda$.

Let us turn to an examination of the standard set of positive roots:

Firstly, the symmetry of H forces $\lambda_n = \lambda\, \delta_{n,0}$. Secondly, we have that for all $k, n \in \mathbb{N}$

$$(4.7) \quad \begin{aligned} &H\big((z^{-n} f)(z^{-2} e)^k v_\lambda, (z^{-n} f)(z^{-2} e)^k v_\lambda\big) = \\ &\varepsilon_1 (\varepsilon_2)^n (\lambda + 2k + nc)\, H\big((z^{-2} e)^k v_\lambda, (z^{-2} e)^k v_\lambda\big). \end{aligned}$$

It is no loss of generality to assume that $(z^{-2} e)^k v_\lambda \neq 0$ since we may otherwise just replace it with some $(z^{-j} e)^k v_\lambda$, $j \in \mathbb{N}$. Thus if the module is unitarizable and non-trivial, then $\varepsilon_1 = \varepsilon_2 = 1$. This is the compact antilinear anti-involution, and hence by the sl_2-theory, $L(\lambda)$ is an integrable representation (observe that $\lambda, c \in \mathbb{Z}_+$, and $c \geq \lambda$).

In the natural case we have

$$(4.8) \quad H(z^n f\, v_\lambda, z^n f\, v_\lambda) = \varepsilon_1 \varepsilon_2^n (-nc + \lambda) \text{ for all } n \in \mathbb{Z}.$$

Thus, positivity of H implies that c = 0 and $\varepsilon_2 = 1$. Further we observe that for a general $a = \sum a_n z^n \in \mathbb{C}[z, z^{-1}]$

$$(4.9) \quad H(a f v_\lambda, a f v_\lambda) = \varepsilon_1 \lambda (a a^*)$$

where $a^* = \sum \bar{a}_n z^{-n}$. Thus, $\varepsilon_1 \lambda$ is a positive linear functional on $\mathbb{C}[z, z^{-1}]$. It is then natural to represent $\mathbb{C}[z, z^{-1}]$ as the set of functions on S^1,

$\mathbb{C}[z, z^{-1}] = \{ \sum_{n=-\infty}^{\infty} a_n e^{in\theta} \mid$ all but a finite number of the a_n's are non-zero $\}$

The positivity of $\varepsilon_1 \lambda$ implies continuity and thus it extends to a positive Radon measure μ (i.e. locally finite) on S^1. It follows from the general result; Proposition 6.2 below, that the module $L_{\mathfrak{b}^{nat},\omega}(\lambda)$ is irreducible except in the case where

supp (μ) is contained in a finite number of points.
Consider then the case $\varepsilon_1 = 1$. This corresponds to a compact (su(2)) situation. Thus there must exist an integer $i_0 \in \mathbb{N}$ and a non-trivial invariant subspace S_{i_0} such that $\forall i \geq i_0 : f^i \in S_{i_0}$. Thus, the measure μ must be finitely supported. It is straightforward that in this case we do have unitarity.
Finally, as we shall see below, the case $\varepsilon_1 = -1$ leads to unitarity.
Let us summarize :

Lemma 4.1. For $\hat{L}(sl_2)$ only the following situations may lead to unitarity.

(4.10) $\mathfrak{b}^{st}$: $\omega(e) = f$, $\omega(f) = e$, $\omega(z\,h) = z^{-1}h$ *and* $c \geq |\lambda|$

and $\mathfrak{p}^{nat}$: $\omega(e) = \varepsilon f$, $\omega(f) = \varepsilon e$, $\omega(z\,h) = z^{-1}h$

and $\lambda(z^n n) = \lambda_n = \varepsilon \int_{S^1} e^{in\theta}\, d\mu(\theta)$

where if $\varepsilon = 1$, μ *is finitely supported, and always,* $\varepsilon\mu$ *is a positive Radon measure.* □

An immediate consequence of this lemma for an arbitrary affine Lie algebra is that for the standard system of positive roots only the compact involution may lead to unitarity (and hence $L(\lambda)$ must be an integrable representation).Turning now to the natural or partly natural ("non-standard") situation we will assume that the measures involved are not finitely supported since this case is easily dealt with (cf. §3). It follows from Lemma 4.1 that now c=0, and $\omega(e_\alpha) = -e_{-\alpha}$ ("non-compact") on the non-standard part.
Let $\dot{\mathfrak{g}}$ denote a simple complex Lie algebra of finite diemnsion. It is well-known [2] and quite straight forward to see that besides conjugation in a compact real form, the only situations that lead to unitarizable highest weight modules are those where $\dot{\mathfrak{g}}$ has a real form corresponding to a Hermitian symmetric space, and where ω is conjugation with respect to this.
Specifically, let $\dot{\mathfrak{g}} = (\dot{\mathfrak{g}}_0)^{\mathbb{C}}$ and let $\dot{\mathfrak{g}}_0 = \dot{\mathfrak{k}} \oplus \dot{\mathfrak{p}}$ be a Cartan decomposition of the real Lie algebra $\dot{\mathfrak{g}}_0$. Then $\dot{\mathfrak{k}}$ has a one-dimensional center $\mathfrak{z} = \mathbb{R}h_0$, where h_0 is chosen such that its eigenvalues under the adjoint action on $\dot{\mathfrak{p}}^{\mathbb{C}}$ are $\pm i$. Let

(4.11) $$\dot{\mathfrak{p}}^+ = \{ X \in \dot{\mathfrak{p}}^{\mathbb{C}} \mid [h_0, X] = iX \}$$

and $$\dot{\mathfrak{p}}^- = \{ X \in \dot{\mathfrak{p}}^{\mathbb{C}} \mid [h_0, X] = -iX \}$$

Let $\dot{\mathfrak{k}}_1 = [\dot{\mathfrak{k}}, \dot{\mathfrak{k}}]$ and let $i\dot{\mathfrak{h}}$ be a maximal abelian subalgebra of $\dot{\mathfrak{k}}$. Then $\dot{\mathfrak{k}} = \dot{\mathfrak{k}}_1 \oplus \mathbb{R}h_0$, $i\dot{\mathfrak{h}} = (i\dot{\mathfrak{h}} \cap \dot{\mathfrak{k}}_1) \oplus \mathbb{R}h_0$, $(i\dot{\mathfrak{h}} \cap \dot{\mathfrak{k}}_1)^{\mathbb{C}}$ is a Cartan subalgebra of $\dot{\mathfrak{k}}_1^{\mathbb{C}}$, and $\dot{\mathfrak{h}} = (i\dot{\mathfrak{h}})^{\mathbb{C}}$ is a Cartan subalgebra of $\dot{\mathfrak{g}}$. The sets of compact and non-compact roots of $\dot{\mathfrak{g}}$ relative to $\dot{\mathfrak{h}}$ are denoted Δ_c and Δ_n, respectively ; $\Delta = \Delta_c \cup \Delta_n$. We choose an ordering of Δ such that $$\dot{\mathfrak{p}}^{+} = \sum_{\alpha\in\Delta_n^+} \dot{\mathfrak{g}}_\alpha$$

For $\gamma \in \Delta$ let H_γ be the unique element of $\dot{\mathfrak{h}} \cap [\dot{\mathfrak{g}}_\gamma, \dot{\mathfrak{g}}_{-\gamma}]$ for which $\gamma(H_\gamma) = 2$.
For $\alpha \in \Delta_n^+$ choose $e_\alpha \in \dot{\mathfrak{g}}_\alpha$ such that

$$[e_\alpha , \omega(e_\alpha)] = H_\alpha \,, \tag{4.12}$$

and let $e_{-\alpha} = \omega(e_\alpha)$.

Since in any non-standard situation c = 0, and since by the assumptions on the measures we are looking at irreducible modules (cf. §6), it follows from the preceding analysis that $\mathcal{U}(\mathbb{C}[z, z^{-1}] \otimes \dot{\mathfrak{k}}_1^{\mathbb{C}}) \cdot v_\lambda$ must be in the Kernel of the Hermitian form. Thus our module is "scalar", that is, of the form

$$M(\lambda) = \mathcal{U}(\mathbb{C}[z, z^{-1}] \otimes \dot{\mathfrak{p}}^{-}) \cdot v_\lambda \tag{4.13}$$

where $(\mathbb{C}[z, z^{-1}] \otimes \dot{\mathfrak{p}}^{+}) \cdot v_\lambda = 0$, $(\mathbb{C}[z, z^{-1}] \otimes \dot{\mathfrak{k}}_1) \cdot v_\lambda = 0$

and $(z^n h) \cdot v_\lambda = -\left(\int_{S^1} e^{in\theta} d\mu(\theta) \right) v_\lambda$

for some positive measure μ which is not supported by a finite number of points. Observe that

$$\lambda_0 = -\int_{S^1} d\mu(\theta) < 0 \,. \tag{4.14}$$

We now recall a result about the scalar modules $\mathcal{U}(\dot{\mathfrak{p}}^{-}) \cdot v_\lambda$ with v_λ as above ([6],[7]). *If there are at least two perpendicular non-compact roots (i.e. the real rank is greater than one) then for (a finite number of) critical values* $c_1, c_2, \ldots$ $\ldots$, $0 > c_1 > c_2 > \ldots$, *the module* $\mathcal{U}(\dot{\mathfrak{p}}^{-}) \cdot v_\lambda$ *with* $\lambda_0 = c_i$, $i = 1, 2, \ldots$ *is reducible. In fact, the Hermitian form restricted to this space is degenerate.* Thus, when λ_0 equals one of these critical values, by the irreducibility of $M(\lambda)$ (Proposition 6.2) there can be no unitarity. Finally, for any $\lambda_0 < 0$, it is easy to deform the measure μ , without destroying the irreducibility, into a measure $\tilde{\mu}$ which yields a λ_0 among these critical values. We may thus state:

Proposition 4.2. In case the real rank of $\dot{\mathfrak{g}}$ *is greater than one there can be no unitarizable module based on a natural parabolic subalgebra and on infinitely supported measures.* □

Remark : Hence only $\dot{\mathfrak{g}} = su(n,1)$ remains. □

To bring our analysis to a conclusion we now turn to the "twisted" affine Lie algebras

$$\mathfrak{g} = \mathfrak{g}_0 \oplus \mathfrak{g}_1 \,, \quad \text{or} \quad \mathfrak{g} = \mathfrak{g}_0 \oplus \mathfrak{g}_1 \oplus \mathfrak{g}_2 \tag{4.15}$$

We maintain the assumption that the measures are not finitely supported. In a non-standard situation, c = 0, and thus, restricted to $\mathfrak{g}_0$ one must be in a non-standard or trivial situation. However, if the module is trivial on $\mathfrak{g}_0$, it is trivial on $\mathfrak{g}$. Thus, by the preceding results, the only case that needs to be considered is $A_2^{(2)}$ with a natural parabolic subalgebra in $\mathfrak{g}_0$:

(4.16) $$A_2^{(2)} = \mathfrak{g}_0 \oplus \mathfrak{g}_1 \qquad \text{and}$$

$$\dot{\mathfrak{g}}_0 = sl_2 = \operatorname{span}\{e, f, h\} \quad \text{as in (3.1).}$$

Observe that the non-standardness of $\mathfrak{g}_0$ forces a non-standardness on the full algebra. Inside $\mathfrak{g}_1$ one can easily find elements u_1^+, u_1^-, u_2^+ and u_2^- such that

(4.17) $$\operatorname{span}\{u_1^+, u_1^-, h\} \cong \operatorname{span}\{u_2^+, u_2^-, h\} \cong sl_2$$

as Lie algebras, and such that

$$[e, u_1^+] = -u_2^+, \qquad \text{and} \qquad [f, u_1^-] = u_2^-.$$

Since we must have $\omega(e) = -f$ and $\omega(u_1^+) = -u_1^-$ it follows that $\omega(u_2^+) = u_2^-$, and this is impossible (it is compact). Thus at this level, there are no unitarizable modules.

Thus, the only highest weight representations which may be unitarizable are those listed in Theorem 3.1.

5. Unitarity.

Let R denote a (non-commutative) associative algebra over $\mathbb{C}$ and let φ be a trace on R, i.e. a linear map of R into $\mathbb{C}$ which satisfies

(5.1) $$\varphi(ab) = \varphi(ba) \qquad \text{for all } a,b \in R.$$

We define the Lie algebra

(5.2) $$sl_2(R,\varphi) = \left\{ \begin{pmatrix} a_1 & a_2 \\ a_3 & a_4 \end{pmatrix} \;\middle|\; a_i \in R, \ i = 1,2,3,4 \ \text{and} \ \varphi(a_1+a_4) = 0 \right\},$$

and we consider the Verma module $M(\varphi)$ defined by the property that there exists a non-zero vector v_φ such that

(5.3) $$\begin{pmatrix} a_1 & a_2 \\ 0 & a_4 \end{pmatrix} v_\varphi = \varphi(a_1) \cdot v_\varphi$$

Let $i \in \mathbb{N}$. We say that $\gamma = (\gamma_1, \ldots, \gamma_s) \in \mathbb{N}^s$ is an *s-partition of* i if

(5.4) $$i = \gamma_1 + \gamma_2 + \ldots\ldots + \gamma_s, \qquad \text{and}$$

$$\gamma_1 \geq \gamma_2 \geq \cdots\cdots \geq \gamma_s > 0.$$

We let $Par_s(i)$ denote the set of all such s-tuples. Further, if $\gamma \in Par_s(i)$ we let $D_\gamma(i)$ denote the set of distributions of i objects in the Young diagram of γ. If $(d_1, \ldots, d_i) \in R^i$ and if $t_\gamma \in D_\gamma(i)$ we let, for $j = 1, \ldots s$, $d(t_\gamma)_j$ denote the product of the elements in the jth row.

Let $\gamma \in Par_s(N)$. Utilizing the fact that φ is a trace, we will say that $\pi_1^1 \times \pi_2^1 \in S_N \times S_N$ is equivalent to $\pi_1 \times \pi_2 \in S_N \times S_N$, where S_N denotes the group of permutations of N letters, if for all $z_1, \ldots, z_i$, $\omega_1, \ldots \omega_i \in R$,

$$(5.5)\ \varphi(z_{\pi_1^1(1)}\omega_{\pi_2^1(1)} \cdots z_{\pi_1^1(\gamma_1)}\omega_{\pi_2^1(\gamma_1)}) \cdots\cdots \varphi(z_{\pi_1^1(\gamma_1+\cdots+\gamma_{s-1}+1)}\omega_{\pi_2^1(\gamma_1+\cdots+\gamma_{s-1}+1)}\cdots)$$

can be obtained from the analogous expression for $\pi_1 \times \pi_2$ by a permutation of the s factors $\varphi(\cdots)$ and/or by cyclic permutation of the variables (e.q. $\varphi(z_3\omega_3\, z_1\omega_1\, z_2\omega_2) = \varphi(z_2\omega_2\, z_3\omega_3\, z_1\omega_1)$.

The set of equivalence classes is denoted by $(S_N \times S_N)(\gamma)$.

Lemma 5.1. *Let* $z_1, \ldots, z_N, \omega_1, \ldots, \omega_N \in R$. *Then in* $M(\varphi)$,

$$(-1)^N \begin{pmatrix} 0 & z_1 \\ 0 & 0 \end{pmatrix} \cdots \begin{pmatrix} 0 & z_N \\ 0 & 0 \end{pmatrix}\begin{pmatrix} 0 & 0 \\ \omega_1 & 0 \end{pmatrix} \cdots \begin{pmatrix} 0 & 0 \\ \omega_N & 0 \end{pmatrix} \cdot v_\varphi =$$

$$(5.6)\ \sum_{s=1}^{N} \sum_{\gamma \in Par_s(N)} \sum_{[\pi_1 \times \pi_2] \in (S_N \times S_N)(\gamma)} (-1)^{\gamma_1} \varphi(z_{\pi_1(1)}\omega_{\pi_2(1)} \cdots z_{\pi_1(\gamma_1)}\omega_{\pi_2(\gamma_1)}) \cdots (-1)^{\gamma_s} \varphi(\cdots z_{\pi_1(N)}\omega_{\pi_2(N)}) \cdot v_\varphi$$

Proof: We proceed by induction. N = 1 is trivial, so assume (5.6) is true up to N. We have, by (5.3),

$$(5.7)\ (-1)^{N+1}\begin{pmatrix} 0 & z_1 \\ 0 & 0 \end{pmatrix} \cdots \begin{pmatrix} 0 & z_{N+1} \\ 0 & 0 \end{pmatrix}\begin{pmatrix} 0 & 0 \\ \omega_1 & 0 \end{pmatrix} \cdots\cdots \begin{pmatrix} 0 & 0 \\ \omega_{N+1} & 0 \end{pmatrix} \cdot v_\varphi =$$

$$= \sum_{i=1}^{N+1} (-1)^N \begin{pmatrix} 0 & z_1 \\ 0 & 0 \end{pmatrix} \cdots \begin{pmatrix} 0 & z_N \\ 0 & 0 \end{pmatrix}\begin{pmatrix} 0 & 0 \\ \omega_1 & 0 \end{pmatrix} \cdots \widehat{\begin{pmatrix} 0 & 0 \\ \omega_i & 0 \end{pmatrix}} \cdots \begin{pmatrix} 0 & 0 \\ \omega_{N+1} & 0 \end{pmatrix} (-1)\varphi(z_{N+1}\omega_i) \cdot v_\varphi +$$

$$+ (-1)^{N+1} \sum_{j>i} \begin{pmatrix} 0 & z_1 \\ 0 & 0 \end{pmatrix} \cdots \begin{pmatrix} 0 & z_N \\ 0 & 0 \end{pmatrix}\begin{pmatrix} 0 & 0 \\ \omega_1 & 0 \end{pmatrix} \cdots \widehat{\begin{pmatrix} 0 & 0 \\ \omega_i & 0 \end{pmatrix}} \cdots \begin{pmatrix} 0 & 0 \\ -\omega_i z_{N+1}\omega_j - \omega_j z_{N+1}\omega_i & 0 \end{pmatrix} \cdots$$

$$\cdots \begin{pmatrix} 0 & 0 \\ \omega_{N+1} & 0 \end{pmatrix} \cdot v_\varphi .$$

This, then, can be evaluated using (5.6), and as a result one will get an expression analogous to this. To treat the constants rigorously it is, by symmetry, enough to examine terms of the form

$$(5.8)\ (-1)^{\gamma_1}\varphi(z_1\omega_1 \cdots z_{\gamma_1}\omega_{\gamma_1}) \cdots (-1)^{\gamma_s}\varphi(z_{(\gamma_1+\cdots\gamma_{s-1}+1)} \cdots z_{N+1}\,\omega_{N+1})$$

Assuming $\gamma_s > 1$ ($\gamma_s = 1$ is trivial), the only way in which such a term can emerge is clearly by replacing ω_N by $-\omega_N z_{N+1}\omega_{N+1}$ in the analogous expression

$$(-1)^{\gamma_1}\varphi(z_1 w_1 \dots z_{\gamma_1} w_{\gamma_1}) \cdots (-1)^{\gamma_s - 1}\varphi(z_{\gamma_1+\dots+\gamma_{s-1}+1} \cdots z_N w_N)$$

in (5.6). □

As a first application of this result we get the unitarizability for su(r, 1) : Let R denote the algebra of polynomial functions from the circle into $gl(r,\mathbb{C})$. Let tr denote the usual trace on $gl(r,\mathbb{C})$ and define

$$(5.9) \qquad \varphi_\mu(p(\theta)) = -\int_0^{2\pi} tr(p(\theta))\, d\mu$$

for some positive Radon measure μ . Observe that $-\varphi_\mu(p_1(\theta)p_2(\theta)^*)$ defines a positive defintie inner product on R. Take the z's of (5.6) to be of the form

$$(5.10) \qquad z = \begin{bmatrix} c_1 e^{in_1\theta} & c_2 e^{in_2\theta} & \dots & c_r e^{in_r\theta} \\ 0 & 0 & \dots & 0 \\ \vdots & \vdots & & \vdots \\ 0 & 0 & \dots & 0 \end{bmatrix}$$

and the w's to be of the form z*, and observe that any product $z_1 \cdot z_2^*$ gives a matrix with at most one non-zero entry, namely the upper left hand corner. It follows that any expression $z_1 z_2^* z_3 z_4^* \dots z_{2N-1} z_{2N}^*$ can be written as a product $z_a \cdot z_b^*$ of just two elements of R. Thus (5.6) expresses the Hermitian form on the module $M(\lambda)$ (with $\lambda = \varphi_\mu$, see (1.5)) as a sum of tensor products of positive definite Hermitian forms, and thus as something positive. This observation completes the proof of Theorem 3.1.

Consider now the case where R is a commutative algebra over $\mathbb{C}$ and assume that $a \to a^*$ is an antilinear involution of R such that

$$(5.11) \qquad \varphi(a^*) = \overline{\varphi(a)} \qquad \text{for all } a \in R$$

Define an antilinear anti-involution ω of $sl_2(R,\varphi)$ by

$$(5.12) \qquad \omega\begin{pmatrix} a & b \\ c & d \end{pmatrix} = \begin{pmatrix} a^* & -c^* \\ -b^* & d^* \end{pmatrix}$$

and identify elements $\begin{pmatrix} 0 & 0 \\ c & 0 \end{pmatrix}$ with the entry $c \in R$. An easy consequence of Lemma 5.1 is then,

Corollary 5.2 : *For* $a, b \in R$ *put* $(a,b) = -\varphi(b^* a)$. *Then*

$$(5.13) \qquad H_\varphi(c_1 \dots c_N . v_\varphi, c'_1 \dots c'_{N'} . v_\varphi) = \delta_{NN'} \sum_{s=1}^{N} \sum_{\gamma \in Par_s(N)} \sum_{t_\gamma, t_{\gamma'} \in D_\gamma(N)} K_\gamma \cdot \prod_{i=1}^{s} (\gamma_i)^{-1} \cdot (d(t_\gamma)_i, d(t'_\gamma)_i)$$

defines the contravariant Hermitian form on $M(\varphi)$, *where, if* $\gamma_1 = \dots = \gamma_{n_1} > \gamma_{n_1+1} = \dots = \gamma_{n_1+n_2} > \gamma_{n_1+n_2+1} = \dots$, $K_\gamma = ((n_1!)(n_2!)\dots)^{-1}$. □

Corollary 5.3 : H_φ is positive semi-definite if and only if the form

$$(a, b) = -\varphi(b^* a) \tag{5.14}$$

is positive semi-definite on R. □

Remark 1. In the case R = $\mathbb{C}[z, z^{-1}]$ with $\varphi \equiv$ evaluation at 1, the Hermitian form has a (large) kernel. Thus, in particular M (φ) need not be irreducible (cf §3 and below). □

Remark 2. Let R be an arbitrary commutative algebra with a basis $\{a_\beta\}_{\beta\in B}$. Define the structure constants $C^\gamma_{\alpha\beta}$ by

$$a_\alpha a_\beta = C^\gamma_{\alpha\beta} \; a_\gamma$$

where the usual summation convention is used. The elements of the form

$$(a_{\alpha_1} f) \; \dots \; (a_{\alpha_i} f) \cdot v_\varphi ,$$

where $af = \begin{pmatrix} 0 & 0 \\ a & 0 \end{pmatrix}$, form a basis of M ($\varphi$), and $(a_\alpha . h) . v_\varphi = \varphi(a_\alpha) v = \lambda_\alpha . v$, and $(ae) v_\varphi = \begin{pmatrix} 0 & a \\ 0 & 0 \end{pmatrix} v_\varphi = 0$.

Due to the commutativity of the af's it is natural to represent M (φ) as a space of polynomials $\mathbb{C}[X_\beta ; \beta \in B]$, and it follows easily that the action by left multiplication on M(φ) is transformed into the following action on $\mathbb{C}[X_\beta ; \beta\in B]$:

$a_{\alpha_0} f$: Multiplication by X_{α_0}

$a_{\alpha_0} h$: $\lambda_{\alpha_0} - 2\, C^\gamma_{\alpha_0 \alpha} \; X_\gamma \; \frac{\partial}{\partial X_\alpha}$

$a_{\alpha_0} e$: $\lambda_\gamma \; C^\gamma_{\alpha_0\alpha} \; \frac{\partial}{\partial X^\alpha} - X_\delta \; C^\gamma_{\alpha_0\alpha} \; C^\delta_{\gamma\beta} \; \frac{\partial}{\partial X_\alpha} \; \frac{\partial}{\partial X_\beta}$.

In particular, for $\mathbb{C}[z, z^{-1}]$ we have that $C^k_{n,m} = \delta_{n+m,k}$ and the formulas (0.9) in the introduction thus follow immediately. □

6. Irreducibility.

Let $\dot{\mathfrak{g}}$ be a finite-dimensional simple Lie algebra over $\mathbb{C}$, $\dot{\mathfrak{h}}$ a Cartan subalgebra, Δ the set of roots, $\Pi = \{\alpha_1, \dots, \alpha_r\}$ a basis of Δ, and Δ^+ the corresponding set of positive roots. Consider a subset Y of Π and let

$$\Delta_Y = \Delta \cap \sum_{\alpha_i \in Y} \mathbb{Z} . \alpha_i , \tag{6.1}$$

with Δ^+_Y, Δ^-_Y defined analogously. Further, let

$$\dot{\mathfrak{g}} = \dot{\mathfrak{g}}_{-1} \oplus \dot{\mathfrak{g}}_0 \oplus \dot{\mathfrak{g}}_1 \tag{6.2}$$

be a decomposition of $\dot{\mathfrak{g}}$ such that $\dot{\mathfrak{g}}_0 \oplus \dot{\mathfrak{g}}_1$, is a compatible parabolic. Let $\{e_\alpha\}_{\alpha\in\Delta}$,

$\{H_\alpha\}_{\alpha\in\Pi}$ be a Chevalley basis of $\dot{\mathfrak{g}}$. We assume that $Y \neq \Pi$ and consider an element Λ of $\dot{\mathfrak{h}}^*$ satisfying $\Lambda(H_{\alpha_i}) = \lambda_i = 0$ for $\alpha_i \in Y$ and $\Lambda(H_{\alpha_j}) = \lambda_j \neq 0$ for $\alpha_j \notin Y$. For later use we now prove a lemma about (generalized) Verma modules. Let $M(\Lambda)$ be the left $\mathcal{U}(\dot{\mathfrak{g}})$-module $\mathcal{U}(\dot{\mathfrak{g}}_{-1}).v_\Lambda$ where v_Λ is a non-zero vector in the 1-dimensional space of the $\mathcal{U}(\dot{\mathfrak{g}}_0 \oplus \dot{\mathfrak{g}}_1)$ -module defined by Λ.

Lemma 6.1 Let $v \in M(\Lambda)$ *be a weight vector such that* $e_\alpha . v = 0$ *for all* $\alpha \in \Delta^+$. *Let* $v = u . v_\lambda$ *for some* $u \in \mathcal{U}(\dot{\mathfrak{g}}_{-1})$. *Then, if* u *is expanded on a standard Poincaré-Birkhoff-Witt basis, at least one of the summands yielding non-zero coordinates contains a factor of some* $e_{-\alpha_i}$, $\alpha_i \in \Pi \setminus Y$.

Proof. Let t_0 denote the minimum height of the α's for which $e_{-\alpha}$ occur in u. Suppose $t_0 > 1$ and let $\alpha_i \in \Pi$ be chosen such that $[e_{\alpha_i}, u]$ contains a term with a factor of height $t_0 - 1$. (Choose e.g. the ordering compatible with height). Since $[e_{\alpha_i}, u] = 0$, another term is needed to remove this again, but this can evidently only be done through the negative of the expression that yielded the first term. Thus, $t_0 > 1$ contradicts the non-triviality of u. □

Let R be a commutative algebra and let $\{\varphi_i\}$ be a finite family of linear functionals on R. To avoid instant degeneracy we assume that

(6.3a) $a \in R$, $\varphi_i(Ra) = 0$ *implies* $a = 0$ *for each* i.

(6.3b) *R contains no finite-dimensional non-trivial ideals.*

We let v_0 define a 1-dimensional $\mathcal{U}(\dot{\mathfrak{g}}_0 \oplus \dot{\mathfrak{g}}_1)$-module through a regular element Λ. We now define a linear map $\Lambda_\varphi : R \otimes \dot{\mathfrak{h}} \to \mathbb{C}$ by

(6.4) $$\Lambda_\varphi(a \otimes H_{\alpha_i}) = \varphi_i(a)\lambda_i$$

with the λ_i's as before, and corresponding to this we consider the left $\mathcal{U}(R \otimes \dot{\mathfrak{g}})$ - module

(6.5) $$M(\Lambda_\varphi) = \mathcal{U}(\dot{\mathfrak{g}}_{-1} \otimes R).v_\varphi ,$$

where we now have $(a \otimes h) v_\varphi = \Lambda_\varphi(a \otimes h) \cdot v_\varphi$, $a \in R$, $h \in \dot{\mathfrak{h}}$

Proposition 6.2. $M(\Lambda_\varphi)$ *is irreducible.*

Proof. Suppose that S is a non-trivial invariant subspace. As in the finite-dimensional case there is at least one non-zero element v of S which satisfies

(6.6) $$\forall a \in R, \ \forall \alpha \in \Delta^+ : \ a e_\alpha v = 0 ,$$

and we may write $v = u_0 . v_\varphi$ for a unique $u_0 \in \mathcal{U}(\dot{\mathfrak{g}}_{-1} \otimes R)$. Let $\alpha_0 \in \Pi \setminus Y$.

Choose an ordering of $\Delta^- \setminus \Delta^-_Y$ such that the smallest elements are those α for which $[e_{\alpha_a}, e_\alpha]=0$. In a Poincaré-Birkhoff-Witt basis we write those terms to the left. Further we shall assume, as we may, that at this level, $\alpha_1+\alpha_2$ is bigger than α_1 and α_2. We write monomials in $\mathcal{U}(\mathring{g}_{-1}\otimes R)$ made up entirely of elements whose $\mathring{g}_{-1}$-part e_α commutes with e_{α_a} as $u(o)$. Again, inside a $u(o)$, the smallest elements g_0 to the left. Next in the ordering come those $-\alpha$'s for which $[e_{\alpha_a}, e_{-\alpha}]\neq 0$ and $\alpha=\alpha_a+\sum_{\alpha_i\in\Pi\setminus\{\alpha_a\}} n_i\alpha_i$ has α_a -coefficient 1. These we order in an arbitrary fixed way. Monomials in $\mathcal{U}(\mathring{g}_{-1}\otimes R)$ corresponding entirely to such roots are denoted by $u(1)$. Then we define $u(2),\ldots u(j)$ analogously. For convenience we also allow the $u(j)$'s to be constants. It follows from Lemma 6.1 that there exists at least one $\alpha_a\in\Pi\setminus Y$ which satisfies that when we write u_0 as a sum of expressions $u(o)u(1)\ldots u(j)$, then at least one $u(1)$ contains a factor $a\,e_{-\alpha_a}$ for some $a\in R$. We proceed with a fixed α_a satisfying this and now claim that any of the $u(1)$'s occuring is made up entirely of products of $r_i\,e_{-\alpha_a}$'s and that, furthermore, the $u(j)$'s for j> 1 must be constants. Again we argue by contradiction : Suppose that some $u(1)$ contains a factor $b e_{-\alpha}$ with $\alpha\neq\alpha_a$. Let us describe this by saying that α satisfies (*). Let $\bar\alpha$ be that element of the set of roots satisfying (*) for which the root $\alpha'=\alpha_a-\alpha$ is smallest. Consider a monomial $u(o)u(1)\ldots u(\ell)$ where the full factor in $u(1)$ containing $e_{\bar\alpha}$'s is, say, $(r_1 e_{-\bar\alpha})\cdots(r_n e_{-\bar\alpha})$.

Assume for simplicity that the r_i's are linearly independent. When we compute $[c e_{\alpha_a}, u_0]$ we then get a monomial where, say, the $u(o)$ part contains a $c\,r_1\,e_{-\alpha'}$ and where the $u(1)$ part contains $(r_2 e_{-\bar\alpha})\ldots(r_n e_{-\bar\alpha})$. On the other hand, by (6.6) we must get zero and hence we must remove this expression again. But this can only be done through the adjoint action of $c\,e_{\alpha_a}$ on some $u(j)$ with $j > 1$ or on portions of $u(1)$ not of the form $r\,e_{-\bar\alpha}$, and this does not change the coefficients of the terms in $u(o)$ involving $r e_{-\alpha'}$'s. Hence, for all $c\in R$, $c.r_1$ must be in the span S of the elements s for which $s\,e_{-\alpha'}$ occurs in some $u(o)$ in the decomposition of u_0. In other words, $R.r_1\subset S$ and S is finite-dimensional This contradicts (6.3b). Further, it now follows analogously that the $u(j)$'s are constants for $j > 1$. Thus u_0 is a sum of terms of the form $u(o)\left(\sum_i (a_i f_{-\alpha_a})^N\right)$ for some fixed N, and where the $u(o)$'s are linearly independent. It then follows from (6.6) that there exists a non-zero element $a\in R$, and some i such that $\varphi_i(ac)=0$ for all $c\in R$. Thus, φ_i violates (6.3a), and this is a contradiction. □

Remark . Let X be a compact pathwise connected Hausdorff space and let R be a non-trivial (i.e. $R\neq 0$, $R\neq\mathbb{C}$) subalgebra of C(X). Then R contains no non-trivial finite-dimensional ideals. □

Acknowledgements

The first author was supported by a Niels Bohr stipend. Both authors acknowledge partial support by NSF grant MCS 8203739.

References.

1. T. Enright, R. Howe, and N. Wallach, A classification of unitary highest weight modules, in :"*Representation theory of reductive groups*", P.C. Trombi ed., Progress in Math. 40, Birkhaüser, Boston, 1983.
2. Harish-Chandra, Representations of semi-simple Lie groups IV, V, Amer. J. Math. 77, 743-777 (1955) ; 78, 1-41 (1956).
3. H.P. Jakobsen, Hermitian symmetric spaces and their unitary highest weight modules, J. Funct. Anal. 52 (1983), 385-412.
4. V.G. Kac, *Infinite dimensional Lie algebras*. Progress in Math. 44. Birkhäuser, Boston 1983.
5. F. Levstein, A classification of involutions of affine Kac-Moody algebras, Dissertation, MIT, 1983.
6. H. Rossi and M. Vergne, Analytic continuation of the holomorphic discrete series of a semi-simple Lie group, Acta Math. 136, 1-59 (1976).
7. N. Wallach, Analytic continuation of the discrete series II, Trans. Amer. Math. Soc. 251 (1979), 19-37.

Added in proof : The exceptional representations of sl_n ($\mathbb{C}$ $[z, z^{-1}]$) can be integrated to (projective) unitary representations of the group of polynomial loops on SU(n,1).

FORMAL INTEGRABILITY OF SYSTEMS OF PARTIAL DIFFERENTIAL EQUATIONS

J. GASQUI

Institut Fourier
Laboratoire de Mathématiques
B.P. 74
38402 ST MARTIN D'HERES - France

We review some basic definitions and results of the formal theory of overdetermined systems of partial differential equations. We have tried to present them the most simply possible, in a language for non-experts. As illustration, we treat a familiar example for physicists : the Einstein equations.

1. - FORMAL SOLUTIONS OF DIFFERENTIAL EQUATIONS.

First we recall some classical notations in analysis. Let U be an open set in $\mathbb{R}^n$. We denote by $C^\infty(U, \mathbb{R}^m)$ the space of differentiable functions on U, with values in $\mathbb{R}^m$. If $\alpha = (\alpha_1,\dots,\alpha_n) \in \mathbb{N}^n$ and $x = (x_1,\dots,x_n) \in \mathbb{R}^n$, we set, $|\alpha| = \alpha_1+\dots+\alpha_n$, $x^\alpha = x_1^{\alpha_1}\dots x_n^{\alpha_n}$; for partial derivatives, we shall write

$$D^\alpha f = \frac{\partial^{|\alpha|} f}{\partial x_1^{\alpha_1}\dots \partial x_n^{\alpha_n}} \quad , \text{ if } f \in C^\infty(U, \mathbb{R}) ,$$

and

$$D^\alpha f = (D^\alpha f^1,\dots,D^\alpha f^m) \ , \quad \text{if } f = (f^1,\dots,f^m) \in C^\infty(U,\mathbb{R}^m) .$$

Let x be a point of U, and k a non negative integer. A k-jet at x of function of U in $\mathbb{R}^m$ is the collection of the values at x of a function $f \in C^\infty(U, \mathbb{R}^m)$ and its partial derivatives up to order k. We denote by $J_k(U,\mathbb{R}^m)_x$ the vector space of these jets and by $j_k(f)(x)$ the k-jet at x of a function $f: U \to \mathbb{R}^m$. We can identify $J_k(U,\mathbb{R}^m)_x$ with the space of Taylor polynomials

$$F = \sum_{|\alpha|\le k} F_\alpha (y-x)^\alpha \ , \quad F_\alpha \in \mathbb{R}^m \ ,$$

of order k at x of functions of U into $\mathbb{R}^m$. If $k=0$, we have a linear isomorphism

$$J_0(U,\mathbb{R}^m)_x \simeq \mathbb{R}^m \ .$$

We also consider the set of all k-jets

$$J_k(U,\mathbb{R}^m) = \bigcup_{x\in U} J_k(U,\mathbb{R}^m)_x \ .$$

If we fix a point x^0 in U, we obviously have

$$J_k(U,\mathbb{R}^m) \simeq U \times J_k(U,\mathbb{R}^m)_{x^0} \ .$$

We denote by

$$\pi : J_k(U,\mathbb{R}^m) \longrightarrow U$$

the source mapping sending $F = \sum_{|\alpha|\le k} F_\alpha (y-x)^\alpha$ on x (or equivalently $j_k(f)(x)$ on x). For $\ell \ge 0$, we have a natural projection

$$\pi_k : J_{k+\ell}(U,\mathbb{R}^m) \longrightarrow J_k(U,\mathbb{R}^m)$$

which sends $F = \sum_{|\alpha|\le k+\ell} F_\alpha (y-x)^\alpha$ on $\sum_{|\alpha|\le k} F_\alpha (y-x)^\alpha$ (or equivalently a $(k+\ell)$-jet $j_{k+\ell}(f)(x)$ on $j_k(f)(x)$). The mapping

$$\pi_0 : J_k(U,\mathbb{R}^m) \longrightarrow \mathbb{R}^m$$

is called tarjet projection.

For $x \in U$, it is easily seen that

$$\mathrm{Ker}\left(\pi_0 : J_1(U,\mathbb{R}^m)_x \longrightarrow \mathbb{R}^m\right) \simeq \mathrm{Hom}(\mathbb{R}^n,\mathbb{R}^m) \ .$$

More generally,

$$\mathrm{Ker}\left(\pi_0 : J_k(U,\mathbb{R}^m)_x \longrightarrow \mathbb{R}^m\right)$$

is the space of Taylor polynomials at x, without constant term. On an other hand,

$$\mathrm{Ker}\left(\pi_{k-1} : J_k(U,\mathbb{R}^m)_x \rightarrow J_{k-1}(U,\mathbb{R}^m)_x\right)$$

is the space $\mathcal{P}^k_x(\mathbb{R}^n,\mathbb{R}^m)$ of homogeneous polynomials $F = \sum_{|\alpha|=k} F_\alpha(y-x)^\alpha$, of degree k on $\mathbb{R}^n$, with values in $\mathbb{R}^m$. Thus we have the exact sequence of vector spaces

$$(1.1)\quad 0 \rightarrow \mathcal{P}^k_x(\mathbb{R}^n,\mathbb{R}^m) \rightarrow J_k(U,\mathbb{R}^m)_x \xrightarrow{\pi_{k-1}} J_{k-1}(U,\mathbb{R}^m)_x \rightarrow 0 .$$

In the following, we shall set

$$\mathcal{P}^k(U,\mathbb{R}^m) = \bigcup_{x\in U} \mathcal{P}^k_x(\mathbb{R}^n,\mathbb{R}^m) .$$

Now we consider a differential operator

$$P : C^\infty(U,\mathbb{R}^m) \rightarrow C^\infty(U,\mathbb{R}^p)$$

of order $k \geq 1$. We have

$$(1.2)\quad (Pf)(x) = \Phi\left(x, \{(D^\alpha f)(x)\}_{|\alpha|\leq k}\right)$$

for $x \in U$ and $f \in C^\infty(U,\mathbb{R}^m)$, where Φ is some differentiable function with values in $\mathbb{R}^p$.

For $x \in U$, we introduce the subset

$$R_{k,x} = \left\{F = \sum_{|\alpha|\leq k} F_\alpha(y-x)^\alpha \ ;\ (PF)(x) = 0\right\}$$

of $J_k(U,\mathbb{R}^m)_x$, which is called the set of <u>formal solutions of order</u> k of P at x. More generally, for $\ell \geq 0$, the set $R_{k+\ell,x}$ of <u>formal solutions of order</u> $k+\ell$ of P at x is the subset of $J_{k+\ell}(U,\mathbb{R}^m)_x$ defined as follows : a jet $F = \sum_{|\alpha|\leq k+\ell} F_\alpha(y-x)^\alpha$ is in $R_{k+\ell,x}$ if the Taylor expansion at x of PF has no term of order $\leq \ell$. In other words, for $f \in C^\infty(U,\mathbb{R}^m)$, the jet $j_{k+\ell}(f)(x)$ belongs to $R_{k+\ell,x}$ if and only if $j_\ell(Pf)(x) = 0$, or equivalently

$$(D^\beta Pf)(x) = 0 , \text{ for } |\beta| \leq \ell .$$

According to (1.2), we remark that $R_{k,x}$ and $R_{k+\ell,x}$ are well-defined. We remark also that, in the non linear case, they can be empty. We set

$$R_k = \bigcup_{x\in U} R_{k,x} , \quad R_{k+\ell} = \bigcup_{x\in U} R_{k+\ell,x} .$$

In the classical terminology of the formal theory, R_k is the <u>differential equation asso-</u>

<u>ciated to</u> P , and $R_{k+\ell}$ the <u>ℓ^{th}-prolongation of</u> R_k .

For $\ell \geq 0$, we define a mapping

$$p_\ell(P) : J_{k+\ell}(U, \mathbb{R}^m) \longrightarrow J_\ell(U, \mathbb{R}^p)$$

sending $F = \sum_{|\alpha| \leq k+\ell} F_\alpha (y-x)^\alpha$ on the Taylor polynomial of order ℓ at x of PF (or, equivalently, a $(k+\ell)$-jet $j_{k+\ell}(f)(x)$ on $j_\ell(Pf)(x))$. It is clear that

$$R_{k+\ell} = \{F \in J_{k+\ell}(U, \mathbb{R}^m) \ ; \ p_\ell(P)F = 0\} \ .$$

Moreover, for $\ell \geq 0$, it is easily seen that the diagram

$$(1.3) \qquad \begin{array}{ccc} J_{k+\ell+1}(U, \mathbb{R}^m) & \xrightarrow{\ p_{\ell+1}(P)\ } & J_{\ell+1}(U, \mathbb{R}^p) \\ \Big\downarrow \pi_{k+\ell} & & \Big\downarrow \pi_\ell \\ J_{k+\ell}(U, \mathbb{R}^m) & \xrightarrow{\ p_\ell(P)\ } & J_\ell(U, \mathbb{R}^p) \end{array}$$

is commutative.

By definition, for $\ell \geq 0$, the natural projection

$$(1.4)_\ell \quad \pi_{k+\ell} : J_{k+\ell+1}(U, \mathbb{R}^m) \longrightarrow J_{k+\ell}(U, \mathbb{R}^m)$$

maps $R_{k+\ell+1}$ in $R_{k+\ell}$. But if the mappings $(1.4)_\ell$ are always surjective, this is not in general the case for the mappings

$$(1.5)_\ell \quad \pi_{k+\ell} : R_{k+\ell+1} \longrightarrow R_{k+\ell} \ .$$

Now we give the

DEFINITION 1.1. - <u>The operator</u> P <u>is formally integrable if the mapping</u> $(1.5)_\ell$ <u>is surjective, for all</u> $\ell \geq 0$.

If $F \in R_{k+\ell}$ is of the form

$$\pi_{k+\ell} G \ , \quad \text{with } G \in R_{k+\ell+1} \ ,$$

we say that the formal solution F of order $k+\ell$ can be extended to a formal solution of order $k+\ell+1$. Thus we see that P is formally integrable iff, for every $\ell \geq 0$, every formal solution of order $k+\ell$ of P can be extended to a formal solution of order $k+\ell+1$ of P.

For $x \in U$, we denote by $J_\infty(U,\mathbb{R}^m)_x$ the space of formal power series

$$\sum_{|\alpha|=0}^{+\infty} F_\alpha (y-x)^\alpha , \quad F_\alpha \in \mathbb{R}^m .$$

A formal power series solution of P at x, is an element F of $J_\infty(U,\mathbb{R}^m)_x$ such that PF vanishes to infinite order at x. The following obvious lemma says that formally integrable operators have many formal power series solutions.

LEMMA 1.1. - *Suppose that* P *is formally integrable, and that, for* $x \in U$, $F = \sum_{|\alpha| \leq k} F_\alpha (y-x)^\alpha$ *is a formal solution of order* k *of* P *at* x. *Then there exists a formal power series solution* $G = \sum_{|\alpha|=0}^{\infty} G_\alpha (y-x)^\alpha$ *of* P *at* x, *such that* $G_\alpha = F_\alpha$, *for* $|\alpha| \leq k$.

The operator P is said to be real analytic if the function Φ in (1.2) is real analytic. In this case, the following proposition guarantees the existence of genuine local solutions :

PROPOSITION 1.1. - *Suppose that* P *is a real analytic, formally integrable operator. For all* $x \in U$ *and* $F \in R_{k,x}$, *there exists a real analytic function* $f : U \to \mathbb{R}^m$, *defined in a neighborhood of* x, *such that*

$Pf = 0$ *and* $j_k(f)(x) = F$.

Remark. In the non-linear case, the operator P is not in general defined on the whole space $C^\infty(U,\mathbb{R}^m)$, but only on functions whose partial derivatives up to an order $k_0 \leq k$ satisfy an open condition (as typical example, we mention the "open" set of maximal rank functions). This situation doesn't change our presentation : everywhere, we have only to restrict our attention to the jets or Taylor polynomials satisfying the open condition.

2. - THE LINEAR CASE.

If we only use definition 1.1, the formal integrability involves an infinity of verifications. In this paragraph (for the linear case) and in the next (for the non-linear case), we present the Cartan-Kähler criterion which permits us to prove the formal integrability in a finite number of steps.

Now, we suppose that

$$P : C^\infty(U, \mathbb{R}^m) \longrightarrow C^\infty(U, \mathbb{R}^p)$$

is a linear differential operator of order k. Then, for $x \in U$ and $\ell \geq 0$, the mapping

$$p_\ell(P) : J_{k+\ell}(U, \mathbb{R}^m)_x \longrightarrow J_\ell(U, \mathbb{R}^p)_x$$

is linear and its kernel is $R_{k+\ell, x}$.

The symbol of P at $x \in U$ is the vector space

$$g_{k,x} = (R_{k,x}) \cap S^k_x(\mathbb{R}^n, \mathbb{R}^m) .$$

Thus an element of $g_{k,x}$ is just a formal solution $j_k(f)(x)$ of order k of P at x such that the partial derivatives $(D^\alpha f)(x)$ vanish for $|\alpha| < k$. For $\ell \geq 0$, the ℓth-prolongation of $g_{k,x}$ is

$$g_{k+\ell,x} = (R_{k+\ell,x}) \cap S^{k+\ell}_x(\mathbb{R}^n, \mathbb{R}^m) .$$

We set

$$g_k = \bigcup_{x \in U} g_{k,x} \quad , \quad g_{k+\ell} = \bigcup_{x \in U} g_{k+\ell,x} .$$

We introduce the symbol mapping

$$\sigma(P) : S^k(U, \mathbb{R}^m) \longrightarrow \mathbb{R}^p$$

of P, which is the restriction of

$$p_0(P) : J_k(U, \mathbb{R}^m) \longrightarrow \mathbb{R}^p .$$

For $\ell \geq 0$, we can also consider the restriction $\sigma_\ell(P)$ of $p_\ell(P)$ to $S^{k+\ell}(U, \mathbb{R}^m)$. Thanks to the commutativity of (1.3), $\sigma_\ell(P)$ takes its values in $S^\ell(\mathbb{R}^n, \mathbb{R}^p)$. The mapping $\sigma_\ell(P)$ is the ℓth-prolongation of $\sigma(P) = \sigma_0(P)$. Obviously, we have

$$g_{k+\ell,x} = \mathrm{Ker}\left(\sigma_\ell(P) : S^{k+\ell}_x(\mathbb{R}^n, \mathbb{R}^m) \longrightarrow S^\ell_x(\mathbb{R}^n, \mathbb{R}^p)\right) ,$$

for all $x \in U$.

Let $\{e_1, \ldots, e_n\}$ be a basis of $\mathbb{R}^n$ and $u_1, \ldots, u_n$ the associated linear coordinates. For $x \in U$ and $1 \leq j \leq n-1$, we denote by $S^k_x(\mathbb{R}^n, \mathbb{R}^m)_{\{e_1, \ldots, e_j\}}$ the subspaces of polynomials F in $S^k_x(\mathbb{R}^n, \mathbb{R}^m)$ which satisfy

$$\frac{\partial F}{\partial u_1} = \dots = \frac{\partial F}{\partial u_j} = 0 .$$

Then we set

$$(g_{k,x})_{\{e_1,\dots,e_j\}} = (g_{k,x}) \cap \rho^k_x(\mathbb{R}^n, \mathbb{R}^m)_{\{e_1,\dots,e_j\}} .$$

We say that P is <u>involutive</u> at a point x of U if there exists a basis $\{e_1,\dots,e_n\}$ of $\mathbb{R}^n$ such that

$$(2.1) \quad \dim g_{k+1,x} = \dim g_{k,x} + \sum_{j=1}^{n-1} \dim (g_{k,x})_{\{e_1,\dots,e_j\}} .$$

A basis such that (2.1) holds is called <u>quasi-regular</u> for $g_{k,x}$. We say that P is involutive if it is involutive at every point of U.

If P satisfies some regularity conditions, we can state the following classical criterion :

THEOREM 2.1. - <u>Suppose that</u>

(i) <u>the map</u> $\pi_k : R_{k+1} \to R_k$ <u>is surjective</u> ;

(ii) P <u>is involutive.</u>

<u>Then</u> P <u>is formally integrable.</u>

Therefore, if we can extend every formal solution or order k of P to a formal solution of order $k+1$, the involutivness insures that there is no more obstruction to extend formal solutions of higher orders. Although this question is very classical, it would be too long to explain here how the involutivness is related to the surjectivity of mappings $(1.5)_\ell$. It is more instructive to give a method for proving (i). For example, in geometric situations, the study of condition (ii) is that of a "flat" case and, on an other hand, (i) strongly depends of the geometric objects attached to the problem.

Let x be a point of U. We have the following exact and commutative diagram

$$
\begin{array}{ccccccccc}
 & & & & \rho_x^{k+1}(\mathbb{R}^n, \mathbb{R}^m) & \xrightarrow{\sigma_1(P)} & \mathrm{Hom}(\mathbb{R}^n, \mathbb{R}^p) & \xrightarrow{\tau} & K_x \longrightarrow 0 \\
 & & & & \downarrow & & \| & & \\
0 & \longrightarrow & R_{k+1,x} & \longrightarrow & J_{k+1}(U, \mathbb{R}^m)_x & \xrightarrow{p_1(P)} & J_1(U, \mathbb{R}^p)_x & & \\
 & & \downarrow \pi_k & & \downarrow \pi_k & & \downarrow \pi_0 & & \\
0 & \longrightarrow & R_{k,x} & \longrightarrow & J_k(U, \mathbb{R}^m)_x & \xrightarrow{p_0(P)} & \mathbb{R}^p & & \\
 & & & & \downarrow & & \downarrow & & \\
 & & & & 0 & & 0 & &
\end{array}
\tag{2.2}
$$

where, in the top row, K_x is the cokernel of $\sigma_1(P)$ and τ the canonical projection. By a standard diagram-chasing argument, we obtain the exact sequence

$$(2.3)\quad R_{k+1,x} \xrightarrow{\pi_k} R_{k,x} \xrightarrow{\Omega} K_x ,$$

where Ω is the linear map constructed by following the dotted path in (2.2). More precisely, if $F = j_k(f)(x)$ is a formal solution of order k of P, we consider the linear map $\psi : \mathbb{R}^n \to \mathbb{R}^p$ given by

$$\psi(e_i^0) = \frac{\partial}{\partial x^i}(Pf)(x) , \quad 1 \le i \le n ,$$

where $\{e_1^0, \ldots, e_n^0\}$ is the canonical basis of $\mathbb{R}^n$. Then we have

$$(2.4)\quad \Omega(F) = \tau(\psi) ,$$

and Ω is well-defined by (2.4).

Now, the mapping $\pi_k : R_{k+1,x} \to R_{k,x}$ is onto if and only if Ω vanishes identically. In the examples, the main point of this part of the theory is to obtain a "good interpretation" of the cokernel K_x and of τ.

3. - THE NON-LINEAR CASE.

We can easily obtain an analogue of Theorem 2.1 in the non-linear case.

We suppose that P is non-linear and defined on $C^\infty(U, \mathbb{R}^m)$. For $f \in C^\infty(U, \mathbb{R}^m)$, let

$$P_f' : C^\infty(U, \mathbb{R}^m) \longrightarrow C^\infty(U, \mathbb{R}^p)$$

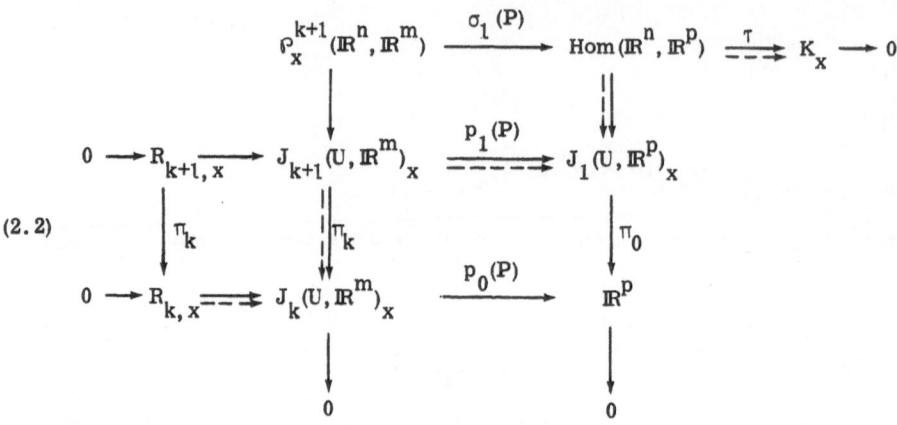

be the linearization of P along f : it is the linear differential operator of order k, defined by

$$P'_f(u) = \frac{d}{dt} P(f+tu)\big|_{t=0} ,$$

for all $u \in C^\infty(U, \mathbb{R}^m)$.

Let $x \in U$ and $f \in C^\infty(U, \mathbb{R}^m)$ such that $F = j_k(f)(x)$ is a formal solution of order k of P at x . By definition, the symbol $g_{k,F}$ of P at F is the symbol at x of the linearization P'_f of P along f . We say that $g_{k,F}$ is involutive if P'_f is involutive at x . We remark that, for $\ell \geq 0$, the mapping

$$\sigma_\ell(P'_f) : \mathcal{P}_x^{k+\ell}(\mathbb{R}^n, \mathbb{R}^m) \longrightarrow \mathcal{P}_x^{\ell}(\mathbb{R}^n, \mathbb{R}^p)$$

depends only of the k-jet of f at x , and, therefore, the notion of involutivness is well-defined in the non-linear case.

Now, P is said to be involutive if $g_{k,F}$ is involutive for all formal solution $F \in R_k$. With this new definition, the theorem 2.1 remains valid in the non-linear case : an involutive non-linear differential operator of order k , all of whose formal solutions of order k can be extended to formal solutions of order $k+1$, is formally integrable. Here also, to be rigorous, we need some obvious supplementary regularity conditions on P .

To complete this paragraph, we define non-linear analogues of (2.3) and (2.4). Let $x \in U$ and $f \in C^\infty(U, \mathbb{R}^m)$ such that $F = j_k(f)(x) \in R_{k,x}$. We have the following commutative diagram

$$\begin{array}{ccccc}
\mathcal{P}_x^{k+1}(\mathbb{R}^n, \mathbb{R}^m) & \xrightarrow{\sigma_1(P'_f)} & \mathrm{Hom}(\mathbb{R}^n, \mathbb{R}^m) & \xrightarrow{\tau_F} & K_F \longrightarrow 0 \\
 & & \downarrow & & \\
J_{k+1}(U, \mathbb{R}^m)_x & \xrightarrow{p_1(P)} & J_1(U, \mathbb{R}^p)_x & & \\
\downarrow \pi_k & & \downarrow \pi_0 & & \\
J_k(U, \mathbb{R}^m)_x & \xrightarrow{p_0(P)} & \mathbb{R}^p & , &
\end{array}$$

where, in the upper row, K_F is the cokernel of $\sigma_1(P'_f)$ and τ_F the canonical projection. By the same sort of methods that in the linear case, an element $\tilde{\Omega}(F)$ of K_F is well-defined by setting

$$\widetilde{\Omega}(F) = \tau_F(\psi) ,$$

where $\psi : \mathbb{R}^n \to \mathbb{R}^m$ is the linear map sending e_i^0 on $\frac{\partial P_f}{\partial x^i}(x)$, $i = 1,\dots,n$. Then we have the

PROPOSITION 3.1. - <u>Let</u> $F \in R_k$. <u>The following conditions are equivalent</u> :

(i) F <u>extends to a formal solution of order</u> $k+1$ <u>of</u> P ;

(ii) $\widetilde{\Omega}(F) = 0$.

When the operator P is quasi-linear, the vector space K_F depends only of the source of F . Then we have a vector space K_x and a mapping $\widetilde{\Omega} : R_{k,x} \to K_x$ such that the sequence

$$(3.1) \quad R_{k+1,x} \xrightarrow{\pi_k} R_{k,x} \underset{0}{\overset{\widetilde{\Omega}}{\rightrightarrows}} K_x$$

is exact in the sense that

$$\pi_k(R_{k+1,x}) = \{F \in R_{k,x} ; \widetilde{\Omega}(F) = 0\} .$$

In the linear case, $\widetilde{\Omega}$ is just the linear map Ω and (3.1) is the exact sequence (2.3).

4. - <u>THE INTRINSIC FORMALISM</u>.

In the classical examples furnished by Geometry, Mechanics or Physics, differential operators are more naturally defined on sections of bundles over manifolds, instead of vector-valued functions on an open set of $\mathbb{R}^n$. Hence, we rapidly traduce the preceding paragraphs in this intrinsic language.

Let X be a differentiable manifold of dimension n and E be a vector bundle of rank m over X . We denote by E_x the stalk of E at $x \in X$, and by $C^\infty(E)$ the space of sections of E over X . By using local coordinates, we easily define jets of functions between differentiable manifolds. In particular, for $k \geq 0$, we can consider the set of k-jets of sections of E , which is a new vector bundle $J_k(E)$ over X . We denote by $j_k(s)(x)$ the k-jet of s at x . For $\ell \geq 0$, let

$$\pi_k : J_{k+\ell}(E) \to J_k(E)$$

be the natural projection sending $j_{k+\ell}(s)(x)$ on $j_k(s)(x)$.

If T and T^* are respectively the tangent and cotangent bundles of X, we denote by S^kT^* the kth-symmetric product of T^*. Thus the sections of the vector bundle $S^kT^*\otimes E$ are the symmetric k-forms on X, with values in E (at a point $x \in X$, $(S^kT^*\otimes E)_x = S^kT^*_x\otimes E_x$ is just the space of homogeneous polynomials of degree k on T_x, with values in E_x). There is a canonical monomorphism of vector bundles

$$\varepsilon : S^kT^*\otimes E \longrightarrow J_k(E)$$

such that the sequence of vector bundles

$$(4.1) \quad 0 \longrightarrow S^kT^*\otimes E \xrightarrow{\varepsilon} J_k(E) \xrightarrow{\pi_{k-1}} J_{k-1}(E) \longrightarrow 0$$

is exact. We recall that ε is well-defined in the following way : for $x \in X$, $s \in C^\infty(E)$ and functions $f_1,\dots,f_k$ of X in $\mathbb{R}$ and vanishing at x, we have

$$\varepsilon\left(df_{1,x}\cdot\ldots\cdot df_{k,x}\otimes s(x)\right) = j_k(f_1\ldots f_k s)(x) .$$

In the left member of the above equality, $df_{1,x}\cdot\ldots\cdot df_{k,x}$.

If $U\subset X$ is an open set of coordinates (identified with a subset of $\mathbb{R}^n$) and trivializing for the bundle E, then we have isomorphisms

$$J_k(E)|_U \simeq J_k(U,\mathbb{R}^m) ,$$
$$P^k_x(\mathbb{R}^n,\mathbb{R}^m) \simeq S^kT^*_x\otimes E_x , \quad x \in U .$$

Moreover, via these isomorphisms, the sequences (1.1) and (4.1) at $x\in U$ are the same.

Let F be another vector bundle of rank p over X, and

$$P : C^\infty(E) \longrightarrow C^\infty(F)$$

be a linear differential operator of order k. For $\ell \geq 0$, we consider the morphism of vector bundles

$$p_\ell(P) : J_{k+\ell}(E) \longrightarrow J_\ell(F)$$

sending $j_{k+\ell}(s)(x)$ on $j_\ell(Ps)(x)$, for all $x \in X$ and $s \in C^\infty(E)$. We set

$$R_k = \text{Ker } p_0(P) \quad \text{and} \quad R_{k+\ell} = \text{Ker } p_\ell(P) .$$

We say that R_k is the linear differential equation of order k associated to P and $R_{k+\ell}$ its ℓth-prolongation.

The <u>symbol mapping</u> of P is the morphism

$$\sigma(P) : S^k T^* \otimes E \longrightarrow F$$

of vector bundles obtained by restriction of $p_0(P)$. In the same way, for $\ell \geq 0$, the ℓth-prolongation of $\sigma(P)$ is the morphism

$$\sigma_\ell(P) : S^{k+\ell} T^* \otimes E \longrightarrow S^\ell T^* \otimes F$$

obtained by restriction of $p_\ell(P)$. The kernel Ker $\sigma(P)$ is the <u>symbol</u> g_k of R_k and $g_{k+\ell}$ = Ker $\sigma_\ell(P)$ its ℓth-prolongation.

If we restrict our attention to an open set of coordinates in X , trivializing for E and F , we recover all the objects introduced in the paragraph 2.

Now, we can develop the formal theory as in paragraph 2. For example, the diagram (2.2) becomes the following exact and commutative diagram

$$(4.2)\qquad \begin{array}{ccccccc}
 & & S^{k+1}T^*\otimes F & \xrightarrow{\sigma_1(P)} & T^*\otimes F & \xrightarrow{\tau} & K \longrightarrow 0 \\
 & & \downarrow \epsilon & & \downarrow \epsilon & & \\
R_{k+1} & \longrightarrow & J_{k+1}(E) & \xrightarrow{p_1(P)} & J_1(F) & & \\
\downarrow \pi_k & & \downarrow \pi_k & & \downarrow \pi_0 & & \\
R_k & \longrightarrow & J_k(E) & \xrightarrow{p_0(P)} & F & &
\end{array}$$

of vector bundles, where K = Coker $\sigma_1(P)$ and τ is the canonical projection. We want to compute the morphism $\Omega : R_k \rightarrow K$, given by diagram-chasing and such that the sequence

$$R_{k+1} \xrightarrow{\pi_k} R_k \xrightarrow{\Omega} K$$

is exact. We begin by the

LEMMA 4.1. - <u>Every connection</u> ∇ <u>on the vector bundle</u> F <u>defines a splitting of the sequence</u>

$$0 \longrightarrow T^* \otimes F \xrightarrow{\epsilon} J_1(F) \xrightarrow{\pi_0} F \longrightarrow 0 .$$

This splitting is nothing but $p_0(\nabla) : J_1(F) \rightarrow T^*\otimes F$. This lemma is very useful, because, in examples, there is often a natural connection on F .

Let ∇ be any given connection in F . If $x \in X$ and $s \in C^{\infty}(E)$ verify $(Ps)(x) = 0$, we easily deduce from Lemma 4.1 that

(4.3) $\Omega(j_k(s)(x)) = \tau(\nabla Ps)(x)$.

For involutivness, we simply notice that a quasi-regular basis for the symbol $g_{k,x}$ at $x \in X$, is a basis of the tangent space T_x and no more of a fixed vector space as in paragraph 2. With this remark, all the definitions can be easily transposed.

Finally, we can also easily give an intrinsic formalism for a non linear operator $P : C^{\infty}(E) \to C^{\infty}(F)$. We simply notice that the bundles E and F are only fibrations, so the notations can be more complicated. For instance, we have to consider the vertical bundles $V(E)$ and $V(F)$ of E and F .

5. - THE EINSTEIN EQUATIONS.

We suppose that X is manifold of dimension $n \geq 3$. If q is a Riemannian metric on X , we denote by $\mathrm{Ric}(q)$ its Ricci curvature, which is a section of the bundle S^2T^* of quadratic forms on X . The operator $q \mapsto \mathrm{Ric}(q)$ is a second order non-linear differential operator, and we want to study the formal integrability of the linearization

$$\mathrm{Ric}'_g : C^{\infty}(S^2T^*) \longrightarrow C^{\infty}(S^2T^*)$$

of this operator along a fixed Riemannian metric g on X . We have

$$\mathrm{Ric}'_g(h) = \frac{d}{dt}\mathrm{Ric}(g+th)\big|_{t=0} ,$$

for all $h \in C^{\infty}(S^2T^*)$.

We consider the diagram (4.2), with $E = F = S^2T^*$, $k = 2$ and $P = \mathrm{Ric}'_g$. The first step is to determine K and τ . Let $\nabla^g = \nabla$ be the Levi-Civita connection of g . We introduce the "divergence of symmetric tensors" which is the first order linear differential operator

$$\mathrm{Div}_g : C^{\infty}(S^2T^*) \longrightarrow C^{\infty}(T^*) ,$$

given in local coordinates by

$$(\mathrm{Div}_g h)_i = g^{jk}(\nabla_j h)_{ik} .$$

By using elementary representation theory of the orthogonal group, one can identify K

with T^* and

$$\tau : T^* \otimes S^2T^* \longrightarrow T^*$$

with the symbol $\sigma(\mathrm{Div}_g)$ of the divergence.

Let $h \in C^\infty(S^2T^*)$ and $x \in X$ such that $\mathrm{Ric}'_g(h)(x) = 0$. According to Lemma 4.1 and (4.3), we obtain

(5.1) $\Omega(j_2(h)(x)) = (\mathrm{Div}_g \mathrm{Ric}'_g h)(x)$.

Now we recall the Bianchi identity

(5.2) $\mathrm{Div}_q \mathrm{Ric}(q) = 0$,

which holds for all Riemannian metric q on X. If we "linearize" the Bianchi identity, we easily find that $\Omega = 0$ if and only if $\mathrm{Ric}(g) = 0$. Thus, we can extend formal solutions of order 2 of Ric'_g to formal solutions of order 3, if and only if the metric g is Ricci-flat.

Next, one can prove that, for $x \in X$, every orthonormal basis of T_x, with respect to g, is quasi-regular for the symbol $g_{2,x}$ of Ric'_g at x. The computations for involutivness at x depend only of the value of g at x, and therefore are independent of the choice of the metric. In fact, with local coordinates, they can be done in the standard Euclidean space.

Finally Ric'_g is formally integrable if and only if $\mathrm{Ric}(g) = 0$. More generally, we can consider, for $\lambda \in \mathbb{R}$, the operator

$$P_g^\lambda : C^\infty(S^2T^*) \longrightarrow C^\infty(S^2T^*)$$
$$h \longmapsto \mathrm{Ric}'_g(h) - \lambda h ,$$

which is the linearization along g of the non-linear Einstein operator

$$q \longmapsto \mathrm{Ric}(q) - \lambda q .$$

Here also we find that P_g^λ is formally integrable if and only if g satisfies the Einstein equations

$$\mathrm{Ric}(g) = \lambda g .$$

Moreover, in this case, Div_g is the compatibility condition of P_g^λ.

Thanks to the Bianchi identity for the Ricci tensor, the non-linear Ricci curvature (or

Einstein) operator is itself formally integrable. From this observation, we can obtain the following result.

Let g_0 be a scalar product on $\mathbb{R}^n$ and R^0 be a (0,4)-tensor at the origin, and having the symmetrics of a curvature tensor. Suppose that

(5.3) $g_0^{ik} R^0_{ijk\ell} = \lambda (g_0)_{j\ell}$,

for $\lambda \in \mathbb{R}$. Then there exists an Einstein metric g , defined in a neighborhood of the origin such that

a) $Ric(g) = \lambda g$;

b) $g(0) = g_0$;

c) the Riemann curvature tensor of g at the origin is R^0 .

In other words, there is no obstruction to extend the algebraic Einstein condition (5.3) to germs. For instance, when $n \geq 4$, Ricci flat metric are generically non flat.

Remark. The fact that the operator Ric is formally integrable means only that the equation $Ric(g) = 0$ is locally solvable and not that $Ric(g) = S$ is solvable for any given S . D. De Turck has proved that this last equation is solvable in a neighborhood of a point x of X , if and only if $S(x)$ is a non-degenerate quadratic form.

M. Dubois-Violette has etablished the same sort of relations between the Yang-Mills equations and their linearizations. One can also prove that the non-linear Yang-Mills equations are formally integrable.

6. - BIBLIOGRAPHY.

The reader can consult :

For linear equations :

H. GOLDSCHMIDT, Existence theorems for analytic linear partial differential equations, Ann. of Math. (1967), 246-270.

D.C. SPENCER, Overdetermined systems of linear partial differential equations, Bull. A.M.S. 75 (1969), 179-239.

For non-linear equations :

H. GOLDSCHMIDT, Integrability criteria for systems of non linear partial differential equations, J. Differential Geometry, 1 (1967), 269-307.

For the proof of existence of local solutions in the analytic case :

B. MALGRANGE, Equations de Lie II, J. Differential Geometry, 7 (1972), 117-141.

For examples :

M. DUBOIS-VIOLETTE, The theory of overdetermined linear systems and its applications to non-linear field equations, (Lectures at the Banach Center, Warsaw, October 83).

J. GASQUI, Sur l'existence locale d'immersions à courbure scalaire donnée, Math. Annalen, 219 (1976), 123-126.

J. GASQUI, Sur l'existence locale de certaines métriques riemanniennes plates, Duke Math. J. 46 (1979), 109-118.

For general applications in Geometry, the introduction of :

J. GASQUI - H. GOLDSCHMIDT, Déformations infinitésimales des structures conformes plates, Progress in Math., Birkhaüser (1984).

For Einstein equations or Ricci curvature :

M. DUBOIS-VIOLETTE, Remarks on the local structure of Yang-Mills and Einstein equations, Phys. Lett. 131 B (1983) 323-326.

J. GASQUI, Sur la résolubilité locale des équations d'Einstein, Compositio Math. 47 (1982), 43-69.

D. DE TURCK, Existence of Metrics with prescribed Ricci curvature : Local theory, Inventiones Math. 65 (1983), 327-349.

QUANTUM INTEGRABILITY AND CLASSICAL INTEGRABILITY

Jarmo Hietarinta
Wihuri Physical Laboratory and Department of Physical Sciences
University of Turku
20500 Turku 50
FINLAND

Abstract: *Quantum integrability and classical integrability are studied from an algebraic viewpoint. In particular we discuss the problem of constructing a quantum integrable system when a classically integrable system is given.*

1 INTRODUCTION

Integrable dynamical systems have been studied quite actively in the recents years. Most of the time the system has been a classical one. Here we will discuss the differences of classical and quantum integrability from an algebraic viewpoint. (For more details see [1].)

Let us start by defining the type of systems that will be studied. The concepts that will be introduced will work for any discrete N-dimensional Hamiltonian systems. However, most of our examples will be from two-dimensional systems, and typically of the form

$$H = \tfrac{1}{2}(p_x{}^2 + p_y{}^2) + V(x,y) \tag{1}$$

In the literature there are many definitions for integrability so let us next agree on the definition that will be used:

> Definition A Hamiltonian system of N degrees of freedom is called *integrable* if there a N-1 globally defined, independent functions, which commute with each other and with the Hamiltonian with respect to the Poisson bracket, i.e
>
> $\{H,I_i\}_{PB}=0,\ i=2,\ldots,N$
> $\{I_i,I_j\}_{PB}=0,\ i,j=2,\ldots,N$

The I_i's of the above definition are variously called *constants of motion, invariants, or integrals of motion*. The term constant of motion might be most appropriate as it stresses the fact that I_i can take any constant value, determined by the initial conditions, and that its value will stay constant on the trajectories.

The above definition was for classical mechanics. What about quantum mechanics? Of course classical and quantum mechanics differ in many ways and some concepts cannot be used in both. In the algebraic sense they are quite close to each other so that the above algebraic definition of integrability can be transported to quantum mechanics without any difficulty:

> Definition: A quantum mechanical Hamiltonian system is called *quantum integrable* if there are N-1 independent, globally defined operators, commuting with each other and the Hamiltonian, i.e.
>
> $[H,I_i]=0, \quad i=2,\ldots,N$
>
> $[I_i,I_j]=0, \quad i,j=2,\ldots,N$

Suppose then that we are given a classically integrable system, for example a Hamiltonian of the type (1), and the required second invariant I. We next make the usual operator substitutions for the momenta and get operators H and I (for I we could have ordering ambiguities!). Do we now have $[H,I]=0$? Not always, although that has been claimed. In the following we shall discuss the various problems that we are faced with when we try to construct a quantum integrable system from a classically integrable one.

2 QUANTUM MECHANICS WITH C-NUMBERS

In the following we will mostly be doing quantum mechanics with c-number functions rather than with operators. There are two reasons for doing it here: 1) Since we are comparing classical and quantum mechanics it is useful to have similar objects in both, 2) for computer algebra systems, where many of the necessary computations were done, it is more convenient to have commuting objects.

We are here only interested in algebraic aspects of quantum mechanics and therefore we only need a replacement for the objects (operators) and for the only algebraic operation between them (commutator). There are many isomorphisms that could be used (corresponding loosely speaking to various ordering rules) of which we will only use the Weyl-Wigner rule. The same result is obtained using star-products.

In any transformation rule (defined by a function F(x,y), see [2]) a c-number function A(p,q) is translated into an operator $\boldsymbol{A}(\boldsymbol{p},\boldsymbol{q})$ using the (formal) integral

$$\mathbf{A}(\mathbf{p},\mathbf{q}) = \int d^n p\, d^n q\, d^n x\, d^n y\, (2\pi\hbar)^{-2n} F(x,y) A(p,q) \exp\left(\frac{i}{\hbar}(x\cdot(\mathbf{p}-p)+y\cdot(\mathbf{q}-q))\right) \qquad (2)$$

In the Weyl rule F(x,y)=1 and for the standard ordering rule ($\boldsymbol{p}$'s to the right) $F(x,y)=\exp(-ixy/2\hbar)$.

The easiest way to obtain the inverse transformation is first to write the operator in the standard ordering and then replace all operators $\boldsymbol{p}$, $\boldsymbol{q}$ by c-numbers p, q. The c-number function As obtained this way is the "standard ordering rule" representation of the given operator. To obtain the "Weyl rule" version Aw of the same operator one uses the transformation

$$A_W(p,q) = \exp\left(\frac{i\hbar}{2}\sum_i \frac{\partial^2}{\partial p_i \partial q_i}\right) A_S(p,q)$$

The major reason for using the Weyl rule is that the associated replacement for the commutator, the Moyal bracket, is quite convenient. It is defined by

$$\{A,B\}_{MB} = A(p,q)\, \frac{2}{\hbar} \sin\left\{\frac{\hbar}{2}\,(\overleftarrow{\partial}_q\cdot\overrightarrow{\partial}_p - \overleftarrow{\partial}_p\cdot\overrightarrow{\partial}_q)\right\}\, B(p,q) \qquad (3)$$

First of all note that the Moyal bracket reduces to the Poisson bracket when $\hbar \to 0$. Another property of considerable importance is that both brackets respect time reversal parity. If the Hamiltonian is even in momenta (e.g. (1)) then this parity conservation implies that both the classical and quantum c-number invariants will also have a definite parity. i.e. they will be either even or odd in p.

3 AUTOMATIC EXTENSION FROM CM TO QM

The Moyal and Poisson brackets give different results only if the higher order terms in (3) contribute. If the Hamiltonian is of type

(1) then in the higher order terms all x-derivatives must operate on V and the p-derivatives on I. In particular if the invariant is at most second order in momenta the higher order terms in the Moyal bracket do not contribute at all, and in that case classical integrability does imply quantum integrability. For example the following classically integrable system is also quantum integrable and the c-number representative of the quantum second invariant is identical to the classical one.

$$H = \tfrac{1}{2}(p_x^{\,2} + p_y^{\,2}) + 2x^3 + xy^2$$

$$I = p_y(yp_x - xp_y) + x^2y^2 + \tfrac{1}{4}y^4$$

The classical invariant can be used as the c-number quantum invariant also in some other cases. For example in systems whose classical integrability follows from the existence of a Lax pair it turns out that the variables appear in such combinations that also quantum integrability can be obtained [3].

As an example consider the Toda lattice whose Hamiltonian in three dimensions reads

$$H = \tfrac{1}{2}(p_x^{\,2} + p_y^{\,2} + p_z^{\,2}) + e^{x-y} + e^{y-z} + e^{z-x} \tag{4}$$

Its invariants from a Lax pair are

$$\begin{aligned} I_1 &= p_x + p_y + p_z \\ I_2 &= p_xp_y + p_yp_z + p_zp_x - e^{x-y} - e^{y-z} - e^{z-x} \\ I_3 &= p_xp_yp_z - p_xe^{y-z} - p_ye^{z-x} - p_ze^{x-y} \end{aligned} \tag{5}$$

The linear and quadratic invariants (which can be combined to make the Hamiltonian) work also in quantum mechanics due to the above mentioned general arguments. In this case also the cubic invariant commutes with the others in the sense of Moyal. This time the p term does not contribute due to its special form (there are no correcponding product terms xyz in the potential).

For many systems it is therefore true that classical integrability implies quantum integrability without any problems. However, this is not always true and in the following sections we will give examples where the situation is not so simple.

4 QUANTUM CORRECTIONS TO THE SECOND INVARIANT

When the invariant is sufficiently complicated the higher order terms in the Moyal bracket do contribute. As an example consider the following Hamiltonian

$$H = \tfrac{1}{2}(p_x^{\,2} + p_y^{\,2}) + \frac{16}{3}\,y^3 + xy^2 + \mu x^{-2} \tag{6}$$

It is classically integrable and its second invariant is

$$I_{cl} = p_x^{\,4} + 4(x^2y + \mu x^{-2})p_x^{\,2} - \tfrac{4}{3}x^3 p_x p_y - \tfrac{4}{3}x^4 y^2 + \tfrac{8}{3}\mu y - \tfrac{2}{9}x^6 + 4\mu^2 x^{-4} \tag{7}$$

If we now calculate the Moyal bracket between (6) and (7) it does not vanish but an $\hbar^2$-term will remain. To correct the situation let us try a modified invariant

$$I_{qu} = I_{cl} + \hbar^2 D(x,y) \tag{8}$$

The equations for D that follow from $\{H, I_q\}_{MB} = 0$ are simply

$$\begin{aligned} D_y &= 0, \\ D_x &= 24x^{-5}\mu \end{aligned} \tag{9}$$

Thus we can solve for D and obtain the c-number quantum invariant

$$I_{qu} = I_{cl} - \hbar^2 6\mu x^{-4} \tag{10}$$

A similar situation appears with many other invariants. Usually the correction terms are necessary only for the lower order terms in p. For further details see [1].

There are several methods by which one can try to change the $\hbar^2$-terms in the invariant. For example one could try to look at the same operator in different ordering rules. Another method is adding powers of the Hamiltonian to the invariant. Note that the c-number representative of H^2 is not H^2 but there will be additional $\hbar^2$ corrections. Both of these will change the correction term D discussed above, however, there is no unique ordering rule where all c-number representatives are without such quantum corrections.

5 DEFORMATIONS NEEDED IN THE HAMILTONIAN

Even these corrections to the second invariant are not always sufficient to make a classically integrable system quantum integrable. As the by now standard example let us take the Holt Hamiltonian

$$H_{cl} = \tfrac{1}{2}(p_x{}^2 + p_y{}^2) + \tfrac{3}{4}\, x^{4/3} + y^2 x^{-2/3} \tag{11}$$

which is classically integrable with the second invariant[4]

$$I_{cl} = p_y{}^3 + \tfrac{3}{2} p_y p_x{}^2 + (\tfrac{9}{2} x^{4/3} + 3x^{-2/3} y^2) p_y + 9x^{1/3} y p_x \tag{12}$$

If we now try a second invariant with the same leading part it turns out that the new system of equations has no solution.

Let us next try $\hbar^2$ corrections for both the second invariant *and* the Hamiltonian, i.e.

$$\begin{aligned} H_{qu} &= H_{cl} + \hbar^2\, A(x,y) \\ I_{qu} &= I_{cl} + \hbar^2 (\, B(x,y) p_x + C(x,y) p_y\,) \end{aligned} \tag{13}$$

The equations for A,B and D following from $\{H,I\}_{MB}=0$ are now

$$2B_x = 3A_y\,,$$

$$C_x + B_y = 3A_x,$$

$$C_y = 3A_y,$$

$$3A_{xxy} + 2A_{yyy} - 8A_x B - 8A_y C = 0,$$

$$9xyA_x + (\tfrac{9}{2}x^2 + 3y^2)A_y + 2yC + (x - \tfrac{2}{3}x^{-1}y^2)B - \tfrac{5}{6}x^{-2}y = 0$$

and they can indeed be solved with the final result

$$H_{qu} = H_{cl} - \frac{5}{72}\hbar^2 x^{-2},$$
$$I_{qu} = I_{cl} - \frac{5}{72}\hbar^2 x^{-2} p_y. \qquad (14)$$

Calculations similar to the above can be carried out with the other Holt potentials whose invariants are of order 4 and 6 in p. Also in these cases we have found that the correction term to the Hamiltonian is the same, $-5/72\,\hbar^2 x^{-2}$ [5].

Another example of this type is the Fokas-Lagerstrom potential

$$H_{cl} = \tfrac{1}{2}(p_x{}^2 + p_y{}^2) + (xy)^{-2/3} \qquad (15)$$

which is classically integrable [6]. For quantum integrability the potential must be deformed by [7]

$$H_{qu} = H_{cl} - \frac{5}{72}\hbar^2(x^{-2} + y^{-2}) \qquad (16)$$

and then the second invariant is

$$I_{qu} = p_x p_y (xp_y - yp_x) + 2(xy)^{-2/3}(xp_x - yp_y) - \frac{5}{36}\hbar^2(xy^{-2}p_x - yx^{-2}p_y) \qquad (17)$$

It is quite surprising that the deformations (14) and (16) are of the same type, including numerical factors. Also the correction seems to be associated with the variable that appears with a fractional power in the rest of the potential. This led us to try, without success, to eliminate the correction term by a canonical point transformations [1] which are also known to produce $\hbar^2$ corrections.

Recently we have been able to solve this intriguing question by applying a canonical transformation together with *a change in the time variable* [8]. It turns out such a combined transformation eliminates the quantum deformation and turns the system into another one which is integrable without any quantum deformations in the potential. What is really surprising is that the Holt potentials are this way transformed into the well know integrable Henon-Heiles potentials!

6 CONCLUSIONS

We have here seen that quantum integrability is by no means a trivial consequence of classical integrability. We have observed in examples three basic types of behavior: 1) The classical invariant can be taken as the c-number quantum invariant, 2) the classical invariant must be modified by $\hbar^2$ terms, 3) also the Hamiltonian must be deformed. We would like to emphasize that there is no way that the quantum deformations could all be explained by an ordering rule (including the star product).

It is easy to imagine many other types of behavior when a classical system is quantized, however, all known models with polynomial (in momenta) invariants fit into the three categories above. Recently some classical models with rational or transcendental invariants have been found [9], but it is still open whether a quantum version of these models can be constructed.

REFERENCES

[1] J. Hietarinta, J. Math. Phys. **25**,1833(1984).

[2] J. Shewell, Am. J. Phys. **27**,16(1959); J. Hietarinta, Phys. Rev. D**25**,2103(1982).

[3] P. Calogero, O. Ragnisco and C. Marchioro, Lett. Nuovo Cimento **13**,383(1975); A.M. Perelomov, Sov. J. Part. Nucl. 10,336(1979).

[4] C.R. Holt, J. Math. Phys. **23**,1037(1982).

[5] J. Hietarinta, Phys. Rev. A**28**,3670(1983).

[6] A.S. Fokas and P. Lagerstrom, J. Math. Anal. Appl. **74**,325(1980).

[7] J. Hietarinta, B. Dorizzi, B. Grammaticos and A. Ramani, C. R. Acad. Sci. Paris Serie II, **297**,791(1983).

[8] J. Hietarinta, B. Grammaticos, B. Dorizzi and A. Ramani, "Coupling constant metamorphosis and duality between integrable Hamiltonian systems", preprint.

[9] J. Hietarinta, Phys. Rev. Lett. **52**,1057(1984).

BÄCKLUND TRANSFORMATIONS FOR NONLINEAR FIELD EQUATIONS*

B. Kent Harrison
Department of Physics and Astronomy
Brigham Young University
Provo, Utah 84602

1. Introduction.

Bäcklund transformations, first discovered a century ago by A. V. Bäcklund [1], have been found to be of renewed interest in recent years due to the discovery of solitons ([23]; see also [19] for example). However, finding such transformations is difficult. Only one even partly systematic method for finding Bäcklund transformations is known, that due to Wahlquist and Estabrook [20, 21]. This method will be discussed here, along with various generalizations.

A Bäcklund transformation (BT), or strictly speaking an auto-Bäcklund transformation, will be taken here simply to mean a set of auxiliary equations, associated with a field equation (or set of field equations), such that one solution of the field equation may be used to generate a second solution, usually by means of integration or solving a simple differential equation. (See [3] for example.) A well-known example is the BT for the sine-Gordon (s-G) equation [15],

$$\phi_{uv} = \sin\phi \tag{1}$$

which is

$$\begin{aligned} \phi_u' &= \phi_u + 2k^{-1} \sin \tfrac{1}{2}(\phi' + \phi) \\ \phi_v' &= -\phi_v + 2k \sin \tfrac{1}{2}(\phi' - \phi) \end{aligned} \tag{2}$$

where ϕ and ϕ' are solutions of Eq. (1), k is a constant, and subscripts mean partial derivatives. It is clear that, if ϕ is known, then it is in principle possible to solve these equations for ϕ'.

While BT's are known for only a small number of field equations, and probably do not exist for most partial differential equations, they still have attracted much interest. Field equations admitting BT's are often nonlinear, with no other systematic means of solution; they may occur in a rather wide variety of contexts; and they may serve as simple approximations for other equations. Thus BT's are important for their study.

*This paper is based on work supported by the National Science Foundation under grant PHY-8308055.

2. Differential equations as differential forms.

The Wahlquist-Estabrook (WE) method [20, 21] requires one to find a function (or functions), called a pseudopotential. This is then used as an auxiliary function in equations for a new solution of the given field equation(s), in terms of the old one, and provides just the freedom needed in that search. Finding the pseudopotential may be done with traditional mathematics, but it is useful and instructive to work in differential forms. Accordingly, we first discuss writing differential equations as differential forms. Since this is discussed elsewhere [14, 12], we merely write the s-G equation in this manner.

In Eq. (1), we put $r = \phi_u$. Then $r_v = \sin\phi$. We now write two 2-forms in a space with variables u, v, r, and ϕ:

$$\alpha = d\phi\wedge dv - r du\wedge dv \tag{3}$$

$$\beta = dr\wedge du - \sin\phi \, dv\wedge du$$

If we restrict these forms to a solution of Eq. (1), then we have $r = r(u, v)$ and $\phi = \phi(u, v)$ (restriction to a space with only parameters u, v), and

$$\alpha = (\phi_u - r) \, du\wedge dv$$

$$\beta = (r_v - \sin\phi) \, dv\wedge du$$

which vanish by the field equations. Thus Eq. (1) is equivalent to requiring $\alpha = \beta = 0$ ("annulling" α and β) when restricted to uv-space.

Alternatively, we could define a set of 1-forms ξ_i:

$$\xi_1 = du \qquad \xi_2 = r du \qquad \xi_3 = \sin\phi \, dv \qquad \xi_4 = \cos\phi \, dv \tag{4}$$

We note that

$$\xi_1\wedge\xi_2 = \xi_3\wedge\xi_4 = 0 \tag{5}$$

identically, and that $d\xi_1 = 0$, $d\xi_2 = \beta + \xi_3\wedge\xi_4$, $d\xi_3 = \alpha\cos\phi + \xi_2\wedge\xi_4$, and $d\xi_4 = -\alpha\sin\phi - \xi_2\wedge\xi_3$. Thus, when α and β are annulled, we have

$$d\xi_1 = 0 \qquad d\xi_2 - \xi_3\wedge\xi_1 = 0 \tag{6}$$

$$d\xi_3 - \xi_2\wedge\xi_4 = 0 \qquad d\xi_4 + \xi_2\wedge\xi_3 = 0.$$

Eqs. (5) and (6) form an alternate set of 2-forms, which may be used to recover Eq. (1).

We denote the set of forms which are equivalent to the field equations by I (an "ideal" of forms). In the above case, I is alternatively taken to be α and β, or to be the forms in (5) and (6). If I may be constructed from a set of 1-forms ξ_i, i = 1 to K, with constant coefficients ("CC"), as in (5) and (6) above, we may call it a "CC ideal" [11].

The KdV equation

$$u_t + 12uu_x + u_{xxx} = 0 \tag{7}$$

can be written as the vanishing of three 2-forms

$$\alpha_1 = du \wedge dt - z dx \wedge dt$$
$$\alpha_2 = dz \wedge dt - p dx \wedge dt \tag{8}$$
$$\alpha_3 = -du \wedge dx + dp \wedge dt + 12uz\, dx \wedge dt$$

($z = u_x$, $p = z_x$, $u_t = -12uz - p_x$), or as the vanishing of a set of thirteen 2-forms constructed from a set of seven 1-forms η_i [5, 21].

The equation of general relativity [4, 7, 9, 11]

$$(\text{ReE})\ \nabla^2 E = (\nabla E)^2 \tag{9}$$

(Ernst equation when $\lambda = -1$, cylindrical wave equation when $\lambda = +1$), where

$$\nabla^2 E = E_{rr} + S^{-1}(S_r E_r + S_z E_z) + \lambda E_{zz},$$
$$(\nabla E)^2 = E_r{}^2 + \lambda E_z{}^2,$$

and S satisfies

$$S_{rr} + \lambda S_{zz} = 0, \tag{10}$$

may be written

$$d(Sf^{-1} * df) + Sf^{-1}\, d\phi \wedge * \, d\phi = 0 \tag{11}$$
$$d(Sf^{-2} * d\phi) = 0$$

and (10) is

$$d(*dS) = 0 \tag{12}$$

where $E = f + i\phi$ and $*$ is a linear Hodge operator defined by $*dr = dz$, $*dz = \lambda dr$. f and ω are metric coeffients and ϕ is defined by $*d\phi = S^{-1}f^2 d\omega$. We define R and η by the equations $* \, dS = dR$ and $* \, df = S^{-1}f(d\eta + \omega d\phi)$. Then define a set of 1-forms ξ_i:

$$\xi_1 = f^{-1}d\phi \qquad \xi_4 = f^{-1}df$$
$$\xi_2 = S^{-1}f d\omega \qquad \xi_5 = S^{-1}dS \tag{13}$$
$$\xi_3 = S^{-1}(d\eta + \omega d\phi) \qquad \xi_6 = S^{-1}dR$$

We can write a set of equations for the $d\xi_i$ ($d\xi_1 = \xi_1 \wedge \xi_4$, etc.) and a set which are to be annulled ($\xi_3 \wedge \xi_1 - \xi_2 \wedge \xi_4 = 0$, etc.) but it is perhaps more interesting to define 1-forms η_i as follows, where $k = \sqrt{\lambda}$ (= 1 or i)

$$\eta_{1(2)} = \xi_4 \pm i\xi_1 + k\xi_3 \pm ik\xi_2$$
$$\eta_{3(4)} = \xi_4 \mp i\xi_1 - k\xi_3 \pm ik\xi_2 \tag{14}$$
$$\eta_{5(6)} = \xi_5 \pm k\xi_6$$

where the subscript in parentheses goes with the lower sign. Then, identically [11]:

$$4d\eta_1 = \eta_1 \wedge (\eta_3 + \eta_6 - \eta_4) - \eta_4 \wedge \eta_5$$
$$4d\eta_2 = \eta_2 \wedge (\eta_4 + \eta_6 - \eta_3) - \eta_3 \wedge \eta_5$$
$$4d\eta_3 = \eta_3 \wedge (\eta_1 + \eta_5 - \eta_2) - \eta_2 \wedge \eta_6 \tag{15}$$
$$4d\eta_4 = \eta_4 \wedge (\eta_2 + \eta_5 - \eta_1) - \eta_1 \wedge \eta_6$$
$$2d\eta_5 = -2d\eta_6 = \eta_5 \wedge \eta_6$$

and the annulling of the following 2-forms yields the field equations:

$$\eta_1 \wedge \eta_2, \quad \eta_1 \wedge \eta_5, \quad \eta_2 \wedge \eta_5 \quad (= 0) \tag{16}$$
$$\eta_3 \wedge \eta_4, \quad \eta_3 \wedge \eta_6, \quad \eta_4 \wedge \eta_6 \quad (= 0)$$

3. Search for a pseudopotential.

If a particular 1-form is an exact differential, given that a set of field equations is satisfied, we say that it defines a potential. For example, for the s-G equation, the quantity ϕ, where $\phi_u = r^2$ $(= \phi_u{}^2)$ and $\phi_v = -2\cos\phi$, is a potential, since $(\phi_u{}^2)_v = (-2\cos\phi)_u$ by means of the field equation (1). In terms of forms, we would write

$$\sigma = r^2 du - 2\cos\phi dv \tag{17}$$

and note that

$$d\sigma = 2r\beta + 2\sin\phi\ \alpha$$

so that, since $I = \{\alpha, \beta\}$,

$$d\sigma = 0 \ (\mathrm{mod}\ I) \tag{18}$$

Thus we could write $\sigma = d\phi$--which defines the potential ϕ. Eq. (17) would become

$$\omega = -d\phi + r^2 du - 2\cos\phi dv \tag{19}$$

where $\omega = 0$ would define ϕ.

If we wanted to search for a potential, we would write

$$\sigma = F(r, \phi)\, du + G(r, \phi) dv \tag{20}$$

and require (18). That would give

$$d\sigma = (F_r dr + F_\phi d\phi) \wedge du + (G_r dr + G_\phi d\phi) \wedge dv$$
$$= F_r(\beta + \sin\phi dv \wedge du) + F_\phi d\phi \wedge du$$
$$+ G_r dr \wedge dv + G_\phi (\alpha + r du \wedge dv)$$

and $d\sigma = 0$ (mod α, β) would yield

$$G_r = F_\phi = rG_\phi - \sin\phi F_r = 0. \tag{21}$$

Solution gives Eq. (17), after we drop trivial constants.

Wahlquist and Estabrook generalized the notion of a potential to allow F and G in (20) to be functions of the potential itself [21]. If we write this new pseudopotential as y, we write, a la (19),

$$\omega = -dy + F(r, \phi, y) dv + G(r, \phi, y) dv \tag{22}$$

and require

$$d\omega = 0 \ \mathrm{mod}\ (I, \omega) \tag{23}$$

where now we include the modulo ω requirement to allow for dy where it occurs. Eqs. (21) now generalize to [12]

$$F_\phi = G_r = 0 \tag{24}$$
$$-F_r \sin\phi + rG_\phi + FG_y - GF_y = 0.$$

As noted elsewere, the solution of these equations is

$$F = rg + b$$
$$G = c\sin\phi + e\cos\phi \tag{25}$$

where g, b, c and e are functions of y satisfying

$$c + ge' - eg' = 0 \qquad -g + bc' - cb' = 0 \tag{26}$$
$$-e + gc' - cg' = 0 \qquad be' - eb' = 0.$$

$g = 0$ gives a trivial solution, so assume $g \neq 0$. Then F and G—and ω—can be divided by g and a new y can be defined by $\int g^{-1}\, dy$. Equivalently, we can require g to be a nonzero constant. We choose $g = -1$, and then we get $c = e'$, $e = -c'$, $bc' - cb' = -1$, and $be' - eb' = 0$. Thus $c = k^{-1} \cos(y + \delta)$ and $e = -k^{-1}\sin(y + \delta)$, where k and δ are constants. We set $\delta = 0$ without loss of generality, note that we must have $b = \lambda e$, and find easily that $\lambda = k^2$. Putting this all together, we see that (22) becomes

$$\omega = -dy + (k\sin y - r)du + k^{-1}\sin(y + \phi)dv. \tag{27}$$

We generalize [12] by making y, F, and G into column vectors (components y^μ, etc.), of unspecified dimension N. The nonlinear term in eq. (24) then becomes a Lie bracket. If then we take F and G to be linear in the y^μ,

$$F = (rA + B)y \tag{28}$$
$$G = (C\sin\phi + E\cos\phi)y$$

where A, B, C and E are matrices, we find the relations

$$[A, C] = -E \qquad [B, C] = -A \tag{29}$$
$$[A, E] = C \qquad [B, E] = 0.$$

Such a set of equations is called a prolongation structure by Wahlquist and Estabrook. It forms an incomplete Lie algebra but presumably can be expressed in terms of a Kac-Moody algebra. By making an Ansatz we may close the Lie algebra. The obvious one here is $B = \lambda E$; the resulting algebra is $s\ell(2, R)$. By taking a particular representation for the B^i and by writing $y^2 = y^1(\tan y/2)$, we get the 1-form (27) for y.

The prolongation structure may be obtained directly (and elegantly) in the CC case by writing ω in terms of the 1-forms (4):

$$\omega = -dy + (B^i \xi_i)y \qquad (i \text{ summed, } 1 \text{ to } 4) \tag{30}$$

(In the general CC case, the sum would be from 1 to K.) Then Eq. (23), where I = {2-forms in (5), (6)}, yields Eq. (29) immediately ($B^1 = B$, $B^2 = A$, $B^3 = C$, $B^4 = E$.)

This suggests that Eq. (30) might be tried in other problems. Wahlquist has shown [22] that this approach, with the η_i mentioned above following Eq. (8), yields the prolongation structure and the pseudopotential equations for the KdV.

If one writes the same thing

$$\omega = -dq + (B^k \eta_k)q \qquad (k \text{ summed, } 1 \text{ to } 6) \tag{31}$$

for the Ernst equation, one also gets a prolongation structure [11]. However, we must now let the B^k be functions of a variable ($k = \sqrt{\lambda}$)

$$\zeta = [k(R + \ell) - S]^{1/2} \; [k(R + \ell) + S]^{-1/2}, \tag{32}$$

where ℓ is a parameter. (ζ may be found by looking for variables invariant under scale changes and translations which leave Eqs. (9) and (10) invariant.) We may set $B^5 = B^6 = 0$. Application of the procedure yields matrix differential equations for the $B^k(\zeta)$, which are easily solved to yield

$$\begin{aligned} B^1 &= a\zeta + b \qquad & B^2 &= c\zeta + d \\ B^3 &= c\zeta^{-1} + d \qquad & B^4 &= a\zeta^{-1} + b \end{aligned} \tag{33}$$

where a, b, c, d are constant matrices satisfying

$$\begin{aligned} &4[a, d] = 4[b, a] = a \\ &4[c, b] = 4[d, c] = c \\ &4[a, c] + 4[b, d] = b - d \end{aligned} \tag{34}$$

Closure once again yields the $s\ell(2, R)$ algebra.

We now choose a representation (the simplest is 2×2, $N = 2$) and evaluate the B^i from (33). Then Eq. (31), with $\omega = 0$, may be integrated to obtain pseudopotentials q^μ. In 2×2 case, a single potential $q = q^2/q^1$ may be defined; since Eq. (31) is linear, the equation for q is a Riccati equation [7, 17]. We may reduce the number of dependent variables from N to N - 1, in the general case, by considering just the ratios of the q^μ in this manner.

4. Search for a Bäcklund transformation.

Our knowledge of the pseudopotentials (once again, denote them by y^μ) now provides us with extra freedom in the search for a Bäcklund transformation. We express new dependent variables as functions of the old ones and of the y^μ and require them to satisfy the ideal I. In the CC case, when all forms may be expressed in terms of 1-forms ξ_i, we simply write (sum from 1 to K) [12]

$$\xi_i' = A_i^{\ k} \, \xi_k \tag{35}$$

where the $A_i^{\ k}$ are functions of the y^μ, and require the ideal I to be annulled for the ξ_i' as well as the ξ_k. Solving for the $A_i^{\ k}$ then gives us the BT.

We will need to evaluate the $d\xi_i'$:

$$d\xi_i' = dA_i^{\ k} \wedge \xi_k + A_i^{\ k} d\xi_k .$$

We have, by (30)

$$\begin{aligned} dA_i^{\ k} &= A^k_{i,\ \mu} dy^\mu \\ &= A^k_{i,\ \mu} (B^\ell)^\mu{}_\nu \, \xi_\ell y^\nu , \end{aligned}$$

where sums on Greek indices run from 1 to N and those on Latin indices run from 1 to K. Thus

$$d\xi_i' = v^\ell(A_i^{\ k}) \, \xi_\ell \wedge \xi_k + A_i^{\ k} \, d\xi_k \tag{36}$$

where $v^{\ell} = y^{\nu}(B^{\ell})^{\mu}{}_{\nu}\ \partial/\partial y^{\mu}$. (37)

(We can show, under the correspondence (37) between the matrix B^{ℓ} and the operator v^{ℓ}, that $[B^{\ell}, B^{m}] \to -[v^{\ell}, v^{m}]$.)

For the s-G equation, with I given by (5) and (6), (35) becomes $\xi_1' = a\xi_1 + b\xi_2$, $\xi_2' = c\xi_1 + e\xi_2$, $\xi_3' = f\xi_3 + g\xi_4$, and $\xi_4 = h\xi_3 + \ell\xi_4$, where a, . . . ℓ are functions of the y^{μ}. The expressions on the right hand sides of these equations were chosen so that $\xi_1' \wedge \xi_2' = \xi_3' \wedge \xi_4' = 0$ would follow from $\xi_1 \wedge \xi_2 = \xi_3 \wedge \xi_4 = 0$. Then we require (from (6), for the ξ_i'):

$$d\xi_1' = v^{\ell}(a)\xi_{\ell} \wedge \xi_1 + a d\xi_1 + v^{\ell}(b)\ \xi_{\ell} \wedge \xi_2 + b d\xi_2 = 0.$$

We use (5) and (6) and set the coefficients of the independent 2-forms equal to zero. That gives

$$v^4(a) = v^4(b) = v^3(b) = 0,\ v^3(a) = -b. \tag{38}$$

We have also

$$d\xi_2' = v^{\ell}(c)\ \xi_{\ell} \wedge \xi_1 + c\ d\xi_1 + v^{\ell}(e)\ \xi_{\ell} \wedge \xi_2 + e\ d\xi_2$$
$$= \xi_3' \wedge \xi_1' = (f\xi_3 + g\xi_4) \wedge (a\xi_1 + b\xi_2),$$

yielding

$$v^4(c) = ga,\ v^3(c) = -e + fa,\ v^3(e) = fb,\ v^4(e) = gb. \tag{39}$$

Two other sets of equations also follow. We now choose a representation for the B^{ℓ}, thus yielding explicit operators v^{ℓ}. The sets of equations (38), (39), etc. now yield equations for the coefficients. It may be useful to assume that the coefficients are functions of only $y = y^2/y^1$ (in the N = 2 case, for example).

It may be necessary to let the coefficients A_i^k be functions of another variable, such as ζ in the Ernst case. One can generalize even further by including other variables; use of the variable ϕ/f in the Ernst case produces apparently more general results, although they turn out to be compositions of known transformations [10].

The BT's for the s-G, KdV, and Ernst equation (and presumably others) may be found directly by this approach. Since they have been discussed thoroughly in the literature they will not be presented here [3, 7, 11, 17, 20, 21].

5. Generalization to any number of independent variables.

The equations discussed above have two independent variables, so that the ideal I is made up of 2-forms. The prolongation form ω (see (22), (23) above, for example) is then a 1-form, so that $d\omega = 0(\text{mod } I, \omega)$ makes sense. For n independent variables, we expect I to contain n-forms (although exceptions occur; see the next section.) In this case, we follow a lead from Morris [16]. We write, if I includes forms up to the nth degree,

$$\omega = \phi \wedge dy + \mu y \qquad (40)$$

where ω is a vector of (n-1)-forms, ϕ is a matrix of (n-2)-forms, y is a vector of 0-forms, and μ is a matrix of (n-1)-forms. (Other choices are possible; for example, we could make the y^μ 1-forms.) We take the analog to Eq. (23) to be linear in ω:

$$d\omega = \rho \wedge \omega \quad (\text{mod } I) \qquad (41)$$

where ρ is a matrix 1-form. We substitute (40) into (41) and equate coefficients of y and dy to 0:

$$(-1)^{n-1} \mu = \rho \wedge \phi - d\phi \qquad (\text{mod } I) \qquad (42)$$
$$d\mu = \rho \wedge \mu$$

We can eliminate μ (if desired) to get

$$(d\rho - \rho \wedge \rho) \wedge \phi = 0 \qquad (\text{mod } I) \qquad (43)$$

We now guess ρ and ϕ (μ too, if desired) to be linear combinations of the basic forms that occur in I. Then (42) or (43) gives equations for the coefficients in these linear combinations.

6. General relativity.

The Ernst equation is found when one specializes the relativity equations to two, instead of four, independent variables. In other words, one assumes the existence of two commuting Killing vectors. If one of these is spacelike and one is timelike, one gets the Ernst equation; if they are both spacelike, then we get a cylindrical wave equation. Solution of the equation solves the whole problem, since the remaining metric coefficient is determined by quadratures.

If there is only one Killing vector (assume nonnull), things are much more complicated. One can introduce an Ernst complex potential as before, but now one must also include equations for the other metric coefficients. It is most convenient to introduce three basis 1-forms θ^A and their (three) associated connection coefficients $\omega^A{}_B$; then one constructs equations for these, including 3-form equations. This problem is not solved but has been reported in the literature [8, 12].

If there are no symmetries, then it is most convenient to use the already existing Cartan formalism [12, 13]. We write the metric as

$$ds^2 = g_{ik}\, \omega^i \otimes \omega^k \qquad (44)$$

where the ω^i are basis 1-forms. The sums go from 0 to 3. We take the g_{ik} to be constants and raise and lower indices with the metric and its inverse. The connection forms $\Omega_i{}^k$ are defined uniquely by

$$d\omega^k = \omega^i \wedge \Omega_i{}^k \qquad (45)$$

and $\Omega_{ki} = -\Omega_{ik}$. (46)

(The last equation must be generalized if the g_{ik} are not constant.) The curvature forms

$$\theta_k{}^i = d\Omega_k{}^i - \Omega_k{}^\ell \wedge \Omega_\ell{}^i \tag{47}$$

have components of the Riemann tensor as coefficients:

$$\theta_{ki} = \tfrac{1}{2} R_{kimn} \omega^m \wedge \omega^n. \tag{48}$$

Integrability of (45) requires that

$$\omega^k \wedge \theta_k{}^i = 0. \tag{49}$$

In order to construct the Einstein equations, one needs the Ricci tensor, defined by

$$R_{in} = R^m{}_{imn}. \tag{50}$$

Thus one needs to be able to extract the components of the Riemann tensor from (48). Eq. (48) contains six terms for each k and i, but one can eliminate all but one by multiplying with two selected 1-forms, say ω^i and ω^ℓ:

$$\theta_{ki} \wedge \omega^j \wedge \omega^\ell = \tfrac{1}{2} R_{kimn}\, \omega^m \wedge \omega^n \wedge \omega^j \wedge \omega^\ell$$
$$= \tfrac{1}{2}\, \varepsilon^{mnj\ell} R_{kinm}\, \sigma \tag{51}$$

where $\varepsilon^{mnj\ell}$ is the alternating symbol and $\sigma = \omega^0 \wedge \omega^1 \wedge \omega^2 \wedge \omega^3$ is the volume 4-form. Inversion leads to

$$R_{k\ell mn}\, \sigma = \tfrac{1}{2}\, \varepsilon_{ijmn}\, \theta_{k\ell} \wedge \omega^i \wedge \omega^j. \tag{52}$$

The vacuum equations $R_{\ell n} = 0$, from (50), are now the 4-form equations

$$g^{mk}\, \varepsilon_{ijmn}\, \theta_{k\ell} \wedge \omega^i \wedge \omega^j = 0. \tag{53}$$

We note that $\theta_k{}^i$, from Eq. (47), resembles Eq. (43) if the ρ is made up of the $\Omega_k{}^i$. So we put

$$\rho = B^{ik}\, \Omega_{ik} \tag{54}$$

and

$$\phi = C_{ik}\, \omega^i \wedge \omega^k \tag{55}$$

in an attempt to reproduce Eq. (43). We must use Eqs. (45), (46), (47), (49), and (53) for our ideal I. We assume that

$$(d\rho - \rho \wedge \rho) \wedge \phi = \lambda_{\ell i}\, \omega^\ell \wedge \omega^k \wedge \theta_k{}^i$$
$$+ \tau^{\ell n}\, \varepsilon_{ijmn}\, g^{km}\, \theta_{k\ell} \wedge \omega^i \wedge \omega^j \tag{56}$$

where the $\lambda_{\ell i}$ and $\tau^{\ell n}$ are arbitrary multipliers, to be eliminated. We find

$$(B^{km} g^{i\ell} - B^{ki} B^{\ell m})\, \Omega_{ki} \wedge \Omega_{\ell m} = 0 \tag{57}$$

which leads to commutators for the B^{ki}, showing that they are generators of the Lorentz group. We also have

$$B^{ki} C_{mn} = \lambda_{[n}{}^{[k} \delta_{m]}{}^{i]} + \varepsilon_{j\ell mn}(\tau^{j[k} g^{|\ell| i]}) \tag{58}$$

where $|\ell|$ is not included in the antisymmetrization denoted by [ ]. These, when solved, are to be used in the equation for the pseudopotentials y:

$$0 = [C_{mn} dy + (B^{ik} C_{mn} + 2C^{[i}{}_{[m} \delta^{k]}{}_{n]})\, \Omega_{ik} y] \wedge \omega^m \wedge \omega^n \tag{59}$$

Choice of a representation for the B^{ki} and solution of (58) for the C_{mn} leads to an equation (59) for y. The work done so far indicates that (59) does not have a nontrivial solution for y. It may be that a solution exists provided we relax some of the conditions assumed for the procedure.

Since this paper was presented in November 1983, it has been shown by F. Chinea [2] and M. Gürses [6] that the vacuum Einstein equations may be written as matrix 3-form equations instead of 4-form equations (reminiscent of Maxwell's equations). Then it is possible to write equations whose integrability conditions are Einstein's vacuum equations, including a BT, and to use them to find solutions of the Einstein equations. The present author is pursuing research in this direction [13].

Sanchez [18] has shown that the self dual Einstein equations with no symmetries, which may be written

$$\partial_{\bar{y}} (J^{-1} \partial_y J) + \partial_{\bar{z}} (J^{-1} \partial_z J) = 0$$

where J is a 4 × 4 complex, nonsingular matrix satisfying $J_{\alpha\beta',\mu} = J_{\alpha\mu,\beta}$, admit a BT with a complex parameter γ:

$$J^{-1} \partial_y J - J'^{-1} \partial_y J' = \gamma \partial_{\bar{z}} (J^{-1} J')$$
$$J^{-1} \partial_z J - J'^{-1} \partial_z J' = -\gamma \partial_{\bar{y}} (J^{-1} J')$$
$$J'_{\alpha\beta,\mu} = J'_{\alpha\mu,\beta}$$

In summary, we know of a family of BT's for the Ernst equation (two Killing vectors); the problem for one Killing vector has not been solved; and we know of a BT for the general vacuum case (but no parameter appears) and for the self dual vacuum, but otherwise general, case (in which a parameter does appear). Whether the BT's for these cases are related or not is not known, but should make for interesting future investigation. Of perhaps greater interest is the question: How broad a family of solutions to Einstein's equations can be found using these BT's?

References

1. A. V. Bäcklund, Einiges uber Curven und Flachentransformationen, Lund Universitets Arsskrift 10 (1875).
2. F. J. Chinea, Einstein equations in vacuum as integrability conditions, Phys. Rev. Lett. 52, 322-324 (1984).
3. A. Dold and B. Eckmann, eds., Bäcklund Transformations, Springer-Verlag, Berlin-Heidelberg, 1976.
4. F. J. Ernst, New formulation of the axially symmetric gravitational field problem, Phys. Rev. 167, 1175-1178 (1968).
5. F. B. Estabrook, Moving frames and prolongation algebras, J. Math. Phys. 23, 2071-2076 (1982).
6. M. Gürses, Prolongation structure and a Bäcklund transformation for vacuum Einstein's field equations, Phys. Lett. A 101A, 388-390 (1984).
7. B. K. Harrison, Bäcklund transformation for the Ernst equation of general relativity, Phys. Rev. Lett. 41, 1197-1200 (1978).
8. B. K. Harrison, Search for inverse scattering formulation of Einstein's vacuum equations with one nonnull Killing vector, paper given at the Ninth International Conference on General relativity and Gravitation, Jena, DDR, 1980.
9. B. K. Harrison, Study of the Ernst equation using a Bäcklund transformation, in Proceedings of the Second Marcel Grossmann meeting on General Relativity, R. Ruffini, ed., North-Holland, 1982.
10. B. K. Harrison, Ernst equation Bäcklund transformations revisited: new approaches and results, in Proceedings of the Third Marcel Grossmann Meeting on General Relativity, Hu Ning, ed., Science Press and North-Holland, 1983.
11. B. K. Harrison, Unification of Ernst-equation Bäcklund transformations using a modified Wahlquist-Estabrook technique, J. Math. Phys. 24, 2178-2187, (1983).
12. B. K. Harrison, Prolongation structures and differential forms, Proceedings of the Workshop on Exact Solutions of Einstein's equations, Retzbach, West Germany, November 1983, Springer-Verlag, to be published.
13. B. K. Harrison, Integrable systems in general relativity, Proceedings of the Nonlinear Studies Summer Seminar, Santa Fe, New Mexico, July 1983, Lectures in Applied Mathematics, American Mathematical Society, to be published.
14. B. K. Harrison and F. B. Estabrook, Geometric approach to invariance groups and solution of partial differential systems, J. Math. Phys. 12, 653-666 (1971).
15. G. L. Lamb, Jr., Bäcklund transformations at the turn of the century, in reference 3, 69-79.
16. H. C. Morris, Prolongation structure and nonlinear evolution equations in two spatial dimensions, J. Math. Phys. 17, 1870-1872 (1976).
17. G. Neugebauer, Bäcklund transformations of axially symmetric stationary gravitational fields, J. Phys. A: Math. Gen. 12, L67-L70 (1979).
18. N. Sanchez, Einstein equations, self-dual Yang-Mills fields and non-linear sigma models, preprint, 1983.
19. A. C. Scott, F. Y. F. Chu, and D. W. McLaughlin, The soliton: a new concept in applied science, Proc. IEEE 61, 1443-1483 (1973).
20. H. D. Wahlquist and F. B. Estabrook, Bäcklund transformation for solutions of the Korteweg-deVries equation, Phys Rev. Lett. 31, 1386-1390 (1973).

21. H. D. Wahlquist and F. B. Estabrook, Prolongation structures of nonlinear evolution equations, J. Math. Phys. 16, 1-7 (1975).
22. H. D. Wahlquist, private communication.
23. N. J. Zabusky and M. D. Kruskal, Interaction of solitons in a collisionless plasma and the recurrence of initial states, Phys. Rev. Lett. 15, 240-243 (1965).

SPECTRAL TRANSFORM APPROACH TO BÄCKLUND TRANSFORMATIONS

A. DEGASPERIS

Dipartimento di Fisica, Università di Roma, 00185 Roma, Italy
Istituto Nazionale di Fisica Nucleare, Sezione di Roma.

INTRODUCTION

The aim of these notes is to provide a brief introduction to the subject; moreover we limit ourselves to point out the main ideas and results on Bäcklund transformations in the particular context of the method based on the spectral transform. Indeed the number of papers produced so far on Bäcklund transformations is incredibly large, but not surprisingly so since these transformations have attracted the attention of both mathematicians and physicists, their relevance being well established in algebra and differential geometry as well as in nonlinear optics, gravitation, gas dynamics and other fields of application. Only few references are quoted below, but these are sufficient for the interested reader to locate proofs, additional results and technical details.

Examples are always a good starting point; consider the following three nonlinear partial differential equations:

(1) $W_t + W_{xxx} + 3W_x^2 = 0, \quad W = W(x,t),$

(2) $V_t + V_{xxx} + 2V_x^3 = 0, \quad V = V(x,t),$

(3) $\varphi_{xt} = (1/2)\sin 2\varphi, \quad \varphi = \varphi(x,t),$

(subscripted variables denote partial differentiation). The remarkable mathematical properties, as well as the relevance in applications, of these evolution equations are very well known (see, for instance, [1, 2, 3]); we merely recall here that i) a large class of solutions can be analyzed, by means of an appropriate spectral transform, in terms of normal modes (i.e. decoupled harmonic oscillators) and solitonic modes (i.e. free particle motion), ii) there exists an infinite set of independent conservation laws, iii) there exists a one-parameter family of (Bäcklund) transformations that map a given solution into a new solution. Of course these three properties are strictly connected to each other as they usually come together; however, in many cases, it seems convenient to use the first one to derive the other two. It should be also plain that infinitely many other nonlinear evolution equations have been found with the same properties of equations (1), (2) and (3), and that we are focussing our attention on these three equations for a mere sake of definiteness.

Consider now the Bäcklund transformations associated with (1), (2) and (3); they read

$$(1a)\quad w_x^{(1)} + w_x^{(2)} = -\frac{1}{2}(w^{(1)} - w^{(2)})(4p + w^{(1)} - w^{(2)}),$$

$$(1b)\quad w_t^{(1)} + w_t^{(2)} = (w_{xx}^{(1)} - w_{xx}^{(2)})(2p + w^{(1)} - w^{(2)}) - 2\Big[(w_x^{(1)})^2 + (w_x^{(2)})^2 + w_x^{(1)} w_x^{(2)}\Big],$$

for the Korteweg-de Vries (KdV) equation (1),

$$(2a)\quad v_w^{(1)} + v_x^{(2)} = -2p \sin(v^{(1)} - v^{(2)}),$$

$$(2b)\quad v_t^{(1)} + v_t^{(2)} = 2p\Big\{(v_{xx}^{(1)} - v_{xx}^{(2)})\cos(v^{(1)} - v^{(2)}) + \Big[(v_x^{(1)})^2 + (v_x^{(2)})^2\Big]\sin(v^{(1)} - v^{(2)})\Big\},$$

for the modified Korteweg-de Vries (mKdV) equation (2), and

$$(3a)\quad \varphi_x^{(1)} + \varphi_x^{(2)} = -2p \sin(\varphi^{(1)} - \varphi^{(2)}),$$

$$(3b)\quad \varphi_t^{(1)} - \varphi_t^{(2)} = -\frac{1}{2p}\sin(\varphi^{(1)} + \varphi^{(2)}),$$

for the Sine-Gordon (SG) equation (3); each of them takes the form of a system of two coupled differential equations for two unknown functions, and all imply that if one of the two functions is a solution of the associated evolution equation then also the second function solves the same evolution equation (this can be easily verified by cross-differentiation). Therefore if, for instance, $w^{(2)}$ is a given solution of the KdV equation, then integrating (1a,b) yields a new solution $w^{(1)}$ of the KdV equation (and note that the order of (1a,b) is lower than the order of the KdV equation itself).

The transformation (3a,b) was first derived by Bäcklund (1873), and then, a century later, (1a,b) was derived by Wahlquist and Estabrook (1973), both in a geometrical context (see, for instance, [4,5], and, for a jet-bundle formulation of the Bäcklund problem, [6]). Before proceeding to the spectral transform approach [2], two facts should be noted about the formulae written above. In the "first half", namely (1a), (2a) and (3a), of the Bäcklund transformations the time variable t enters only parametrically, and therefore one half of the Bäcklund transformation is in fact an ordinary (rather than partial) differential equation (in the variable x). In the second place compare (1) and (1a) with (2) and (2a), respectively; although the KdV and mKdV equations are very similar, the "first half" of the corresponding Bäcklund transformations (1a) and (2a) are rather different from each other. On the contrary, the mKdV and SG equations are very different from each other, and, nevertheless, the "first half" of their corresponding Bäcklund transformation (2a) and (3a) are identical! As shown below, these peculiar features of the evolution equations considered here are neither accidental nor particular properties of these equations, but are a reasonable consequence of the general scheme based on the spectral transform [2].

As for the use of Bäcklund transformations to construct solutions of evolution equations, and to derive an infinite set of conserved quantities, the reader is referred to [2] and the references quoted there. For a recent application to potential scattering theory, see [7].

THE SPECTRAL TRANSFORM

The basic step here is to introduce a time dependent linear differential (with respect to x) operator $H(t,k)$ such that the nonlinear evolution equation of interest results as the compatibility condition between the eigenvalue problem

(4) $H\,(t,k)\Psi = 0,$

and the evolution equation

(5) $\Psi_t = M\,(t,k)\Psi\,,$

where k is the eigenvalue, or spectral parameter, $M(t,k)$ is a linear differential operator and the dependence of H and M on k is assumed to be rational.

The two equations (4) and (5) are known as Lax pair, and for instance, those associated with the evolution equations considered above are

(1c) $H(t,k) = \partial_x^2 + w_x(x,t) + k^2\,,$

(1d) $M(t,k) = 2\left[2k^2 - w_x(x,t)\right]\partial_x + w_{xx}(x,t)\,,$

for the KdV equation (1),

(2c) $H(t,k) = \partial_x - i\, V_x(x,t)\, \sigma_2 + ik\sigma_3$,

(2d) $M(t,k) = 2ik\,[V_x^2\,(x,t) - 2k^2]\sigma_3 + 2ikV_{xx}(x,t)\,\sigma_1 - i[V_{xxx}(x,t) +$

$+ 2V_x^3\,(x,t) - 4k^2\,V_x(x,t)]\,\sigma_2$,

for the mKdV equation (2), and

(3c) $H(t,k) = \partial_x - i\,\varphi_x(x,t)\,\sigma_2 + ik\,\sigma_3$,

(3d) $M(t,k) = (i/4k)\left\{\cos\,[2\,\varphi(x,t)]\,\sigma_3 - \sin\,[2\,\varphi(x,t)]\,\sigma_1\right\}$,

for the SG equation (3) (here $\sigma_1 = \begin{pmatrix} 0 & 1 \\ 1 & 0 \end{pmatrix}$, $\sigma_2 = \begin{pmatrix} 0 & -i \\ i & 0 \end{pmatrix}$ and $\sigma_3 = \begin{pmatrix} 1 & 0 \\ 0 & -1 \end{pmatrix}$ are the Pauli matrices).

These formulae explicitly show that the variable t enters parametrically in the eigenvalue problem (4) (just as in the "first half" of the Bäcklund transformations (1a), (2a) and (3a)), and that the operators H(t,k) associated with the mKdV equation (2) and SG equation (3) coincide. This suggests, in connection with the two remarks made in the previous section, that in fact the "first half" of a Bäcklund transformation has nothing to do with the time evolution, and is related only to the eigenvalue equation (4). That this is indeed the case will be clear below; for the moment we first consider the eigenvalue problem (4) in detail by leaving out the time dependence. This will lead us to define the spectral transform.

To this aim we choose to consider the operator

(6) $H(k) = \partial_x - Q(x) + ik\sigma_3$,

with

(7) $Q(x) = \begin{pmatrix} 0 & q^{(-)}(x) \\ q^{(+)}(x) & 0 \end{pmatrix}$,

a generalized version of (2c) (or, equivalenty, (3c)), known as the generalized Zakharov-Shabat operator; for a treatment analogous to the following, but for the operator (1c), see [2]. The explicit expression of the eigenvalue equation (4), with (6), reads

(8) $\Psi_x = [-ik\sigma_3 + Q(x)]\Psi$, $\Psi = \Psi(x,k)$,

where we assume that the condition (see (7))

$$(9)\quad \int_{-\infty}^{+\infty} dx \ \left| q^{(\pm)}(x) \right| < \infty$$

holds for the formulae below to make sense. The spectrum of the eigenvalue problem (8) has a real component, i.e. all real values of k, and possibly (but not necessarly) a discrete component, i.e. a finite number of complex values of k. For the real spectrum the solution of (8) is the 2x2 matrix $\Psi(x,k)$ defined by the asymptotic conditions

$$(10a)\quad \Psi(x,k) \xrightarrow[x \to +\infty]{} \begin{pmatrix} e^{-ikx} & \alpha^{(-)}(k)\ e^{-ikx} \\ \alpha^{(+)}(k)\ e^{ikx} & e^{ikx} \end{pmatrix},$$

$$(10b)\quad \Psi(x,k) \xrightarrow[x \to -\infty]{} \begin{pmatrix} \beta^{(+)}(k)\ e^{-ikx} & 0 \\ 0 & \beta^{(-)}(k)\ e^{ikx} \end{pmatrix},$$

that uniquely define the reflection matrix

$$(11)\quad \alpha(k) = \begin{pmatrix} 0 & \alpha^{(-)}(k) \\ \alpha^{(+)}(k) & 0 \end{pmatrix},$$

as a function of the real variable k. For the discrete spectrum (if any), we choose, merely for notational convenience, to indicate with $k_n^{(+)}$, $n = 1,2,\ldots,N^{(+)}$ and $k_n^{(-)}$, $n = 1,2,\ldots,N^{(-)}$, the discrete eigenvalues that lie in the upper and, respectively, lower half of the complex k-plane, i.e.

$$(12)\quad \pm \ \mathrm{Im}\ k_n^{(\pm)} > 0 , \qquad n = 1,2,\ldots,N^{(\pm)} .$$

Thus the vector solution $\varphi_n^{(\pm)}(x)$, corresponding to the eigenvalue $k_n^{(\pm)}$, vanishes as $x \to \pm\infty$ and is defined by the equations

$$(13)\quad \varphi_{nx}^{(\pm)} = \left[- ik_n^{(\pm)} \sigma_3 + Q(x)\right] \varphi_n^{(\pm)} , \quad n = 1,2,\ldots,N^{(\pm)} ,$$

$$(14)\quad \int_{-\infty}^{+\infty} dx \ \left[\varphi_n^{(\pm)}(x)\right]^T \sigma_1 \varphi_n^{(\pm)}(x) = 1 ,$$

where V^T is the transpose of the vector (i.e. one-column matrix) V. Through the asymptotic behaviours

$$(15)\quad \varphi_n^{(+)}(x) \xrightarrow[x \to +\infty]{} \gamma_n^{(+)}\ e^{ik_n^{(+)}x} \begin{pmatrix} 0 \\ 1 \end{pmatrix},$$

$$(15b)\quad \varphi_n^{(-)}(x) \xrightarrow[x \to +\infty]{} \gamma_n^{(-)}\ e^{-ik_n^{(-)}x} \begin{pmatrix} 1 \\ 0 \end{pmatrix},$$

we finally define the spectral quantities

(16) $\rho_n^{(\pm)} = [\gamma_n^{(\pm)}]^2$.

We are now in the position to introduce the spectral transform $S[Q(x)]$ of the off-diagonal matrix $Q(x)$, see (7), by the following definition (see (11) and (16)):

(17) $$S[Q(x)] = \{\alpha(k);\ k_n^{(\pm)},\ \rho_n^{(\pm)},\ n = 1,2,\ldots,N^{(\pm)}\};$$

we also refer to the problem of computing the spectral transform of a given matrix $Q(x)$ as the direct problem. Note that solving the direct problem $Q(x) \rightarrow S[Q(x)]$ ammounts to integrating the linear ordinary differential equation (8).

The rationale for the definition (17) is that the knowledge of the quantities in the r.h.s. of (17) is necessary and sufficient to reconstruct the matrix $Q(x)$, namely that the mapping $Q(x) \rightarrow S[Q(x)]$ can be inverted. The reconstruction of $Q(x)$ from its spectral transform is known as the inverse problem, and its solution is obtained via the three following steps [2,8] :

(18) $$M(z) = \frac{1}{2\pi}\int_{-\infty}^{+\infty} dk\ \alpha(k)\ \exp(ikz\,\sigma_3) + \begin{pmatrix} 0 & \sum_{n=1}^{N^{(-)}} \rho_n^{(-)} e^{-ik_n^{(-)}z} \\ \sum_{n=1}^{N^{(+)}} \rho_n^{(+)} e^{ik_n^{(+)}z} & 0 \end{pmatrix},$$

(19) $$K(x,y) + M(x+y) + \int_x^{+\infty} dz\ K(x,z)\ M(z+y) = 0\ , \quad x \leq y$$

(20) $$Q(x) = -2\begin{pmatrix} 0 & K_{12}(x,x) \\ K_{21}(x,x) & 0 \end{pmatrix} .$$

This shows that, in order to solve the inverse problem, one has to solve the linear matrix integral equation (19) for the unknown (2x2 matrix) $K(x,y)$; in this equation both the Kernel and the nonhomogeneous term are given in terms of the off-diagonal matrix $M(z)$ whose explicit expression (18) obtains as a Fourier integral over the continuous spectrum plus a finite sum over the discrete spectrum. The formula (20) finally yields the explicit solution of the inverse problem. It is worth noticing here that the mapping $S[Q(x)] \rightarrow Q(x)$ defined by (17), (18), (19) and (20) is of course nonlinear, and that its linearized approximation coincides with the usual Fourier transform, namely ($N^{(\pm)} = 0$)

(21a) $$\alpha(k) \simeq \int_{-\infty}^{+\infty} dx \exp(2ikx\,\sigma_3)\, Q(x) \; ,$$

(21b) $$Q(x) \simeq \frac{1}{\pi} \int_{-\infty}^{+\infty} dk \exp(-2ikx\,\sigma_3)\, \alpha(k) \; .$$

BÄCKLUND AND DARBOUX TRANSFORMATIONS

Here we sketch a technique to derive the Bäcklund transformations (or, more precisely, the "first half" of them) associated with the generalized Zakharov-Shabat operator (6). The following approach has been extensively applied to many other differential operators, and, for a recent treatment including its extension to the multi-dimensional case, see [9] and the references quoted there.

Consider the generalized Zakharov-Shabat equation (8) with a matrix $Q^{(2)}(x)$,

(22) $$\psi_x^{(2)} = \left[-ik\,\sigma_3 + Q^{(2)}(x)\right] \psi^{(2)} \quad ,$$

and let $T(x,k)$ be a matrix transforming a matrix solution $\psi^{(2)}$ of (22) into a matrix solution of the generalized Zakharov-Shabat (8), but with a different matrix $Q^{(1)}(x)$, namely

(23) $$\psi^{(1)}(x,k) = T(x,k)\, \psi^{(2)}(x,k) \quad ,$$

(24) $$\psi_x^{(1)} = \left[-ik\,\sigma_3 + Q^{(1)}(x)\right] \psi^{(1)} \; .$$

This implies that the matrix $T(x,k)$ satisfies the matrix differential equation

(25) $$T_x + ik\left[\sigma_3, T\right] + T\,Q^{(2)} - Q^{(1)}\,T = 0 \quad ;$$

in order the analyze this equation, it is convenient to write down separately its diagonal and off-diagonal parts. Thus we introduce the splitting

(26) $$T(x,k) = C(x,k) + A(x,k) \quad ,$$

where C is the diagonal (commuting with σ_3) component of T,

$$(27a)\qquad C \equiv \begin{pmatrix} T_{11} & 0 \\ 0 & T_{22} \end{pmatrix},$$

and A is the off-diagonal (anticommuting with σ_3) component of T,

$$(27b)\qquad A \equiv \begin{pmatrix} 0 & T_{12} \\ T_{21} & 0 \end{pmatrix}.$$

By taking into account that $Q^{(j)}$, j=1,2, is off-diagonal (see (7)), the equation (25) is then equivalent to the two coupled equations

$$(28a)\qquad C_x + AQ^{(2)} - Q^{(1)}A = 0 \quad ,$$

$$(28b)\qquad A_x + 2ik\ \sigma_3 A + CQ^{(2)} - Q^{(1)}C = 0 \quad .$$

Integrating the first equation,

$$(29)\quad C(x,k) = -\ \sigma_3\ \Gamma(k) + \int_x^{\infty} dy\ \left[A(y,k)\ Q^{(2)}(y) - Q^{(1)}(y)A(y,k)\right] ,$$

where $\Gamma(k)$ is an arbitrary diagonal x-independent matrix, and substituting this expression into the second equation, the system (28) reduces to the integrodifferential equation

$$(30)\quad (\Lambda - 2ik)\ A + \Gamma Q^{(2)} + Q^{(1)}\Gamma = 0 ,$$

where we have introduced the important integrodifferential operator Λ that is defined by the formula

$$(31)\ \Lambda F(x) = -\ \sigma_3\Big\{ F_x(x) + Q^{(1)}(x)\int_x^{\infty} dy\left[Q^{(1)}(y)\ F(y) - F(y)\ Q^{(2)}(y)\right] - \\ - \int_x^{\infty} dy\left[Q^{(1)}(y)\ F(y) - F(y)\ Q^{(2)}(y)\right] Q^{(2)}(x) \quad ,$$

that specifies its action on a generic off-diagonal matrix F(x). The operator Λ, that plays a basic rôle in the spectral transform method, has been introduced first by a different way, namely by generalizing the well known wronskian relations [2].

We now make the strong assumption that the transformation matrix T(x,k), see (23), be a polynomial in the spectral variable k; because of (26) and (29), this ammounts to ask that

$$(32a)\quad \Gamma(k) = \sum_{n=0}^{M} (2ik)^n \, \Gamma^{(n)} \quad ,$$

$$(32b)\quad A(x,k) = \sum_{n=0}^{M} (2ik)^n \, A^{(n)}(x) \quad ,$$

where of course the coefficients $\Gamma^{(n)}$ are arbitrary constant diagonal matrices, and $A^{(n)}(x)$ are off-diagonal matrices that satisfy the recursion relation

$$(33)\quad A^{(n)} = \Lambda A^{(n+1)} + Q^{(1)} \Gamma^{(n+1)} + \Gamma^{(n+1)} Q^{(2)} , \quad n=0,1,2,\dots,M,$$

that is readly implied by (30), together with the initial conditions $A^{(M+1)} = \Gamma^{(M+1)} = 0$. It is easily seen that the equation (30), being polynomial in k, implies M+2 equations for the coefficients; moreover M+1 of these equations can be explicitely solved to yield the matrices $A^{(n)}(x)$,

$$(34)\quad A^{(n)}(x) = \sum_{m=0}^{M-n-1} \Lambda^m \left[Q^{(1)}(x)\, \Gamma^{(n+m+1)} + \Gamma^{(n+m+1)} Q^{(2)}(x) \right] ,$$

$n=0,1,2,\dots,M-1,$

with $A^{(M)}(x) = 0$, while the last equation

$$(35)\quad \sum_{m=0}^{M} \Lambda^m \left[Q^{(1)}(x)\, \Gamma^{(m)} + \Gamma^{(m)} Q^{(2)}(x) \right] = 0 \quad ,$$

gives a relationship between $Q^{(2)}(x)$ and $Q^{(1)}(x)$. It is now convenient to introduce the following notation:

$$(36)\quad \Gamma^{(m)} = g_m \sigma_3 + h_m \quad ,$$

$$(37)\quad g(z) = \sum_{m=0}^{M} g_m z^m \quad , \quad h(z) = \sum_{m=0}^{M} h_m z^m \quad ,$$

$$(38)\quad g^{\#}(z',z) = [g(z') - g(z)]/(z'-z) \quad , \quad h^{\#}(z',z) = [h(z') - h(z)]/(z'-z) \quad ,$$

where the coefficients g_m and h_m are (arbitrary) complex numbers, so that the resulting expression of the matrix $T(x,k)$ reads

$$(39)\quad T(x,k) = -g(2ik) - h(2ik)\sigma_3 - \int_x^{\infty} dy \left[Q^{(1)}(y) A(y,k) - A(y,k) Q^{(2)}(y) \right] +$$

$$+ A(x,k) \quad ,$$

$$(40)\quad A(x,k) = g^{\#}(2ik,\Lambda)\{[Q^{(1)}(x) - Q^{(2)}(x)]\sigma_3\} + h^{\#}(2ik,\Lambda)[Q^{(1)}(x) + Q^{(2)}(x)] \quad .$$

On the other hand, in this notation the equation (35) relating $Q^{(2)}(x)$ to $Q^{(1)}(x)$ takes the compact form

$$(41)\quad g(\Lambda)\{[Q^{(1)}(x) - Q^{(2)}(x)]\sigma_3 + h(\Lambda)[Q^{(1)}(x) + Q^{(2)}(x)] = 0 \quad ,$$

where $g(z)$ and $h(z)$ are the arbitrary polynomials defined by (36) and (37).

The transformation $\psi^{(2)} \rightarrow \psi^{(1)}$ given by (23) with a polynomial dependence of $T(x,k)$ on k is usually referred to as a Darboux transformation (since Darboux introduced this transformation in the context of the Schroedinger operator (1c) and in the particular case in which T is of first degree in k [2]). Moreover the relation (41) between $Q^{(2)}(x)$ and $Q^{(1)}(x)$, that corresponds to the Darboux matrix (39), with (40), will be named Bäcklund transformation (for instance from $Q^{(2)}(x)$ to $Q^{(1)}(x)$); before qualifying this terminology in connection with the content of the introduction, we consider first the relationship between the spectral transforms of $Q^{(2)}(x)$ and $Q^{(1)}(x)$ that corresponds to the Bäcklund transformation (41). However, for the sake of simplicity, we limit our consideration to the continuous component of the spectral transform (for a more detailed discussion see [2]).

Consider then the Darboux transformation that relates the matrix solutions $\psi^{(2)}(x,k)$ and $\psi^{(1)}(x,k)$ that satisfy the asymptotic behaviours (10), namely

$$(42)\quad \psi^{(j)}(x,k) \xrightarrow[x\to+\infty]{} \exp(-ikx\sigma_3)[1 + \alpha^{(j)}(k)] \quad , \quad j=1,2 \quad ,$$

where $\alpha^{(j)}(k)$ is the off-diagonal matrix (see (11) that defines the continuous component of the spectral transform of $Q^{(j)}(x)$ (see (17)). This transformation reads

$$(43)\quad \psi^{(j)}(x,k) = T(x,k)\,\psi^{(2)}(x,k)\,[T(+\infty,k)]^{-1} \quad ,$$

where the appearance of the matrix $T(+\infty,k)$ in the r.h.s. takes into account the appropriate asymptotic conditions (42); indeed one should note that, as implied by (39), (40) and (31),

$$(44)\quad T(+\infty,k) = -g(2ik) - h(2ik)\,\sigma_3$$

is diagonal, and that the normalization condition adopted in (43) by multiplying from the right by $[T(+\infty,k)]^{-1}$ does not affect the results and formulae given above. Combining now the behaviour (42) with the Darboux transformation (43), with (44), one easily obtains the corresponding Bäcklund transformation for the spectral transforms, namely

$$(45)\quad \alpha^{(1)}(k) = \alpha^{(2)}(k)\,[g(2ik) - h(2ik)\,\sigma_3][g(2ik) + h(2ik)\,\sigma_3]^{-1} .$$

These results are our equipment for deriving and investigating nonlinear evolution equations, as well as their associated Bäcklund transformations. In fact, let us consider first a parameter (i.e. time) dependent matrix $Q(x,t)$ in the generalized Zakharov-Shabat equation (8), and let us set

$$(46)\quad Q^{(2)}(x) = Q(x,t) \quad , \quad Q^{(1)}(x) = Q(x,t+\varepsilon) \quad ,$$

in the previous formulae, together with

$$(47)\quad g(z) = 2/\varepsilon \qquad , \quad h(z) = \gamma(z) \qquad ;$$

then, in the limit $\varepsilon \to 0$, the relation (41) yields the (AKNS class [8] of) nonlinear evolution equation

$$(48)\quad Q_t(x,t) = \sigma_3\,\gamma(L)\,Q(x,t) \quad ,$$

where L is of course the integrodifferential operator that obtains from Λ with $Q^{(1)}(x) = Q^{(2)}(x) = Q(x,t)$, namely (see (31))

$$(49)\quad LF(x) = -\sigma_3\Big\{F_x(x) + 2Q(x,t)\int_x^{\infty} dy\,[Q(y,t), F(y)]\Big\} .$$

The corresponding evolution equation for the spectral transform of $Q(x,t)$ is very simple and reads

$$(50a)\quad \alpha_t(k,t) = \gamma(2ik)\,\sigma_3\,\alpha(k,t) \quad ,$$

$$(50b)\quad dk_n^{(\pm)}/dt = 0 \quad ; \quad n=1,2,\ldots,N^{(\pm)} \quad ,$$

$$(50c)\quad d\rho_n^{(\pm)}(t)/dt = \mp\,\gamma(2ik_n^{(\pm)})\,\rho_n^{(\pm)}(t) \quad , \quad n=1,2,\ldots,N^{(\pm)} .$$

Therefore the initial value problem associated with the evolution equation (48) can be solved by means of linear operations only, via the steps $Q(x,o) \longrightarrow S[Q(x,o)] \longrightarrow S[Q(x,t)] \longrightarrow Q(x,t)$.

For instance the mKdV equation (2) obtains by setting in (48) $\gamma(z) = z^3$ and

(2e) $Q(x,t) = iV_x(x,t)\,\sigma_2$,

while the SG equation (3) obtains with $\gamma(z) = -z^{-1}$ and

(3e) $Q(x,t) = i\varphi_x(x,t)\,\sigma_2$.

Let us consider next a solution $Q^{(2)}(x,t)$ of the evolution equation (48) and a matrix $Q^{(1)}(x,t)$ related to it by the equation (41), with the condition that the polynomials $g(z)$ and $h(z)$ be not dependent on time; then also $Q^{(1)}(x,t)$ is a solution of the same evolution equation (48). A direct proof of this statement is certainly far too difficult, while the spectral transform approach provides a proof that, in the case in wich the spectral transforms of $Q^{(1)}$ and $Q^{(2)}$ have no discrete spectrum component, is even trivial. In fact, if the matrix $\alpha^{(2)}(k,t)$ satisfies the (linear) evolution equation (50a), then of course also $\alpha^{(1)}(k,t)$, related to $\alpha^{(2)}(k,t)$ by (45), is a solution of (50a) (in this proof, that is evident by simple inspection, it is clear that the condition that $g(z)$ and $h(z)$ be time indipendent is indeed essential). A more careful analysis is required to prove the statement made above if the spectral transforms of $Q^{(2)}(x,t)$ and/or $Q^{(1)}(x,t)$ possess also a discrete spectrum component; for a general discussion of this point see [2]. These results justify the name of Bäcklund transformation for the equation (41), as this equation defines a mapping of a solution $Q^{(2)}(x,t)$ of the evolution (48) into the new solution $Q^{(1)}(x,t)$ of the same evolution equation. Moreover, it should be pointed out that the Bäcklund transformation (41), characterized by the two polynomials $g(z)$ and $h(z)$, does not depend on the particular evolution equation of the class (48) (i.e. on the polynomial $\gamma(z)$ that characterizes a particular evolution equation). Therefore two different evolution equations of the class (48), for instance the mKdV and SG equation, should have the same (first half of the) Bäcklund transformation. For instance, the Bäcklund transformation (2a) (or (3a)) obtains by setting in (41)

(51) $g(z) = \frac{1}{2}z \quad , \quad h(z) = p \quad ,$

where p is a real parameter, and by using the expression (2e) (or(3e)). In this context, it is clear that the KdV and mKdV equations (1) and (2) should not be expected to have the same Bäcklund transformation since their associated linear operators, namely the Schroedinger and Zakharov-Shabat operators, respectively, are different from each other.

Bäcklund transformations have been frequently used to construct new solutions of a nonlinear evolution equation out of a known one; in particular, the one-soliton solution is obtained by applying a Bäcklund transformation to the vanishing solution. In this respect, quite useful formulae are the so-called superposition formulae; they follow from the remarkable property of the Bäcklund transformations (41) to commute with each other. The proof of the commutativity theorem is rather simple in the spectral transform approach [2]; in fact, in the particular case in which no discrete spectrum is involved, the proof is an obvious consequence of the relation (45). A direct proof of the commutativity theorem for the Bäcklund transformation (3a,b) was first given by L. Bianchi (1896) in his lecture notes on differential geometry.

To give an example of a superposition formulae, let us consider the Bäcklund transformation (1a); this transformation depends on the real parameter p, and therefore one can sequentially apply two Bäcklund transformations, one with parameter p_1 and the second with parameter p_2, to a given solutio $w^{(0)}(x,t)$ of the KdV equation (1): $w^{(0)}(x,t) \xrightarrow{1} w^{(1)}(x,t) \xrightarrow{2} w(x,t)$. On the other hand, the order of the two Bäcklund transformations can be reversed without changing the final outcome because of the commutativity theorem: $w^{(0)}(x,t) \xrightarrow{2} w^{(2)}(x,t) \xrightarrow{1} w(x,t)$. It is then easy to eliminate from all the relevant equations corresponding to these four Bäcklund transformations all derivatives to end up with the following (purely algebraic) superposition formula [2]

$$(52) \quad w = w^{(0)} - 2(p_1 + p_2)\left[w^{(1)} - w^{(2)}\right] / \left[w^{(1)} - w^{(2)} + 2(p_1 - p_2)\right].$$

A different (but equivalent) superposition formula obtains by eliminating instead the starting solution $w^{(0)}(x,t)$, and it reads [10]

$$(53) \quad w = \frac{1}{2}\left(w^{(1)} + w^{(2)}\right) + \left[w_x^{(1)} - w_x^{(2)} - (p_1 + p_2)(w^{(1)} - w^{(2)})\right] / \\ / \left[w^{(1)} - w^{(2)} + 2(p_1 - p_2)\right].$$

Interesting new formulae obtain from these superposition formulae by setting $p_1 = p + \varepsilon$, $p_2 = p$, $w^{(2)}(x,t) = \bar{w}(x,t);p)$, $w^{(1)}(x,t) = \bar{w}(x,t;p+\varepsilon)$

and then by letting ε vanish; for instance, one of the resulting expressions is

$$W(x,t;p) = W^{(0)}(x,t) - 2p\bar{W}_p(x,t;p) \Big/ \left[1 + \frac{1}{2}\bar{W}_p(x,t;p)\right] \quad , \tag{54}$$

and this yields a new solution of the KdV equation (1). In this expression $\bar{W}(x,t;p)$ is of course the solution that obtains from $W^{(0)}$ by a Bäcklund transformation with parameter p. From this formula, a new explicit (singular) solution of the KdV equation (1) is easily derived by taking $W^{(0)}(x,t) = 0$ [10]. Additional results, and the derivation of analogous formulae from the Bäcklund transformations associated with the generalized Zakharov-Shabat problem can be found in [10]; there the algebraic construction of a two-dimensional lattice of explicit solutions is also reported.

REFERENCES

[1] R.K. Bullough, P.J. Caudrey (editors), Solitons, Topics in Current Physics, 17, Springer, Berlin, 1980.

[2] F. Calogero, A. Degasperis, Spectral Transform and Solitons, North-Holland, Amsterdam, Vol. I, 1982; Vol. II, in preparation.

[3] R.K. Dodd, J.C. Eilbeck, J.D. Gibbon, H.C. Morris, Solitons and Nonlinear Waves, Academic Press, New York, 1982.

[4] R.M. Miura (editor), Bäcklund Transformations, Lecture Notes in Mathematics 515, Springer, Berlin, 1976.

[5] C. Rogers, W.F. Shadwick, Bäcklund Transformations and Their Applications, Academic Press, New York, 1982.

[6] F.A.E. Pirani, D.C. Robinson, W.F. Shabwick, Local Jet-bundle Formulation of Bäcklund Transformations, Reidel, Dordrecht, 1979.

[7] P.C. Sabatier, "Rational reflection coefficients in one dimensional inverse scattering and applications", preprint PM/83/5, Montpellier, 1983.

[8] M.J. Ablowith, H. Segur, Solitons and Inverse Scattering Transform. SIAM, Philadelphia, 1981.

[9] M. Boiti, B.G. Konopelchenko, F. Pempinelli, "Bäcklund Transformations via Gauge Transformations in 2+1 Dimensions" Inverse Problems (to be published).

[10] F. Calogero, A. Degasperis, "Elementary Bäcklund Trasformations, nonlinear superposition formulae and algebraic construction of solutions for the nonlinear evolution equations solvable by the Zakharov-Shabat spectral problem", Physica D (to be published).

GAUGE COUPLING OF NON-LINEAR σ-MODEL AND A GENERALIZED MAZUR IDENTITY

B. CARTER
Groupe d'Astrophysique Relativiste
D.A.F., Observatoire de Paris-Meudon

Abstract

An inversion symmetric class of non-linear σ-models is constructed. The original pure model with field values in the coset space of a classical matrix group $\mathcal{G}$ with respect to an isotropy subgroup under the adjoint action is generalized to a minimally gauge coupled model in which the field is a section in a bundle with group $\mathcal{G}$ acting on the coset space as fibre with a non-trivial connection of (for example) Yang-Mills type. It is shown that the gauge coupled models admit a natural generalisation of the identities originally constructed by Mazur for the pure non-linear σ-models whereby the divergence of a quantity whose surface integral vanishes when suitable boundary conditions are satisfied is shown to be equal to a functional of the difference between two sets of field variables that is positive definite in many relevant situation. In such cases, which occur when the base-space metric is positive definite (so that the system is of elliptic type) and the isotropy subgroup is compact, the identities lead directly to uniqueness theorems for the solutions.

1. Introduction

In recent years the rather wide class of field models that would be described by mathematicians as harmonic mappings onto homogeneous spaces has come to be known in physics under the appelation of non-linear σ-models. Our present purpose is to consider a sub-class of such models in which the fields belong to the symmetric coset space generated as the quotient of one of the classical (Lie) matrix groups by an isotropy subgroup of the adjoint action of the group on itself. In particular a considerable body of work originated by Geroch[1] and

Kinnersley[2] and developed more recently by Sanchez[3], Mazur[4] and others has shown that the Ernst formulation of the Einstein equations for a stationary axisymmetric vacuum solution and its electromagnetic generalization can be considered as a model of this type.

Among the many interesting results to emerge from this work has been the developement by Mazur[5] of a systematic and elegant procedure for the construction of identities relating the divergence of a quantity whose surface integral vanishes when suitable boundary conditions are satisfied to the sum of a quantity that vanishes when the field equations of the model are satisfied and a functional of the difference between two sets of the field variables that turns out to be automatically positive definite in many relevant cases. The class of divergence identities constructed in this way includes as special limit cases the identities having the same properties that were developed successively by the present author[6] and by Robinson[7],[8] for the purpose of establishing the uniqueness of solutions to the black hole equilibrium state problem as formulated in terms of a non-linearly coupled elliptic system on a two-dimensional base space by the present author[6] [9] [10]. The common ancestor to all such identities is the well known relation

$$\nabla_\mu(\varphi\nabla^\mu\varphi) = \varphi\nabla_\mu\nabla^\mu\varphi + (\nabla^\mu\varphi)\nabla_\mu\varphi$$

which is used to prove uniqueness of solutions to the ordinary linear Laplace equation

$$\nabla_\mu\nabla^\mu\varphi = 0$$

on any base space with positive definite metric $g_{\mu\nu}$, subject to boundary conditions such that φ must tend to zero somewhere and that $\varphi\nabla_\mu\varphi$ must tend to zero everywhere on the boundary of the base space region under consideration, where φ is the difference between any two allowed solutions. The idea in all the applications of the method is simply to integrate over the base space region under considerations and to use Green's theorem and the boundary conditions to establish that the integral of the positive definite functional must vanish and hence that its argument must be zero throughout, which in most cases is sufficient to immediately establish the identity of the two hypothetical solutions.

At about the same time as Mazur's construction of his divergence identity was carried out, an even more general class of identity relating a divergence to a right hand side with the same kind of positivity property was developed independently by Bunting[11],[12] for the same purpose, namely completing the proof of the uniqueness theorem[9],[7],[10] for electromagnetic black hole equilibrium states. Although they do not have such an elegantly explicit form as the Mazur type identities, the Bunting identities have the advantage that they can be constructed not only for non-linear σ-models but even for more general harmonic mappings in which the image space is non-homogeneous, provided that an appropriate negativity condition is satisfied by the image space curvature. A concise description of the Bunting construction, and an explicit demonstration that it applies in particular to all the non-linear σ-models covered by the original Mazur construction (in so much as they can be shown to satisfy the curvature negativity condition automatically) has been given elsewhere by the present author[13]. This work showed that all though they are necessary for the elegant explicit form of the Mazur identities, the special symmetry properties of the non-linear σ-models to which they apply are not really essential for the unicity theorems that are obtained thereby.

Our present purpose is to show how the Mazur type divergence identities can also be generalized in a quite different manner from that achieved by Bunting. Moreover unlike the Bunting construction, the procedure to be described here depends in an essential way on the existence of the continuous symmetry group action on the image space of the non-linear σ-models. We start by showing how the ordinary non-linear σ-models of the symmetric coset space kind considered by Mazur can be generalized in a uniquely natural way by minimal coupling of the standard kind to a background gauge field belonging to the algebra of the continuous symmetry group, and which might itself (for example) be supposed to obey field equations of the standard (coupled) Yang-Mills type. After a general description of the structure of such gauge-coupled non-linear σ-models, it is shown how appropriate generalized Mazur identities can be constructed and used for proving uniqueness of solutions with a given gauge background field under conditions of the same kind as were necessary when the gauge background field was absent, including in particular the requirement that the base space metric be positive definite (so that the system is of elliptic type) and the requirement that the isotropy subgroup characterising the coset space should be of compact type even though the group as a whole is not.

2. Inversion Symmetry of the Homogeneous Coset Space

We shall be concerned with a set of field variables Φ belonging to a manifold $\mathcal{X}$ that can be identified with an orbit of the adjoint action of a matrix group $\mathcal{G}$ acting on itself. More specifically it will be supposed that the group $\mathcal{G}$ is of the standard type as defined by the invariance requirement for a standard (non-generate) form η (with components $\eta^{a\dot{b}}$) under the natural action of the elements Q (with components $Q^a{}_b$) of $\mathcal{G}$ as specified by

$$Q \in \mathcal{G} \quad \Longleftrightarrow \quad Q\cdot\eta\cdot Q^* = \eta \tag{2.1}$$

where a heavy dot denotes matrix multiplication (contraction of adjacent indices) and the star denotes hermitian conjugation (i.e $Q^*{}_{\dot{b}}{}^{\dot{a}} = \overline{Q^a{}_b}$). The elements Φ of $\mathcal{X}$ are supposed to be hermitian matrices with component $\Phi_{\dot{a}b}$ such that the corresponding matrices $^\#\Phi$ with components $\Phi^a{}_b$,as defined by the operation of index raising by contraction with η , should satisfy the group characterization condition (2.1), i.e.

$$^\#\Phi \equiv \eta\cdot\Phi \in \mathcal{G} \tag{2.2}$$

Since the adjoint group action

$$^\#\Phi \longmapsto Q^{-1}\cdot{}^\#\Phi\cdot Q \tag{2.3}$$

(which is required to be transitive over $\mathcal{X}$) is equivalent to

$$\Phi \longmapsto Q^*\cdot\Phi\cdot Q \tag{2.4}$$

it can be seen that it automatically preserves the hermiticity requirement

$$\Phi^* = \Phi \tag{2.5}$$

A homogeneous space $\mathcal{X}$ constructed in this way can evidently be regarded as a coset space,

$$\mathcal{X} = \mathcal{G} / \mathcal{H} \tag{2.6}$$

where $\mathcal{H}$ is an isotropy subgroup of $\mathcal{G}$ defined to consist of elements $\mathbf{Q} \in \mathcal{G}$ such that some arbitrarily chosen element $\Phi \in \mathcal{X}$ remains invariant under transformations of the form (2.4). Such a space has a natural (Riemannian or pseudo-Riemannian) metric

$$\begin{aligned} ds^2 &= \tfrac{1}{4} \langle \Phi^{-1} \cdot d\Phi , \Phi^{-1} \cdot d\Phi \rangle \\ &= \tfrac{1}{4} \langle {}^{\#}\Phi \cdot d{}^{\#}\Phi , {}^{\#}\Phi \cdot d{}^{\#}\Phi \rangle \end{aligned} \tag{2.7}$$

induced by the corresponding natural metric

$$ds^2 = \tfrac{1}{4} \langle \mathbf{Q}^{-1} \cdot d\mathbf{Q} , \mathbf{Q}^{-1} \cdot d\mathbf{Q} \rangle \tag{2.8}$$

on $\mathcal{G}$ where the bracket product is defined on the Lie algebra $\mathcal{a}$ of $\mathcal{G}$, whose elements $\mathbf{a}$ (with components $a^a{}_b$) are characterised by

$$\mathbf{a} \cdot \eta + \eta \cdot \mathbf{a}^* = 0 \tag{2.9}$$

according to the standard prescription

$$\langle \mathbf{a}, \mathbf{b} \rangle = \tfrac{1}{2}\, tr\{\mathbf{a} \cdot \mathbf{b}\} \tag{2.10}$$

which automatically satisfies the reality condition

$$\langle \mathbf{a}, \mathbf{b} \rangle = \langle \mathbf{a}^*, \mathbf{b}^* \rangle = \overline{\langle \mathbf{a}, \mathbf{b} \rangle} \tag{2.11}$$

as a consequence of (2.11)

It is well known that a homogeneous space $\mathcal{X}$ constructed in this manner is automatically a symmetric space in the sense that there is a tangent space reversing automorphism associated with any arbitrarily chosen fixed element Φ. This automorphism is induced on $\mathcal{X}$ by a corresponding automorphism

$$Q \mapsto Q^{\dagger} = ({}^{\#}\Phi)^{-1} \cdot Q \cdot {}^{\#}\Phi \tag{2.12}$$

on $\mathcal{G}$. This automorphism can be seen to be an inversion, in the sense that

$$(Q^{\dagger})^{\dagger} = Q \tag{2.13},$$

as a consequence of the relation

$$({}^{\#}\Phi)^{-2} = \eta^{*} \cdot \eta^{-1} \tag{2.14}$$

(which expresses the group membership requirement (2.2) subject to the hermiticity condition (2.5)) and of the commutation relation

$$[\eta^{*} \cdot \eta^{-1}, Q] = 0 \tag{2.15}$$

which follows directly from (2.1). To see that this inversion mapping does in fact reverse the tangent space to $\mathcal{X}$ at Φ, one starts by noting that an infinitesimal displacement $d\Phi$ in the tangent space must (by differentiation of 2.3) be expressible in the form

$$d\,{}^{\#}\Phi = [{}^{\#}\Phi, a]\, d\lambda \tag{2.16}$$

where $d\lambda$ is an infinitesimal displacement parameter and a is some element of the algebra $\mathcal{A}$ of $\mathcal{G}$ as characterized by (2.9). The automorphism (2.12) on induces a corresponding automorphism

$$a \mapsto a^{\dagger} = ({}^{\#}\Phi)^{-1} \cdot a \cdot {}^{\#}\Phi \tag{2.17}$$

on the algebra. This automorphism also has the inversion property

$$(a^{\dagger})^{\dagger} = a \tag{2.18}$$

as a consequence of (2.14) and of the analogue,

$$\left[\mathfrak{a}, \eta^{*}\cdot\eta^{-1}\right] = 0 \tag{2.19}$$

of (2.15). The same properties (2.14) and (2.19) also imply

$$\left[{}^{\#}\phi, \mathfrak{a}^{\dagger}\right] = \left[\mathfrak{a}, {}^{\#}\phi\right] \tag{2.20}$$

so that one sees directly that the tangent space elements (2.16) will indeed be reversed by the inversion, i.e.

$$\left(\alpha\,{}^{\#}\phi\right)^{\dagger} = -\alpha\,{}^{\#}\phi \tag{2.21}$$

The automorphism specified according to (2.17) for any arbitrarily chosen fixed point ϕ in $\mathcal{X}$ determines a natural decomposition of the algebra $\mathfrak{a}$ as a direct sum of even and odd subspaces $\mathfrak{a}^{+}$ and $\mathfrak{a}^{-}$: we shall have

$$\mathfrak{a} = \mathfrak{a}^{+} + \mathfrak{a}^{-} \tag{2.22}$$

where the even and odd parts $\mathfrak{a}^{+}$ and $\mathfrak{a}^{-}$ are defined by

$$\mathfrak{a}^{\pm} = \tfrac{1}{2}\left(\mathfrak{a} \pm \mathfrak{a}^{\dagger}\right) \tag{2.23}$$

so that

$$\left(\mathfrak{a}^{\pm}\right)^{\dagger} = \pm\,\mathfrak{a}^{\pm} \tag{2.24}$$

It follows immediately from (2.17) that the commutator of any pair of algebra elements $\mathfrak{a}$ and $\mathfrak{b}$ will satisfy

$$\left[\mathfrak{a}, \mathfrak{b}\right]^{\dagger} = \left[\mathfrak{a}^{\dagger}, \mathfrak{b}^{\dagger}\right] \tag{2.25}$$

The commutators among the odd and even subspaces can therefore be seen to satisfy

$$\left[\mathfrak{a}^{+}, \mathfrak{b}^{+}\right]^{-} = \left[\mathfrak{a}^{-}, \mathfrak{b}^{-}\right]^{-} = \left[\mathfrak{a}^{+}, \mathfrak{b}^{-}\right]^{+} = 0 \tag{2.26}$$

which shows that the odd (but not the even) subspace is a subalgebra. In terms of this notation scheme, the expression (2.16) for a displacement in the tangent space to $\mathcal{X}$ at Φ can be rewritten as

$$d\Phi = 2\,\Phi\cdot a^{-}\,d\lambda \tag{2.27}$$

3. The Gauge Coupling

In pure non-linear σ-models of the standard kind, the elements Φ in the space $\mathcal{X}$ are considered as ordinary physical field variables over the relevant base space $\mathcal{M}$. In terms of the (flat or curved) base space metric

$$ds^2 = g_{\mu\nu}\,dx^\mu dx^\nu \tag{3.1}$$

(where x^μ , $\mu = 1,\dots n$ are local base space coordinates) the standard general form for the Lagrangian L from which the field equations for Φ are to be derived is

$$L = -\tfrac{1}{2}\rho^{-1}\langle \mathcal{J}_\mu, \mathcal{J}^\mu\rangle \tag{3.2}$$

where the base space metric (3.1) is used for raising and lowering of the base space indices μ and for the specification of the base space covariant differentiation operation occurring in the definition

$$\mathcal{J}_\mu = \rho\,\Phi^{-1}\cdot D_\mu \Phi \tag{3.3}$$

of the current matrix. The quantity ρ appearing here is any given (e.g. uniform) weight field on the base space, and the brackets appearing in (3.1) indicate the ordinary trace scalar product on the group algebra $\mathcal{a}$ to which the current matrix must evidently belong.

The purpose of the present work is to consider a rather wider class of models in which the Φ are no longer considered as ordinary physical field variables but as a section[14] of a fibre bundle over the base $\mathcal{M}$ with gauge group $\mathcal{G}$ and adjoint action (2.3) over non-linear fibres of the form $\mathcal{X}$. Such a structure belongs to the general class of non-linear fibre bundles which includes as special cases the familiar examples of the class of principle bundles (in which the fibre is the group itself, subject to its own left action) and the class of ordinary vector bundles (for which the fibre is linear, as in the case of the most commonly used kinds of physical gauge theories such as quantum chromodynamics). The definition of covariant differentiation in any (linear or nonlinear bundle) requires the specification of a differential connection on the associated principle bundle, such a connection being represented in any local gauge patch by a gauge one-form whose components A_μ say are elements of the Lie algebra $\mathcal{a}$ of the group, which in the present case means that they must satisfy

$$A \cdot \eta + \eta^* \cdot A = 0 \tag{3.4}$$

by (2.9). For a non-linear action of the kind specified globally by (2.4) and locally by (2.27) the corresponding operation of covariant derivation on Φ will be given by

$$D_\mu \Phi = \partial_\mu \Phi - \Phi \cdot A^-_\mu \tag{3.5}$$

where ∂_μ denotes ordinary partial differentiation with respect to the local base space coordinates x^μ, and, in accordance with the notation scheme introduced in the previous section,

$$2A^-_\mu = A_\mu + \Phi^{-1} \cdot A^*_\mu \cdot \Phi \,. \tag{3.6}$$

With this definition a local gauge transformation of the form (2.4) will induce a corresponding covariant transformation on $D_\mu \Phi$ provided that A_μ undergoes a corresponding transformation of the appropriate (noncovariant) form, i.e.

$$A_\mu \longmapsto Q^{-1} \cdot (\partial_\mu Q + A_\mu \cdot Q) \,. \tag{3.7}$$

Since the current can be considered globally as a section in the associated (linearly fibred) adjoint algebra bundle, it will itself be subject to an operation of gauge covariant differentiation of the standard kind, namely

$$D_\mu \mathcal{J}_\nu = \nabla_\mu \mathcal{J}_\nu + [A_\mu, \mathcal{J}_\nu] \tag{3.8}$$

where ∇ denotes the ordinary operation of covariant differentiation as defined purely in terms of the metric (3.1) without reference to the gauge field. In terms of this notation the integrability condition on $\mathcal{J}_\mu$ for the existence of Φ satisfying (3.3) is expressible as the identity

$$2 D_{[\mu}\rho^{-1}\mathcal{J}_{\nu]} + \rho^{-2}[\mathcal{J}_\mu, \mathcal{J}_\nu] = -2 \mathbb{F}^-_{\mu\nu} \tag{3.9}$$

where square brackets on indices denote antisymmetrisation and $\mathbb{F}^-_{\mu\nu}$ is the odd part of the covariant gauge curvature field, i.e.

$$2\mathbb{F}^-_{\mu\nu} = \mathbb{F}_{\mu\nu} + \Phi^{-1}\cdot \mathbb{F}^*_{\mu\nu}\cdot \Phi \tag{3.10}$$

where

$$\mathbb{F}_{\mu\nu} = 2\,\partial_{[\mu}A_{\nu]} + [A_\mu, A_\nu] \tag{3.11}$$

As in the case of the ordinary non-linear σ-models for which the gauge field is absent, it can be seen still in this more general case that the hermiticity condition (2.5) on Φ ensures that the current matrix $\mathcal{J}_\mu$ defined by (3.3) and (3.5) will automatically belong to the odd subalgebra $\mathcal{a}^-$, in the sense specified by (2.23), i.e. we shall have

$$\Phi^{-1}\cdot \mathcal{J}^*_\mu \cdot \Phi = \mathcal{J}_\mu \tag{3.12}$$

or equivalently in our condensed notation scheme

$$\mathcal{J}^+_\mu = 0 \tag{3.13}$$

Differentiation of (3.12) leads to the deduction that the even part of the covariant derivative has the antisymmetric form

$$(D_\mu \mathcal{J}_\nu)^+ = \tfrac{1}{2}\rho^{-1}[\mathcal{J}_\nu, \mathcal{J}_\mu] \tag{3.14}$$

and hence that its contraction must vanish, i.e.

$$(D_\mu \mathcal{J}^\mu)^+ = 0 \tag{3.15}$$

thereby establishing the important conclusion that the covariant divergence $D_\mu \mathcal{J}^\mu$ of the current matrix will also automatically belong to the odd subalgebra $\mathcal{A}^-$.

4. Field Equations

We are now in a position to obtain the field equations that are specified by the Lagrangian (3.1), which takes the explicit form

$$L = -\frac{1}{4\rho} g^{\mu\nu} tr\{\Phi^{-1}(D_\mu\Phi)\cdot\Phi^{-1}\cdot D_\nu\Phi\} \tag{4.1}$$

Variation of all the quantities involved gives

$$\begin{aligned} dL = {} & \tfrac{1}{2}(T^{\mu\nu} - L g^{\mu\nu})\, dg_{\mu\nu} + tr\{\mathcal{J}^\mu\cdot dA_\mu\} \\ & + \tfrac{1}{2} tr\{(D_\mu\Phi)\cdot\Phi^{-1}\cdot d\Phi\} \\ & - \nabla_\mu tr\{\tfrac{1}{2}\mathcal{J}^\mu\cdot\Phi^{-1}\cdot d\Phi\} \end{aligned} \tag{4.2}$$

The variation of the gauge field and the metric has been included here in order that the corresponding current and effective energy momentum tensor may be read out. It can be checked that the current matrix $\mathcal{J}^\mu$ appearing in (4.2) does in fact agree precisely with our previous expression (3.3), while the effective energy momentum tensor takes the form

$$T^{\mu\nu} = \frac{1}{2\rho} tr\{\mathcal{J}^\mu\cdot\mathcal{J}^\nu - \tfrac{1}{2} g^{\mu\nu}\mathcal{J}^\rho\cdot\mathcal{J}_\rho\} \tag{4.3}$$

The variational field equations will be specified by the requirement that one should have

$$tr\{(D_\mu \mathcal{J}^\mu)\cdot \Phi^{-1}\cdot d\Phi\} = 0 \tag{4.4}$$

for all allowed variations of the field quantities Φ. If the matrix components of Φ were unconstrained, we should immediately obtain the requirement that the field equations should have the form

$$D_\mu \mathcal{J}^\mu = 0 \tag{4.5}$$

as a consequence of the evident non degeneracy of the trace metric (2.10). However more care is needed in the case under consideration here, for which the field variables are supposed to be constrained by the requirement that $^\#\Phi$ should lie on a single trajectory (characterized by the hermiticity property (2.5)) of the adjoint group action (2.3). Under these conditions the matrix $d\Phi$ is no longer arbitrary, being restricted to have the form (2.27), but we can nevertheless use the fact that the product $\Phi^{-1}\cdot d\Phi$ will be an arbitrary member of the odd subalgebra $\mathcal{a}^-$ as characterized by (2.9) and (2.24). Since we have seen in the previous section that the covariant current divergence $D_\mu \mathcal{J}^\mu$ will also belong automatically to this same odd subalgebra $\mathcal{a}^-$ (by (3.15)), we get back to a set of field equations having the <u>same</u> form (4.5) as in the unconstrained case, provided that the restriction of the trace metric (2.10) to the subalgebra $\mathcal{a}^-$ is known to be non-degenerate, a requirement that can easily be seen to be satisfied in the standard cases for which the conserved form η characterizing the group is herm tean in the complex case or symmetric or antisymmetric in the real case, corresponding to groups of (pseudo) unitary complex matrices or of pseudo (orthogonal) or symplectic real matrices.

An alternative way to derive (4.5) is to work out the Noether identity arising from the invariance of L under a (globally uniform) group action of the form (2.4), which (subject to (3.12)) leads directly to a current divergence condition of the form

$$\nabla_\mu \mathcal{J}^\mu = [\mathcal{J}^\mu, A_\mu] \tag{4.6}$$

which is evidently equivalent to (4.5). The analogous divergence condition on the effective energy momentum tensor that is obtained from the Noether identity arising from the general covariance of the expression (4.1) can be seen to take the form

$$\nabla_\mu T^\mu{}_\nu = tr\{\mathcal{J}^\mu \cdot F_{\mu\nu}\} \qquad (4.7)$$

Exactly divergence free current components and energy momentum tensor could be obtained by adding on appropriate contributions arising purely from the gauge field in the case when the latter obeys dynamic equations of Yang-Mills type in the sense that it is obtained from a total Lagrangian scalar L_T the form

$$L_T = L + L_F \qquad (4.8)$$

where

$$L_F = \frac{1}{8\pi g^2}\langle F_{\mu\nu}, F^{\mu\nu}\rangle \qquad (4.9)$$

for some coupling constant g. The analogue of (4.2) for the Yang-Mills contribution to the total Lagrangian is

$$dL_F = \frac{1}{2}(T_F^{\mu\nu} - L_F g^{\mu\nu})dg_{\mu\nu} + tr\{\mathcal{J}_F^\mu \cdot dA_\mu\}$$
$$- \nabla_\mu tr\{\frac{1}{4\pi g^2} F^{\mu\nu} \cdot dA_\nu\} \qquad (4.10)$$

where the pure Yang-Mills contribution to the total current

$$\mathcal{J}_T^\mu = \mathcal{J}^\mu + \mathcal{J}_F^\mu \qquad (4.11)$$

has the form

$$\mathcal{J}_F^\mu = \frac{1}{4\pi g^2} D_\mu F^{\nu\mu} \qquad (4.12)$$

and the pure Yang-Mills contribution to the total effective energy momentum

$$T_{T}{}^{\mu}{}_{\nu} = T^{\mu}{}_{\nu} + T_{F}{}^{\mu}{}_{\nu} \tag{4.13}$$

has the form

$$T_{F}{}^{\mu}{}_{\nu} = \frac{1}{4\pi g^2} tr\{\mathbb{F}^{\mu\rho}\cdot\mathbb{F}_{\nu\rho} - \frac{1}{4}\mathbb{F}^{\rho\sigma}\cdot\mathbb{F}_{\rho\sigma}\, g^{\mu}{}_{\nu}\} \tag{4.14}$$

Conservation of the total current is trivial since the Yang-Mills field equations evidently reduce to the requirement that it be zero, i.e.

$$\mathcal{J}_{T} = 0 \tag{4.15}$$

which is equivalent to the generalized Maxwell source equation

$$D_{\nu}\mathbb{F}^{\mu\nu} = 4\pi g^2 \mathcal{J}^{\mu} \tag{4.16}$$

for which the field equation (4.5) for Φ is an integrability condition. On the other hand the Noether identity expressing the conservation of the total effective energy momentum tensor has the non-trivial form

$$\nabla_{\mu} T_{T}{}^{\mu\nu} = 0 \tag{4.17}$$

5. Generalized Mazur Identity

Our purpose in this final section is to show that the gauge coupled σ-model set up in the previous sections satisfies a natural generalization of the identity previously constructed by Mazur in the absence of a gauge field for the purpose of proving uniqueness of solutions to wide classes of boundary condition problem subject to appropriate signature conditions.

The situation we wish to consider here is one in which we have <u>two</u> solutions $\Phi_{[0]}$ and $\Phi_{[1]}$ say of the field equations (4.5) as defined with respect to the <u>same</u> background gauge field A_μ. Our purpose is to study a deviation matrix Δ (with components $\Delta_{\dot{a}}{}^{\dot{b}}$) defined by

$$\Phi_{[0]} \cdot \Phi_{[1]}^{-1} = 1 + \Delta \qquad (5.1)$$

so that Δ has the property of vanishing if and only if the two solutions $\Phi_{[0]}$ and $\Phi_{[1]}$ coincide.

The specification (3.5) of the covariant differentiation operation on Φ can be seen to determine a corresponding operation on Δ specified by

$$D_\mu \Delta = \partial_\mu \Delta - [A^*_\mu, \Delta] \qquad (5.2)$$

It is convenient to introduce an abreviated notation in which a bulls eye denotes the difference between any functional of $\Phi_{[1]}$ and the analogous functional of Φ, the functionals themselves being distinguished by the corresponding suffices, so that in particular for the current matrix functional we shall write

$$\mathcal{J}^{\mu\odot} = \mathcal{J}^\mu_{[1]} - \mathcal{J}^\mu_{[0]} \qquad (5.3)$$

In terms of this notation scheme, the covariant derivative (5.2) of Δ can be expressed by

$$\rho D_\mu \Delta = \Phi_{[1]} \cdot \mathcal{J}_\mu^{\odot} \cdot \Phi_{[0]}^{-1} \qquad (5.4)$$

Following lines suggested by the work of Mazur in the absence of the gauge field, we now take the gauge covariant divergence of (5.4), which leads to the identity

$$D_\mu(\rho D^\mu \Delta) =$$

$$\Phi_{[1]} \cdot \{(D_\mu \mathcal{J}^\mu)^\odot + \rho^{-1}(\mathcal{J}^\mu_{[1]} \mathcal{J}_\mu^\odot - \mathcal{J}_\mu^\odot \mathcal{J}^\mu_{[0]})\} \cdot \Phi_{[0]}^{-1} \qquad (5.5)$$

When one takes the trace of this, the fact that the trace of a commutator must vanish causes the gauge covariant derivative on the left hand side to reduce to ordinary covariant derivatives. Thus using the abreviation

$$\Delta = tr\{\boldsymbol{\Delta}\}$$

for the (evidently gauge covariant) trace of the deviation matrix, we obtain the required basic identity in the form

$$\nabla^{\mu}\{\rho\,\partial_{\mu}\Delta\} - tr\{\,\Phi_{[1]}\cdot(D_{\mu}\mathcal{J}^{\mu})^{\odot}\cdot\Phi_{[0]}^{-1}\}$$
$$= \rho^{-1}\,\|\mathcal{J}_{\odot}\|^{2} \qquad (5.7)$$

where we introduce a field dependent norm defined by

$$\|\mathcal{J}_{\odot}\|^{2} = g^{\mu\nu}tr\{\,\mathcal{J}^{*\odot}_{\mu}\cdot\Phi_{[1]}\cdot\mathcal{J}^{\odot}_{\nu}\cdot\Phi_{[0]}^{-1}\} \qquad (5.8)$$

and where use has been made of the oddness properties (3.13) and (3.15) that result from the hermiticity requirement (2.5). The basic identity (5.7) has the property that the left hand side reduces to a pure divergence when the field equations (4.5) one satisfied, since the vanishing of the gauge covariant current associated with the separate solutions evidently implies the vanishing of their difference, i.e.

$$(D_{\mu}\mathcal{J}^{\mu})^{\odot} = 0 \qquad (5.9)$$

Such an identity can be used to establish a rather wide class of uniqueness theorems for cases in which the base metric $g_{\mu\nu}$ and the field matrix Φ have a well defined positive (or negative) definite character, i.e.

$$\xi^{\mu} \neq 0 \;\Rightarrow\; \xi^{\mu} g_{\mu\nu}\xi^{\nu} > 0 \qquad (5.10)$$

and

$$u^a \neq 0 \quad \Rightarrow \quad \bar{u}^{\dot{a}} \Phi_{\dot{a}b} u^b > 0 \qquad (5.11)$$

the latter property being evidently invariant under transformations of the form (2.3). Such a situation arises for coset spaces $\mathcal{G}/\mathcal{H}$ in which the isotropy subgroup $\mathcal{H}$ is the intersection of $\mathcal{G}$ with a strictly unitary group (in the complex case) or a strictly orthogonal group in the real case. Thus in the complex case, for hermitean η with $\mathcal{G}$ of the form $SU(p,q)$ it would be necessary that $\mathcal{H}$ should have the form $S(U(p)\times U(q))$. In the real case, for symmetric η with $\mathcal{G}$ of the form $SO(p,q)$ it would be necessary that $\mathcal{H}$ should have the form $S(O(p)\times O(q))$, while for antisymmetric η (in an even number of dimensions in order to be non-degenerate) with $\mathcal{G}$ of the form $Sp(2q)$ it would be necessary to have $\mathcal{H}$ of the form $OSp(2q)$.

Since a tensor product of positive definite bilinear or sesquilinear forms (in the present instance the tensor with components $g_{\mu\nu}\,\Phi_{[1]\dot{a}b}\,\Phi_{[0]}^{-1\,c\dot{d}}$) is always itself a positive definite form for the tensor product of the corresponding vector spaces (which in the present instance contains the current matrix with components $\mathcal{J}_\nu{}^b{}_c$) one sees that the positivity properties (5.10) and (5.11) imply the corresponding positivity property

$$\mathcal{J}^{\mu\odot} \neq 0 \quad \Longrightarrow \quad \|\mathcal{J}^\odot\|^2 > 0 \qquad (5.12)$$

This property enables one to establish uniqueness theorems for a large class of problems in which Φ is defined over a base space domain Σ say subject to any boundary conditions, on a finite or asymptotic surface S bounding Σ, that are sufficiently stringent to ensure that one has

$$\oint_S \rho(\nabla^\mu\Delta)\, dS_\mu \longrightarrow 0 \qquad (5.13)$$

where dS_μ is the metric normal surface element. Under these circumstances application of Green's theorem to (5.7) gives

$$\int_\Sigma \rho^{-1} \| \mathcal{J}^{\circ} \|^2 d\Sigma = 0 \tag{5.14}$$

where $d\Sigma$ is the metric surface element on the base space. When the conditions (5.10) and (5.11) are satisfied, we can apply (5.12) to (5.14) so as to obtain the conclusion that we must have

$$\mathcal{J}_\mu^{\circ} = 0 \tag{5.15}$$

throughout the domain Σ. By (5.4) and (5.2) this latter result can be seen to be equivalent to the conclusion that the deviation matrix Δ must satisfy

$$\partial_\mu \Delta = [A_\mu^*, \Delta] \tag{5.16}$$

throughout the domain Σ. Hence if the boundary conditions are sufficiently strong to fix the limiting value of Φ uniquely (so that Δ tends to zero there) even at a single point, then we can conclude from the homogeneity of (5.16) that Δ must vanish everywhere in Σ which by (5.1) establishes the required unicity theorem to the effect that

$$\Phi_{[1]} = \Phi_{[0]} \tag{5.17}$$

everywhere.

We can in fact establish such a result for an even more general class of boundary conditions in cases when η is hermitian or real symmetric, so that $\mathcal{G}$ is a pseudo-unitary or pseudo-orthogonal group of type $SU(p,q)$ or $SO(p,q)$, since in these cases it can easily be checked (e.g. by working in a system such that η and $\Phi_{[0]}$ are simultaneously diagonal) that the trace of the deviation matrix itself has the positivity property

$$\Phi^{\circ} \neq 0 \quad \Rightarrow \quad \Delta > 0 \tag{5.18}$$

Under these circumstances by multiplication of (5.7) by Δ^{γ} where γ is an arbitrary positive index, and using (5.9), we obtain a more general divergence relation of the form

$$\nabla^{\mu}\{\rho\Delta^{\gamma}\partial_{\mu}\Delta\}$$
$$= \rho^{-1}\Delta^{\gamma-1}\{\Delta\|\mathcal{J}^{\circ}\|^{2} + \gamma\rho^{2}g^{\mu\nu}(\partial_{\mu}\Delta)\partial_{\nu}\Delta\} \tag{5.19}$$

in which both terms in the bracket on the right will still be non-negative. It follows that even if the boundary, requirement (5.13) is weakened to the more general form

$$\oint_{S}\rho\Delta^{\gamma}(\nabla^{\mu}\Delta)\,dS_{\mu} \longrightarrow 0 \tag{5.20}$$

for any fixed index satisfying

$$\gamma \geq 0 \tag{5.21}$$

then we shall still obtain the conclusion that (5.15) must hold throughout in order to avoid the first term in the bracket on the right hand side of (5.19) becoming strictly positive. The obvious condition

$$\partial_{\mu}\Delta = 0 \tag{5.22}$$

ensuring that the other non-negative term in the bracket should vanish, does not represent an independent restriction but is obtained automatically by taking the trace of the consequence (5.16) of (5.15). Once (5.22) has been obtained in this way it is evident from (5.18) that it can be used directly to establish the unicity property (5.17) whenever the boundary conditions fix the limit of Φ at least at some point.

Acknowledgements

The author wishes to thank G. Gibbons, J. Madore, and N. Sanchez for helpful discussions.

References

(1) R. Geroch (1971) J. Math. Phys. 12, 918.
(2) W. Kinnersley (1973) J. Math. Phys. 14, 651.
(3) N. Sanchez (1982) Phys. Rev. D26, 2589.
(4) P.O. Mazur (1983) Acta Physica Polonica, B14, 219.
(5) P.O. Mazur (1982) J. Phys. A15, 3173.
(6) B. Carter (1971) Phys. Rev. Lett., 26, 331.
(7) D.C. Robinson (1974) Phys. Rev. D10, 458.
(8) D.C. Robinson (1975) Phys. Lett. 34, 905.
(9) B. Carter (1973) in Black Holes ed. B. and C. Dewitt, Gordon and Breach, New York.
(10) B. Carter (1979) in General Relativity ed. S. W. Hawking and W. Israel, Cambridge U.P..
(11) G. Bunting (1981) The Black Hole Uniqueness Theorem in unpublished proceedings of the 2nd Australian Mathematics Convention Sydney, 11-15 May, 1981.
(12) G. Bunting (1983) Proof of the Uniqueness Conjecture for Black Holes, Ph.D. Thesis, Dept of Mathematics, University of New England, Armidale, N.S.W..
(13) B. Carter (1984) preprint, Observatoire de Paris-Meudon.
(14) B. Carter (1979) in Recent Developments in Gravitation (Cargèse, 1978), in Nato Advanced Study Institutes, Series B44, ed. M. Levy and S. Deser Plenum, New York
M. Levy and S. Deser Plenum, New York.

MAGNETIC MONOPOLES

W. Nahm

Physikalisches Institut
der Universität Bonn
Nussallee 12
5300 Bonn 1

1. Introduction

The study of magnetic monopoles exemplifies the new interaction between high energy physics and mathematics. On one hand the discovery of such monopoles would allow deep insights into the basic structure of matter and the early history of our universe. On the other hand, a limiting form of the equations describing the monopoles turns out to be completely integrable, and now is an important member of the slowly increasing set of completely integrable non-linear differential equations, which are much studied by mathematicians both for their intrinsic interest and as a tool in topology and algebraic geometry.

The Maxwell equations in free space are symmetric under interchange of electric and magnetic fields. In matter this symmetry is destroyed, as only the electric field has sources, namely electrically charged matter. Of course it was tempting to introduce magnetic charge into the equations to restore the symmetry, and to look for such charges in exotic places like moon rocks, so far without success. For static charges one obtains

$$B = -\partial\varphi_m , \qquad (1)$$

where B is the magnetic field, and φ_m the magnetic potential, which is related to the magnetic charge density ρ_m by

$$-\partial^2\varphi_m = 4\pi\rho_m \qquad (2)$$

For classical fields this is perfectly consistent.

However, quantization requires a Lagrange formalism, in which the electromagnetic field is the curvature of a U(1) connection, the electromagnetic potential. One component of the Bianchi identities for this curvature is

$$\partial_i B^i = 0, \tag{3}$$

which states that the magnetic field has no sources.

In 1931, Dirac found a way out. In modern terminology, he eliminated a point from space and considered connections on non-trivial U(1) bundles on R^3-pt. These bundles are integer powers of the Hopf bundle. The integers correspond to the magnetic flux through a sphere around the singular point,

$$e\int B\,dS/\hbar = 2\pi n, \quad n \in \mathbb{Z}. \tag{4}$$

Thus one obtains Dirac's quantization condition

$$eg = \tfrac{1}{2} n \hbar \tag{5}$$

for the electric charge e and the magnetic charge

$$g = \int B\,dS/4\pi. \tag{6}$$

Note that this construction only works for the gauge group U(1), not for its non-compact covering. As the unitary representations of U(1) are labeled by the integers, the existence of magnetic charge implies the quantization of electric charge, a conclusion which also is reflected in eq. (5).

The main disadvantage of Dirac's construction is the existence of singular points at the monopole centres, which leads to difficulties. In most of the present work on magnetic monopoles these singularities are regarded just as an idealization which hides the complex structure of the core of real monopoles.

Today most high energy physicists believe that the electromagnetic gauge group $U(1)_{em}$ is a subgroup of a simple or at least semisimple gauge group G, under which the laws of nature are invariant. Such theories are called "grand unified" (GUT), as G should describe electromagnetic, weak, and strong interactions. Dirac's U(1) bundles on R^3-pt then can be replaced by G bundles over all of R^3, which far from the monopole positions reduce to U(1) bundles, in the sense that the components of the connection which are orthogonal to the U(1) generator decrease exponentially with increasing distance from the monopole cores.

The fact that in most physical phenomena only a subgroup H of G is manifest as symmetry group is explained by the lack of G symmetry of the physical vacuum, whose invariance group is H. From experiments we know

$$H \supset (U(1)_{em} \otimes SU(3)_c)/Z_3 \, , \tag{7}$$

where $SU(3)_c$ is the gauge group of the strong interactions. It is not inconceivable that H has further factors corresponding to not yet discovered interactions.

Outside of the monopole cores the gauge fields should belong to a subgroup H_m of H, which can be described as the holonomy group of a connection on a sphere around the monopole core. Dirac's monopoles would have $H_m=U(1)_{em}$, but today one expects an $H_m=U(1)_m$ which projects both to $U(1)_{em}$ and to a subgroup $U(1)_c$ of $SU(3)_c$. Such monopoles also would carry colour magnetic charge. At distances larger than 1fm from the centre, the colour magnetic charge is expected to be screened by vacuum fluctuations, and the monopole should look like the one conceived by Dirac. However, here these quantum effects will not be discussed further.

The physical vacuum should not be described as empty space, but as containing condensed matter of some kind. The simplest description of such matter is by a scalar field $\varphi(x)$, the Higgs field, though this just may be a rough description of a more complex situation. The possible values of $\varphi(x)$ in the vacuum should lie in one orbit of G, and the invariance group of such a $\varphi(x)$ is conjugate to $H \subset G$, such that the vacuum states correspond to the points of G/H.

A two-sphere whose points lie in regions of approximate vacuum contains net charge of magnetic type, if the corresponding element of the homotopy group $\pi_2(G/H)$ is non-trivial. For theories, in which $U(1)_{em}$ is contained in a semisimple GUT group G, the homotopy group $\pi_2(G/H)$ always is non-trivial, and the possible existence of magnetic monopoles is a necessary consequence. However, GUT theories predict monopoles of very large mass, such that only in early stages of the big bang monopole production should have occurred. Unfortunately, not enough is known to predict the present monopole abundance.

In the same way as a non-trivial $\pi_2(G/H)$ predicts point like defects in the physical vacuum, a non-trivial $\pi_1(G/H)$ would yield string like defects, and a non-trivial $\pi_0(G/H)$ would yield domain walls. The latter are excluded on observational grounds, but cosmic strings may have shaped the mass distribution in our universe.

2. Degrees of freedom

The simplest equation for magnetic monopoles arises, if one equates the magnetostatic potential φ_m in eq. (1) with the Higgs field. This is unrealistic, as the real Higgs field is expected to be massive and to yield short range interactions, in contrast to the electromagnetic field. Nevertheless, many interesting properties of magnetic monopoles survive in the limit of vanishing Higgs mass and self-interaction.

Like B, also φ has to belong to the Lie algebra of G, and for its derivation one has to use the covariant derivative

$$\mathcal{D}_i = \partial_i + A_i , \tag{8}$$

such that eq. (1) is replaced by

$$B = -\mathcal{D}\varphi . \tag{9}$$

Outside the core this again yields eq. (1), up to exponentially decrease-ing terms.

Eq. (9) is due to Bogomolny[1]. As we want to obtain the solutions of finite energy, it must be supplemented by the condition

$$\int B^2 d^3x < \infty , \tag{10}$$

where the invariant bilinear form in the Lie algebra of G is not written down explicitly. Eqs. (9-10) imply that far from the monopole core φ is a covariant constant. Moreover, due to the Bianchi identity the energy of the magnetic field can be written as a surface integral,

$$\int B^2 d^3x = -\int B\varphi \, dS , \tag{11}$$

such that non-trivial solutions require a non-vanishing field φ at in-finity. Thus φ indeed has the properties expected of a Higgs field.

Typical solutions of eq. (9) can be visualized as systems of several magnetic monopoles of the same charge in equilibrium. The magneto static repulsion of these monopoles is compensated exactly by the attract ion due to the Higgs field. This effect does not occur for realistic monopoles, for which the Higgs attraction is of short range. Nevertheless the monopoles of the limiting case described by eq. (9) share important properties with realistic ones, in particular their degrees of freedom.

Monopoles have a translational degree of freedom, i.e. a position. They also have an internal degree of freedom, which can be seen in the following way: Consider well separated monopoles and a sphere around one of them. Cut out the interior of S and apply an infinitesimal gauge transformation to it. The new potential inside S is

(12)

If the covariant derivative of α is sufficiently small on S, the new potential can be fitted to the potential outside S after a small adjustment. The condition that α is approximately a covariant constant on S implies that it commutes with the holonomy group H_m. Conversely each generator of the centralizer $C(H_m)$ of H_m yields a possible α. The procedure yields a physically distinct state, if α does not commute with the complete holonomy group inside S. According to eq. (9) one possible choice for α is φ. To complete the fit of the interior and the exterior of S after the transformation, each possible α must commute with φ, such that φ itself yields a U(1) subgroup of the internal symmetry group of the monopole. For eq. (9), φ is a generator of H_m, such that the invariance of φ under action of α need not be imposed separately.

An investigation into the solutions of eq. (9) gave the first hints that the internal degrees of freedom of monopoles are given by $C(M_m)$[2]. Before it was generally assumed that monopoles transform under all of H, like ordinary particles. Only recently it was recognized that a global action of H meets topological obstructions[3,4]. Indeed, consider an H bundle with a global action of $H_e \subset H$ on the fibres. This action can be used to fix the fibre coordinates globally up to an action of $C(H_e)$, i.e. the holonomy group H_m must commute with H_e.

In general one has

$$H \supset H_e' \otimes H_a \otimes H_m' , \qquad (13$$

$$H_e = H_e' \otimes H_a , \qquad (14)$$

$$H_m = H_m' \otimes H_a , \qquad (15)$$

where H_e' and H_m' are non-abelian and H_a is abelian. For GUT monopoles one expects e.g.

$$H_a = U(1)_c \otimes U(1)_{em} \qquad (16)$$

and

$$H_e' = SU(2)_c . \qquad (17)$$

Excitations of the H_e degrees of freedom yield charges of electric type, i.e. the monopoles become dyons. The dyonic degree of freedom may be essential for the monopole phenomenology. I expect the monopoles ground state to have a $U(1)_c$ charge, due to interactions with fermionic matter. Such dyons would be confined. Only bound states of two monopoles and a quark should exist as free particles. This might help to explain the scarcity of monopoles in our universe.

3. Complete integrability

Now let us leave real physics and come back to the solution of eq. (9). Extending the space R^3 to an R^4 with standard euclidean metric and a dummy coordinate x^0 and introducing the R^4 connection

$$A = \varphi\, dx^0 + A_i\, dx^i \tag{18}$$

eq. (9) can be written as the self-duality equation

$$F = *F \tag{19}$$

for the curvature

$$F = dA + A \wedge A. \tag{20}$$

The components of F may be written as commutators of covariant derivatives,

$$F_{\mu\nu} = [\mathcal{D}_\mu \mathcal{D}_\nu]. \tag{21}$$

Eq. (19) has many special properties, which have been collected under the heading "complete integrability", though a unifying theory of completely integrable PDEs (or even ODEs) does not yet exist. In particular, it can be written as compatibility condition for a pair of ordinary differential equations, i.e. a Lax pair.

For this purpose we use the quaternionic structure of R^4 and the notation

$$x = x^\mu q_\mu , \tag{22}$$

where the quaternions q_μ are represented by 2x2 matrices. Rotations of R^4 then take the form

$$x' = a x b, \qquad a, b \in SU(2), \tag{23}$$

which manifestly shows that R^4 vectors transform as (1/2,1/2) representations of $SO(4)=(SO(3)\times SO(3))/Z_2$. Eq. (19) means that F forms a (1,0) representation, whereas arbitrary antisymmetric tensors also have (0,1) components, which change sign under the Hodge duality operation $*$.

Writing

$$D = \mathcal{D}_\mu q^\mu \tag{24}$$

and introducing a new complex variable

$$\pi = \begin{pmatrix} \pi_1 \\ \pi_2 \end{pmatrix} \in CP^1 \tag{25}$$

the Lax pair for eq. (19) is given by

$$(D\pi)_A \, \chi(x,\pi) = 0, \quad A = 1,2. \tag{26}$$

The proof uses the unit antisymmetric 2x2 matrix ε, which is the charge conjugation matrix for the quaternions,

$$q_\mu^* = \varepsilon q_\mu \varepsilon^{-1}, \tag{27}$$

such that

$$(\pi q_\nu)^T \varepsilon q_\mu \pi = \pi^T \varepsilon q_\nu^+ q_\mu \pi. \tag{28}$$

Moreover, $q_\mu^+ q_\nu - q_\nu^+ q_\mu$ is anti-self-dual, due to the fact that according to eq. (23) the expression $x^+ y - y^+ x$ transforms as (0,1) representation of SO(4). The contraction of (0,1) with (1,0) representation vanishes, which completes the proof.

Eq. (26) is a local expression of self-duality. If tha latter is valid in a domain $U \subset R^4$, it is compatible in $U \times CP^1$. Now let us introduce twistor coordinates by

$$\omega = x\pi. \tag{29}$$

Eq. (25) yields

$$x\Pi = \Omega, \tag{30}$$

where

$$\Pi = (\pi, -\varepsilon\pi^*), \quad \Omega = (\omega, -\varepsilon\omega^*), \tag{31}$$

such that x can be recovered from (π,ω). If these twistor coordinates vary over all of CP^3, the corresponding base space is S^4, which compactifies R^4. Otherwise one works within $U\times CP^1 \subset CP^3$.

The projection of $U\times CP^1$ to CP^1 lifts the connection A to a connection in $U\times CP^1$, with trivial components in CP^1. In terms of the twistor coordinates one finds

$$\mathcal{D}_{\pi^*} = -\frac{1}{2}(\pi^+\pi)^{-1} x^+ D\pi + \partial_{\pi^*}, \tag{32}$$

$$\mathcal{D}_{\omega^*} = \frac{1}{2}(\pi^+\pi)^{-1} D\pi. \tag{33}$$

Supplementing eq. (26) by the compatible equation

$$\partial_{\pi^*}\chi(x,\pi) = 0, \tag{34}$$

we see that χ can be described as a holomorphic section of a bundle over a domain in twistor space. If one gauges $A_{\pi}*$ and $A_{\omega}*$ to zero, χ becomes a holomorphic function of the twistor coordinates.

Now take two solutions χ_i of eq. (26) which are holomorphic for $\pi_i=0$, i=1,2 resp. If eq. (26) is formulated in the principal bundle, such that the χ are group elements, one obtains

$$\partial_{\pi^*}h = \partial_{\omega^*}h = 0, \tag{35}$$

where

$$h = \chi_1^{-1}\chi_2 \tag{36}$$

can be interpreted as transition function in the holomorphic bundle over $U\times CP^1$. As the splitting of h into χ_1, χ_2 is unique up to a function which only depends on x, one can recover from $h(\pi,\omega)$ the χ_i and by eq. (26) the connection A up to a gauge transformation. This construction is due to Ward[5)].

If one chooses h arbitrarily, the corresponding solutions of the self-duality equation will have unwanted singularities and belong to connections of the complexified gauge group G_c instead of the compact G. Several procedures have been tried to restrict the form of h to the one which yields monopoles solving eqs. (9-10). One may start with a given solution of eq. (26) and act on it with a Kac-Moody group. This procedure has been developed independently by physicists[6,7] and, in the context of the mKdV hierarchy, by mathematicians[8]. Another possibility is the choice of a suitable ansatz for h, which has been applied successfully in different guises[9,10].

If one wants to implement condition (10) from the start, it is advantageous not to work with D but with its quaternionic adjoint D^+. The Dirac-Weyl equation

$$D^+ \psi(x) = 0 \tag{37}$$

for connections of the desired type has square integrable solutions, and these are the ones we shall use. As the potential A does not change if x^o is translated, we may look for solutions of the form

$$\psi(x) = \exp(i x^o z)\, \psi(x^i). \tag{38}$$

For a given z let $\Psi(z)$ be a vector whose components form a complete set of orthonormal solutions of eq. (37). Now consider a connection T in a new R^4 dual to the original one,

$$T = T_o(z)\, dz + T_k(z)\, dp^k, \tag{39}$$

where

$$T_o(z) = \int \Psi^+ \frac{\partial}{\partial z} \Psi \, d^3x \tag{40}$$

and

$$T_k(z) = -i \int \Psi^+ x^k \Psi \, d^3x. \tag{41}$$

Using the projector

$$\Psi \Psi^+ = 1 - D(D^+D)^{-1} D^+ \tag{42}$$

and the fact that the self-duality implies

$$D^+ D = \mathcal{D}^2 \tag{43}$$

it is easy to see explicitly[11,12] that the curvature F_T of T is itself self-dual, i.e.

$$\frac{dT_i}{dz} + [T_0 T_i] = \varepsilon_{ijk} T_j T_k . \tag{44}$$

The matrix T^0 may be gauged to zero.

From orthonormalized solutions $v(x,z)$ of the dual Dirac-Weyl equation

$$D_T^+ v = 0 \tag{45}$$

one may recover the original connection A by

$$A_k = \int v^+ \partial_k v \, dz \tag{46}$$

$$\varphi = -i \int v^+ z v \, dz . \tag{47}$$

Explicit calculations are easier than it may seem. The z integrations in eqs. (46-47) can be carried out algebraically[13] and the solution of eq. (45) may be reduced to some algebra and an ordinary integration[14]. Most importantly, the non-linear ODE (44) is exactly integrable in terms of Riemannian theta functions, as it can be transformed to a linear flow on the Jacobian of an algebraic curve[15].

To prove these results, one uses the fact that due to its self-duality the connection T satisfies the compatibility condition for the Lax pair

$$(D_T \pi) f(p,z,\pi) = 0 . \tag{48}$$

Using solutions of this equation, one can find the transition matrix h_T for the holomorphic bundle associated to T, namely[16]

$$h_T(\pi,\omega) = \exp\left(\frac{1}{2\pi_1\pi_2} (\pi^T \varepsilon q_k \pi) T^k(z_0) \left(2z_0 - \omega_1/\pi_1 - \omega_2/\pi_2\right)\right) . \tag{49}$$

Different choices of z_0 yield equivalent transition matrices.

On the other hand, one may look for solutions of the form

$$f(p,z,\pi) = \exp(i p_k x^k) f(z,\pi) . \tag{50}$$

Then eq. (48) yields the linear algebraic equation

$$y_k (T^k + i x^k) f = 0 . \tag{51}$$

where

$$y_k = \pi^T \varepsilon q_k \pi . \tag{52}$$

Eq. (51) can be solved, if

$$\det \left(y_k T^k(z) + i y_k x^k \right) = 0 . \tag{53}$$

Due to the self-duality of F_T, the determinant does not depend on z.

Eq. (53) yields an algebraic curve in the complex space with projective coordinates π and

$$\eta = i y_k x^k . \tag{54}$$

Eq. (51) yields a z-dependent line bundle on this curve, i.e. a flow in its Jacobian. From the construction it somehow should be obvious that this flow is linear, but I don't know an easy proof.

The space with coordinates (π, η) is obtained from twistor space by the projection

$$(\pi, \omega) \rightarrow (\pi, i\pi^T \varepsilon \omega) = (\pi, \eta) . \tag{55}$$

It can be interpreted as the space of oriented lines in R^3, given by a direction

$$u^k = -i(\pi^+\pi)^{-1} \pi^+ q^k \pi \tag{56}$$

and a point x mod u. Note that u and y determine each other by the equation

$$\vec{u} \times \vec{y} = i\vec{y} . \tag{57}$$

From eq. (48) one obtains the ODE

$$(u^k D_k + z - i\varphi) f = 0 . \tag{58}$$

Thus the holomorphic bundle on twistor space can be reinterpreted as a holomorphic bundle on the space of lines in R^3, a construction due to Hitchin[17]. He introduced the spectral curve of the monopole as the set of lines for which eq. (57) has a square integrable solution. It turns out that this curve is identical to the one given by eq. (53). Actually this has been proved only for the case where there is a single spectral curve, in particular for G=SU(2). In general eq. (44) is valid apart from some points on the z axis and yields a finite number of spectral curves. In this case the situation is less clear, though partial results have been obtained[18].

The set of lines which constitute the spectral curve has an envelop in R^3 which is the real section of a possibly reducible curve in C^3. This envelop consists of isolated points and closed curves. It has a direct physical significance: In a suitable gauge the Higgs field has the form

$$\varphi(x) = \varphi_{alg}(x) + O(\exp(-c|x|)), \qquad (59)$$

where φ_{alg} is an algebraic function of x. Now the envelop is just the locus of singularities of φ_{alg}. If one continues φ_{alg} around the closed curves of the envelop, it becomes multivalued. The reason is the following: φ_{alg} is constructed by considering all oriented lines of the spectral curve through a given point and selecting just those which run towards the monopole positions. Far away from these, this selection is unique, but it cannot be maintained, when one meets the envelop. The envelop and its relationship to the algebraic part of the Higgs field has been studied both from the point of view presented here[11] and from Hitchin's approach[19].

Finally I would like to mention two results for G=SU(2). Donaldson has considered the moduli space for the corresponding monopoles[20]. It turned out to be the same as the space of rational functions of CP^1. Rouhani has obtained the transition function h for the connection A in terms of T and v, such that a closer comparison to other monopole constructions now may be possible[21].

References

1) E.B.Bogomolny, Sov.J.Nucl.Phys.24 (1976)861.
2) E.Weinberg, Nucl.Phys.B167 (1980) 500, B203 (1982) 445.
3) P.Nelson and A.Manohar, Phys.Rev.Lett.50 (1983) 943.
4) A. Balachandran et al., Phys.Rev.Lett.50 (1983) 1553.
5) R.Ward, Phys.Lett.61A (1977) 81.
6) Ling-Lie Chau and Wu Yong-Shi, Phys.Rev. D26 (1982) 3581.
7) L. Dolan, Phys.Lett. 113B (1982) 387.
8) G.Segal and G.Wilson, Publ.Math. IHES (1984)
G.Wilson, Habillage et fonctions, IHES preprint M-84-28.
9) E.Corrigan and P.Goddard, Comm.Math.Phys. 80 (1981) 575.

10) P.Forgács, Z.Horváth and L.Palla, Phys.Lett. 109B (1982) 200.
11) W.Nahm, in: Group Theoretical Methods in Physics, Trieste 1983 p.189.
12) E.Corrigan and P.Goddard, DAMTP preprint 83-19, to be published in Annals of Physics.
13) H.Panagopoulos, Phys.Rev.D28 (1983) 380.
14) W.Nahm, in: Group Theoretical Methods in Physics, Istanbul 1982, p. 456.
15) P.Griffiths, Linearizing flows and a cohomological interpretation of Lax equations, preprint 1984.
16) P.Wainwright, Durham preprint DTP-83-21.
17) N.Hitchin, Comm.Math.Phys.83 (1982) 579, 89 (1983) 145.
18) M.Murray, Comm.Math.Phys. 90 (1983) 263 and private communication.
19) J.Hurtubise, The asymptotic Higgs field of a monopole, to be published in Comm.Math.Phys.
20) S.Donaldson, Nahm's equation and the classification of monopoles, to be published in Comm.Math.Phys.
21) S.Rouhani, Phys.Rev. D30 (1984) 819.

MULTIMONOPOLES AND THE RIEMANN-HILBERT PROBLEM.

P. Forgács

Central Res. Inst. Phys.
H-1525 Budapest 114, P.O.B. 49
Hungary.

I give an overview on generating self-dual multimonopole solutions applying soliton theoretic methods, and point out other possible uses of these ideas.

1 - Introduction.

It seems that substantial progress in today's physics is more and more linked to the understanding of the nonlinearities of proposing yet another model of the world. There is no hope to understand the strong interactions in a nonperturbative way. Spontaneously broken gauge theories (SBGT) provide us with an attractive framework for the description of electroweak interactions and for the unification of strong and electroweak forces. SBGT's are perturbatively renormalizable however these theories also contain "topological excitations" (vortices, monopoles) at least at the classical level. In other words, there are topologically inequivalent sectors in the space of all classical configurations. Also in four Euclidean dimensions in pure gauge theories there are topologically nontrivial solutions (instantons) which give the vacuum a rich structure. It is of course an open question what is the role of classical solutions in the full quantum theory.
In this talk I shall give an overview of the method our group in Budapest [Z. Horvath, L. Palla, P.F.] has developed to generate fully explicit solutions in SBGT in 3 dimensions corresponding to static multimonopoles (that is monopoles with magnetic charge greater than 1).
These solutions exist only under rather special circumstances. The Higgs field must be in the adjoint representation and there should be no Higgs potential. Although this last condition seems to be rather unphysical,

introducing supersymmetry which is motivated by the Montonen-Olive conjecture, automatically forbids Higgs self-interaction terms.
Up to now one can find only the solutions of the Bogomolny equations [1] although it was shown by Taubes [2] that static, finite energy nontrivial solutions exist which do not satisfy the first order Bogomolny eqs. The key to solve the Bogomolny eqs. is the existence of an associated linear system which can be succesfully tackled by a generalization of the "inverse scattering" method of Zakharov-Shabat and al. [3].
The full second order field equations also have an associated linear system but one has to go to 8 dim. to write it down and no explicit solutions were generated but it seems worthwhile to pursue further these ideas.
I think that understanding the mathematical structure of gauge in a deeper way it will prove relevant for the real world as well.

2 - Connection between monopoles and self-dual gauge fields.

We consider an SU(N) gauge theory with scalar fields in the adjoint representation in the limit of vanishing Higgs potential.
The Lagrangian density is

$$L = -\frac{1}{4} F^a_{\mu\nu} F^{a\mu\nu} \frac{1}{2} (D_\mu \Phi)^a (D^\mu \Phi)^a \tag{1}$$

where

$$F_{\mu\nu}{}^a = \partial_\mu A^a_\nu - \partial_\nu A^a_\mu - f^{abc} A^b_\mu A^c_\nu$$

$$(D_\mu \Phi)^a = \partial_\mu \Phi^a - f^{abc} A^b_\mu \Phi^c$$

The hamiltonian density for static configurations with no electric fields ($A^a_o = 0$) is

$$H = \frac{1}{4} F^a_{ij} F^{aij} + \frac{1}{2} (D_i \phi)^a (D^i \phi)^a \tag{2}$$

$([i,j = 1,2,3 \; ; \; a = 1, \ldots N]$
The field equaitons of this theory are solved by configurations satisfying the Bogomolny equations :

$$F^a{}_{ij} = - \varepsilon_{ijk} (D^k \Phi)^a \tag{3}$$

The engergy, E, can be written using the Bogomolny equations as

$$E = \int_{\mathbb{R}^3} H d^3 x = \frac{1}{2} \int_{\mathbb{R}^3} \Delta |\phi|^2 \, d^3x \tag{4}$$

where

$$|\phi|^2 = \sum_{a=1}^{N} \Phi^a \Phi^a \quad \text{and} \quad \Delta = \partial_i \partial^i$$

The topological charge, n, is given by

$$n = \lim_{r\to\infty} \frac{1}{8\pi v} \int_{r=\text{const.}} dS^i \; \partial_i \; |\phi|^2 \qquad (5)$$

Since the asymptotic boundary condition $\lim_{r\to\infty} |\phi|=v$ is imposed, for a configuration with topological charge n the large distance behaviour of $|\phi|$ is

$$|\phi| \to v - \frac{n}{r} + 0(\frac{1}{r^2}) \qquad \text{as } r\to\infty \qquad (6)$$

One should keep in mind that (6) alone does not guarantee that the topological charge is indeed n due to the presence of possible singularities In the rest of this article I take the vacuum expectation value of the Higgs field, v = 1.

Let us consider now a pure SU(N) gauge theory with Lagrangian

$$L = - \frac{1}{4} \; F_{\mu\nu}^{\;a} \; F^{a\mu\nu}$$

The Euclidean space ($\mathbb{R}^4$) field equations are solved by self-dual (SD) configurations which satisfy

$$F_{\mu\nu}^{\;a} = {}^*F_{\mu\nu}^{\;a} = \frac{1}{2} \; \varepsilon_{\mu\nu\rho\sigma} \; F^{a\rho\sigma}$$

If all the gauge fields A_μ^a are independent of one coordinate ("Euclidean time") the self-duality equations (SDE) reduce to

$$F_{ij}^a \;\; = - \;\; \varepsilon_{ijk} \; (D_k A_o)^a$$

where we recognize the Bogomolny equations (3) reinterpreting A_o^a as the Higgs field.

This formal connection between the SDE and the Bogomolny eqs. has extremely interesting and useful consequences. The spherically symmetric, charge one monopole [4] has a so called instanton chain representation [5,6] that is a special multi-instanton configuration which exactly reproduces the monopole field in the limit when the number of instantons goes to infinity. (Clearly a finite energy configuration in $\mathbb{R}^3$ has infinite action in $\mathbb{R}^4$). This observation has a far reaching generalization, namely any monopole solution has such instanton-chain representation and moreover these chains can be explicitly constructed [7].

As the first step towards quantizations is to study the fluctuation spectrum around the classical solutions it is very important to note that the connection between the "static" Euclidean pure Yang-Mills fields and the monopole model (2) can be easily extended to the small oscillation problem [6]. Therefore using the results of instanton theory one can construct the solutions of the small oscillation problem around a multimonopole configuration if there exists a instanton-chain representation. In view of this remark the results of Chakrabarti [7,8] seem especially important.

Let us now introduce the associated linear system to the SDE. Defining complex coordinates as : $y = x_1 + ix_2$, $z = x_3 + ix_4$, $\bar{y} = x_1 - ix_2$, $\bar{z} = x_3 - ix_4$ the SDE take the form [9]

$$F_{yz} = F_{\bar{y}\bar{z}} = 0$$
$$F_{y\bar{y}} = F_{z\bar{z}} = 0$$

The first two eqs. are solved by

$$Ay = - D_{,y} D^{-1} \ , \ A\bar{y} = D^{+-1} D^{+}_{,\bar{y}} \ , \text{ etc}$$

where $D \in SL(N,\mathbb{C})$. Introducing the gauge invariant quantity

$$\dot{g} = D^{+}D \tag{7}$$

The last equation reduces to [10]

$$(g_{,y}\, g^{-1})_{\bar{y}} + (g_{,z}\, g^{-1})_{\bar{z}} = 0 \tag{8}$$

It is easy to see that g is a hermitian matrix with unit determinant. The key observation is that eq. (8) is the compatibility condition for the following linear system [11]

$$\begin{aligned} (\lambda\, \partial_{\bar{z}} + \partial_y)\psi &= B_y\ \psi \\ (-\lambda\, \partial_{\bar{y}} + \partial_z)\psi &= B_z\ \psi \end{aligned} \tag{9}$$

where $B_y = g_{,y}\, g^{-1}$, $B_z = g_{,z}\, g^{-1}$, $\psi = \psi\,(\lambda, y, \bar{y}, z, \bar{z})$ is an NxN matrix function for which we impose the boundary condition

$$\psi\,(\lambda = 0,\ y,\bar{y},z,\bar{z}) = g$$

The parameter λ is the so called spectral parameter.

3 - A solution generating method for the SDE.

In what follows we show in detail haw can one construct solutions of the linear system (9) which enables us to generate the multimonopoles. We need the following input for our construction :

a) a known solution of (8) denoted by g_o

b) the corresponding solution of the linear system (9) $\psi_o(\lambda)$ satisfying $\psi_o(\lambda=0) = g_o$

Given a) & b) we look for solutions of (9) in the following form

$$\psi(\lambda) = \chi(\lambda)\ \psi_o(\lambda) \tag{10}$$

and the next crucial input is the following ansatz for $\chi(\lambda)$:

$$\chi(\lambda) = 1 + \sum_{k=1}^{n} \frac{Rk}{\lambda - \mu_k} \tag{11}$$

where R_k's are NxN matrices independent of λ . In other words $\chi(\lambda)$ is a meromorphic function in the complex λ plane. Using (11) as we shall see below the construction of the Rk matrices becomes completely algebraic. The equations for $\chi(\lambda)$ are :

$$(\lambda\, \partial_{\bar z} + \partial_y)\, \chi \cdot \chi^{-1} + \chi\, \overset{\circ}{B}_y\, \chi^{-1} = B_y$$
$$(-\lambda\, \partial_{\bar y} + \partial_z)\, \chi \cdot \chi^{-1} + \chi\, \overset{\circ}{B}_z\, \chi^{-1} = B_z \qquad (12)$$

Here we introduced the obvious notation $B_z{}^o = g_{o,z}\, g_o{}^{-1}$, $B_y{}^o = g_{o,y}\, g_o{}^{-1}$.
From (11) one can easily deduce that

$$Rk\ \chi^{-1}(\mu) = 0 \qquad (13)$$

and (13) implies

$$R_k = n^{(k)} \circ m^{(k)}\ ; \quad \chi^{-1}(\mu_k) = q^{(k)} \circ p^{(k)}$$
$$m^{(k)} . q^{(k)} = \sum_{a=1}^{2} m_a^{(k)} q_a^{(k)} = 0 \qquad (14)$$

In fact in (14) we assumed that the gauge group is SU(2). The situation for SU(N) is analogous but the formulae become rather messy. See [12] for details. The LHS of (12) would contain terms with second and first order poles at $\lambda = \mu_k$ whereas the RHS of (12) is analytic in λ . Therefore the residua must vanish at the pole sites :

Absence of second order poles :

$$\mu_k\, \partial_{\bar z}\, \mu_k + \partial_y\, \mu_k = 0$$
$$-\mu_k\, \partial_{\bar y}\, \mu_k + \partial_z\, \mu_k = 0$$

The general solution for the poles, μ_k, is given as

$$h(\mu_k\, y - \bar z,\ \mu_k\, \bar z + y,\ \mu_k) = 0 \qquad (15)$$

where h is an arbitrary (nice) function.

Absence of first order poles :

$$(\mu_k\, R_{k,\bar z} + R_{k,y})\, \chi^{-1}(\mu_k) + R_k \overset{\circ}{B}_y\, \chi^{-1}(\mu_k) = 0$$
$$(-\mu_k\, R_{k,\bar y} + R_{k,z})\, \chi^{-1}(\mu_k) + R_k \overset{\circ}{B}_z\, \chi^{-1}(\mu_k) = 0 \qquad (16)$$

From eqs. (16) we obtain the vectors $m^{(k)}$:

$$m_a^{(k)} = M_b^{(k)}\, [\psi_o^{-1}(\mu_k)]_{ba}$$

where $M^{(k)} = M^{(k)}(\mu_k\, y - \bar z,\ \mu_k \bar z + y,\ \mu_k)$ are otherwise arbitrary vectors. To complete the construction of the matrix R_k we have to determine the $n^{(k)}$ vectors. This can be achieved using the hermiticity of the matrix g. This hermiticity property ensures that our solutions will be in SU(2) indeed. We impose the following condition :

$$g = \chi(\lambda)\, g_o\, \chi^+\left(-\frac{1}{\bar\lambda}\right) \qquad (17)$$

which guarantees the hermiticity of g. As a consequence we see that if $\chi(\lambda)$ has a pole at $\lambda = \mu$ then $\chi^{-1}(\lambda)$ has a pole at $\lambda = -\frac{1}{\mu}$. The $n^{(k)}$ vectors are determined taking the residua at $\lambda = -\frac{1}{\mu_k}$ of eq. (17), we get

$$n^{(k)} = \sum_{\ell=1} \frac{1}{\bar\mu_\ell}\, g_o\, \bar m^{(\ell)}\, (\Gamma^{-1})_{\ell k}$$

where the matrix Γ is defined as

$$(\Gamma)_{k\ell} = \frac{\sum_{d,c} m^{(k)}(g_o)_{dc} m^{(\ell)}_c}{1 + \mu_k \bar{\mu}_\ell} \qquad (18)$$

This way we completed the construction of the R_k matrices and we have a new solution $\psi(\lambda)$ of our linear system. The matrix g satisfying (8) can be written as :

$$g_{ab} = (g_o)_{ab} - \sum_{k,\ell} \frac{1}{\mu_k \bar{\mu}_\ell} \Gamma^{-1}_{\ell k} N^\ell_a N^k_b \qquad (19)$$

where $N^k_a = m^{(k)}_b (g_o)_{ba}$. There is still one problem, the determinant of g as in (19) is not unity.

Although to calculate detg is not entirely trivial here I just quote the result (for details see [12]) :

$$\det g = (-1)^n \left(\prod_{k=1}^{n} \frac{1}{|\mu_k|^2} \right) \det g_o \qquad (20)$$

So defining the normalized g_n as

$$g_n = g \prod_{k=1}^{n} |\mu_k| \qquad (21)$$

it still satisfies (8) and det $g_n = (-1)^n$ det g_o. Therefore if we start with a g_o having det $g_o = 1$, after an even number of steps detg_n = 1 however after an odd number of steps detg_n=-1 which is a problem. In this latter case one has to start with a seed solution g_o where detg_o=-1. This corresponds in general to a complex solution. To generate the monopoles we start with a diagonal g_o and there it is sufficient to multiply g_o with a constant diagonal matrix to change the sign of detg_o.

Now it is very easy to adopt the above results for generating solutions of the Bogomolny eqs, which in our notations look like as

$$(g_{,y}\, g^{-1})_{\bar{y}} + (g_z\, g^{-1})_{z} = 0 \qquad (22)$$

the associated linear system is

$$\begin{aligned} (\lambda\, \partial_z + \partial_y)\psi &= g_{,y}\, g^{-1}\, \psi \\ (-\lambda\, \partial_y + \partial_z)\psi &= g_{,z}\, g^{-1}\, \psi \end{aligned} \qquad (23)$$

where $y = \frac{1}{2}(x_1 + ix_2)$, $z = x_3$

The formulae for g are the same as in eqs (19,21) only the characteristic variables and the poles change a bit :

The implicit equation defining the poles (15) becomes :

$$h(\gamma(\mu), \mu) = 0 \qquad (24)$$

where

$$\gamma(\mu) = y_\mu - \frac{\bar{y}}{\mu} - z$$

Also $M^k = M^k(\gamma(\mu_k), \mu_k)$.
This completes the algebraic solution generating method of the SDE (Bogomolny eqs.). The only analytic work to be done is to solve (23) to get $\psi_o(\lambda)$, which is not very difficult for simple seed solutions.

4 - Construction of monopole solutions.

I proceed now to show how to apply the formalism presented in section3 in practice. First I generate the axially symmetric monopoles and then the general charge n solutions without any symmetry.
The reduction of eqs. (22) & (23) to axial symmetry is rather interesting. Then $g = g(\rho,z)$ $(x_1 + ix_2 = \rho\, e^{i\varphi})$ and assuming that g is a real symmetric 2x2 matrix eqs(22) reduce to the eqs. of an SL(2,$\mathbb{R}$) sigma model (in 2 dim.) which are also equivalent to the Einstein eqs. describing stationary, axially symmetric space-times (in vacuum). These eqs. are known as the Ernst eqs. [13]. When we impose axial symmetry on the linear system (23) there is a subtlety ; namely $\psi(\lambda)$ also depends on the polar angle in the following way :

$$\psi(\lambda,\rho,z,\varphi) = \psi(\lambda e^{i\varphi}, \rho, z)$$

Defining $\tilde{\lambda} = \lambda e^{i\varphi}$ we obtain from (23) the following linear eigenvalue problem :

$$\begin{aligned} \tilde{\lambda}\, \psi_{,z} + \psi_{,\rho} + \frac{\tilde{\lambda}}{\rho}\, \psi_{,\tilde{\lambda}} &= g_{,\rho}\, g^{-1}\psi \\ -\tilde{\lambda}\, \psi_{,\rho} + \frac{\tilde{\lambda}^2}{\rho}\, \psi_{,\tilde{\lambda}} + \psi_{,z} &= g_{,z}\, g^{-1}\, \psi \end{aligned} \qquad (25)$$

It is easy to show that (25) is equivalent to the linear eigenvalue problem for the Ernst eq. "pulled out of a hat" by Belinski, Maison, Zakharov. [14]. It is important to note that in the axially symmetric case the pole eqs. have a unique solution (h is a completely determined function in this case) :

$$\mu = \frac{-w+z \pm \sqrt{(w-z)^2 + \rho^2}}{2\, y} = \frac{-w+z + R(w)}{\rho}\; e^{-i\varphi} \qquad (26)$$

where w is an arbitrary constant. As we shall see the value of these constants is completely fixed from imposing regularity on our solutions. This corresponds to the following choice of the function $h(\gamma,\mu)$ in the pole eq. (24) :

$$h = \gamma(\mu) + w = 0 \qquad (27)$$

To generate the charge one monopole we now have to specify the seed solution g_o, then solve (23) to find $\psi_o(\lambda)$. It is natural to choose a g_o describing a Higgs vacuum, that is $|\phi|^2 = 1$ and $F^a_{ij} = 0$. We made the following simple choice for g_o :

$$g_o = \text{diag}\,(e^z\ ,\ e^{-z}) \qquad (28)$$

However this choice for g_o is clearly not unique, that is there are other g_o's also corresponding to the Higgs vacuum but not all of them gives the same result for the new g_n. It is easy to solve (23) for $\psi_o\,(\lambda)$:

$$\psi_o(\lambda\,) = \text{diag}\,(e^{\,z-\,\lambda y}\ ,\ e^{\,-z\,+\,\lambda y}) \qquad (29)$$

We now apply one step to ψ_o. As we remarked earlier an odd number of steps changes the sign of det g_o, so we multiply g_o with C=diag(1,-1). It is easily verified that g as given in (20) depends only on the ratio $M_1^{(k)}\,M_2^{(k)-1}$, therefore without loss of generality we can take

$$M_a^{(k)} = \left(\exp\,[-f_k(\gamma,\lambda\,)]\,,\ \exp\,[f_k\,(\gamma,\lambda\,)]\right) \qquad (30)$$

In the axially symmetric case the f_k's are just constants. If we make one step only $\Gamma_{k\ell}$ is a 1x1 matrix which as calculated without any problem :

$$\Gamma = \Gamma_{11} = \frac{\rho^2}{z+R}\ \frac{\sinh(R-\alpha)}{R} \qquad (31)$$

where $R = \sqrt{z^2+\rho^2}$, $\alpha = f + \bar{f}$. A necessary condition for the regularity of the solution is the regularity of Γ and the existence of Γ^{-1}. In (31) this implies $\alpha = 0$. From (20), (21) we obtain g_n :

$$\begin{aligned} g_{11} &= \frac{e^2}{\rho}\,(z + R - \frac{Re^R}{\sinh R}\,) \\ g_{22} &= -\frac{e^2}{\rho}\,(z + R + \frac{Re^R}{\sinh R}\,) \\ g_{12} &= g_{21} = \frac{R}{\sinh R} \end{aligned} \qquad (32)$$

One can immediatly verify for (32) that det g = 1.

Any hermitian 2x2 matrix with unit determinant can be parametrized as

$$g = \frac{1}{f}\begin{pmatrix} 1 & -\psi \\ -\bar{\psi} & f^2 + \psi\bar{\psi} \end{pmatrix} \qquad (33)$$

For g given by (32) we get

$$f = \frac{\rho\ e^{-2}}{z-R\ \coth R} \qquad \psi = f\,\frac{R}{\rho\,\sinh\,R} \qquad (34)$$

One can compute from (34) the components of the gauge potentials and verify that we indeed have the spherically symmetric, charge one monopole. I note here that Chadrabarti found a simpler form [7] for the charge one monopole, mainly because his seed solution is still spherically symmetric. Also, I would like to emphasize that the choice of g_o is crucial. For example choosing $g_o = \text{diag}\,(\rho^k e^z, -\frac{1}{\rho^k} e^{-z})$ the resulting new g after one step is a singular solution, very different from the monopole although g_o also corresponds to the Higgs vacuum. It is also clear that the drawbacks of the method I presented are the following : One has to choose parameters to guarantee the regularity of the solution (in other words we generate local solutions) and it is very cumbersome to compute the gauge potentials from the matrix g. When one is within axial symmetry, one gets around this last problem in two ways. Either one applies Bäcklund transformations or introduce the so called superpotentials [15] from which it is relatively easy to get the length of the Higgs field $|\phi|^2$ which virtually contains all the relevant information about the solution. I will say a few words about these later. In the general (non - symmetric) case only the present method works.
Let us now proceed to construct the axially symmetric charge n monopole configurations. Using (29), (30) we get for Γ_{ij} :

$$\Gamma_{ij} = \frac{1}{1+\mu_i \bar{\mu}_j} [\exp\{-z-(f_i+\bar{f}_j)+\mu_i y+\bar{\mu}_j\bar{y}\} +(-1)^n \exp\{z+f_i+\bar{f}_j-\mu_i y-\bar{\mu}_j\bar{y}\}] \quad (35)$$

where we multiplied g_o with $C = \text{diag}(1,(-1)^n)$. The poles, μ_k, satisfy

$$\prod_{k=1}^{n} (\gamma(\mu) - \tilde{W}_k) = 0 \quad (36)$$

Demanding regularity of $|\Phi|^2$ fixes the values of $\tilde{w}_k$'s and restrict the possible values of the constants f_k 's. The simplest choice for the fk's is fk = 0, k = 1,...n.
Then the $\tilde{w}_k$'s are :

$$\tilde{w}_k = i\pi \frac{n+1-2k}{2} \qquad 1 \le k \le n \quad (37)$$

There are other allowed choices for the fk's, however they do not give any free parameter. This is not surprising as it was shown in [16] that axial symmetry strongly reduces the degrees of freedom for finite energy monopole solutions so the only possible configurations are "superimposed" monopoles at a single point. This means that the number of parameters for this case is five. (3 parameters for the location and 2 for the direction). Since we fixed the position of the monopoles at the origin and the symmetry axis is the z axis we expect not to have any free parameters. As it was shown in [17,18] any n-monopole solution belongs to a 4n-1 parameter class, so the next step in our program is to look for these more general configurations. It is essential to find the correct poles.

The only guideline we have is the form of $h(\gamma,\mu)$ in the axially symmetric case, (36). It is natural to take again for $h(\gamma,\mu)$ a polinomial of degree n in γ and to restrict the degree of the equation for μ to 2n. Then the most general h can be written as

$$h(\gamma,\mu) = \sum_{i=1}^{n} a_i(\mu)\,\gamma^i \tag{38}$$

where $a_i(\mu) = \sum_{j=o}^{n-i} (b_j\,\mu^j + c_j\,\mu^{-j})$

$h(\gamma,\mu)$ contains $2(n+1)^2$ -2 free parameters. In (30) the fk's are not constants any more, however the "minimal deformation" of the axially symmetric $M_a^{(k)}$'s is to assume that all the $M_a^{(k)}$'s are described by just one function :

$$fk(\gamma,\lambda) = f(\gamma,\lambda) \quad k = 1, \ldots n$$

In addition we make one last assumption which is quite natural, namely that $f(\gamma,\lambda)$ is analytic at $\lambda = 0$, i.e.

$$f(\gamma,\lambda) = \sum_{i=o}^{\infty} f_i(\lambda)\,(\gamma\lambda)^i$$

As in (20), (21) only $\psi_0^1(\mu_k)$ and $M^k(\mu_k, \gamma(\mu_k))$'s are present we effectively need the values of the arbitrary function, f, at the pole sites, μ_k's. Since the algebraic equation defining the poles is of degree n in γ, we can express all γ^m's when $m > n-1$ in terms of γ^i $(i=0,\ldots n-1)$ at the pole sites μ_k. Consequently one can uniquely associate with any power series in γ a polinomial of degree n-1 (in γ) yielding the same values at all pole sites, μ_k. Therefore without any loss of generality we can take f as a polinomial of degree n-1 in γ.

A necessary condition for the regularity of the solution is the non-singularity of Γ_{ij} when $\mu_i = -\bar{\mu}_j^{-1}$. The expression we get for Γ_{ij} is:

$$\Gamma_{ij} = (1+\mu_i\,\bar{\mu}_j^{\,1})\,[e^{\alpha ij} + (-1)^n e^{-\alpha ij}] \tag{39}$$

where $\alpha ij = \gamma(\mu_i) + \bar{\gamma}(\mu_i^{-1}+\bar{\mu}_j) - f(\mu_i,\gamma(\mu_i)) - \bar{f}(\mu_j,\gamma(\mu_j))$

The regularity conditions imply that

$$\alpha(\mu_i, \gamma(\mu_i)) = i\pi\, w_i^{(n)} \tag{40}$$

where $\alpha(\lambda,\gamma(\lambda)) = \gamma - f(\lambda,\gamma(\lambda)) - \bar{f}(-\bar{\lambda}^{-1}, \gamma(-\bar{\lambda}^{-1}))$.

α is a polinomial of degree n-1 in γ :

$$\alpha = \sum_{i=0}^{n=1} \alpha_i(\lambda)\,\gamma^i$$

Assuming that α_i (λ)'s are analytic in an annular domain, using Cauchy's formula we get for $f(\lambda,\gamma)$:

$$f(\lambda,\gamma) = \frac{1}{2\pi i} \oint_{|\lambda|<|\zeta|} \frac{\alpha(\zeta,\gamma)}{\lambda-\zeta} d\zeta + \gamma \tag{41}$$

The next step is to compute α explicitely. Since α is a polinomial of degree n-1 with its values given at n points (in eqs (40)) it is uniquely determined by Lagrange's interpolation formula

$$\alpha(\lambda, \gamma(\lambda)) = i\pi \sum_{i=1}^{n} W_j^{(n)} \prod_{i\neq j} \frac{\gamma(\lambda) - \gamma_i(\lambda)}{\gamma_j(\lambda) - \gamma_i(\lambda)} \tag{42}$$

where we wrote h as

$$h = \prod_{i=1}^{n} (\gamma - \gamma_i(\lambda)).$$

Since $\alpha(\mu, \gamma(\mu)) = \bar{\alpha}(-\bar{\mu}^{-1}, \gamma(-\bar{\mu}^{-1}))$ by construction this implies extra constraints for h, namely

$$h = (\lambda, \mu) = \bar{h}(\gamma(-\bar{\mu}^{-1}), -\bar{\mu}^{-1})$$

Furthermore in eq. (42) the w_i's should satisfy

$$W_i^{(k)} = -W_j^{(k)} \text{ if } \gamma_i(\lambda) = \bar{\gamma}_j(-\bar{\lambda}^{-1})$$

The $W_i^{(k)}$'s are determined from the axially symmetric case yielding

$$W_i^{( )} = \frac{n+1-2i}{2} \quad 1\leq i\leq n$$

This implies that the number of free parameters would be halved, so h contains only $(n+1)^2-1$ free parameters. From eq. (41) we obtain the following constraints for the $\alpha_i(\lambda)$'s :

$$\frac{\alpha_i(\zeta)}{\zeta^{j+1}} d\zeta = 0 \qquad \begin{matrix} j=-i+1,\dots,i-1 \\ 2\leq i\leq n-1 \end{matrix} \tag{43}$$

$$\frac{1}{2\pi i} \oint \frac{\alpha_1(\zeta)}{\zeta} d\zeta = 1 \tag{44}$$

The constraints (43), (44) express the fact that f is a polinomial of $\lambda\gamma$ and they represent $\sum_{i=2}^{n-1}(Zi-1)+1$ additional relations among the $(n+1)^2-1$ free parameters in h. Consequently our solutions depends on 4n-1 parameters. It reduces to the axially symmetric case by construction when $a_i(\mu)$'s are constants, independent of μ (uniquely determined by w_i's).
It is not very difficult to derive the asymptotic behaviour of $|\phi|$ using $\partial_z \mu_i \approx -r^{-1} \mu_i$:

$$|\phi| \to 1 - \frac{n}{r} \text{ as } r \to \infty$$

Unfortunately it is very hard to get explicit formulae for the length of the Higgs field for a general n monopole solution. One should parametrize the constraints (43),(44) in a clever way and solve the algebraic eqs. (of degree 2n) for the μ's. For the n=2 case we have numerically computed $|\phi|$ and proved that it is nowhere singular [19].
Let us describe the separated two monopole in some detail. In this case $f(\lambda,\gamma)$ can be easily computed. We took h as

$$h = \gamma^2 + A(\mu^2+\mu^{-2}) + B \tag{45}$$

where A,B are real constants. We note that using similar arguments as in [20] h can be always reduced to this form. The only constraint now is (44) which implies the following relation between A and B

$$\sqrt{B} = \frac{1}{\sqrt{1+\beta}}\, K\left(\frac{2\beta}{1+\beta}\right),\quad A = \frac{1}{2}\,\beta B,\ -1 < \beta \leq 0$$

where K(m) is a complete elliptic integral of the first kind with parameter m [21]. For the choice (45) of h f is given by

$$f(\lambda,\gamma) = \frac{\gamma\lambda}{4i} \oint_{|\lambda|<|\zeta|} \frac{d\zeta}{\zeta-\lambda}\,[A(\zeta^4+1) + B\zeta^2]^{-\frac{1}{2}} \tag{46}$$

In the case when β is negative the integrand in (46) has four branch points on the real axis. We deform the contour of integration in (46) to the cuts chosen to run on the real axis from $-\infty$ the smallest branch point and from the largest to ∞. As a result we get

$$f = (\lambda,\gamma) = \frac{\gamma}{\sqrt{-A}\,\tilde{\beta}}\left(K(\tilde{\beta}^{-4}) - \pi(\lambda^2\tilde{\beta}^{-2}\backslash\delta)\right)$$

$$\sin\delta = \tilde{\beta}^{-2} \qquad \tilde{\beta} = \sqrt{-\beta^{-1} + \sqrt{\beta^{-2}-1}}$$

and $\Pi(n\backslash\delta)$ is the complete elliptic integral of the third kind [21]. The only remaining free parameter, β, in our solution determines the location of the zeroes of $|\phi|$ on the x_1 axis as

$$X_1 = \pm\sqrt{-m}\,K(m)\ ,\ m = \frac{2\beta}{1+\beta}$$

One easily sees that the distance between the monopoles goes to zero as $\beta \to 0$. The separation of the monopoles, d, tends to infinity as $\beta \to -1$ $d \sim -\ln(1+\beta)$.
I would like to say a few words about the superpotentials [15] which are very useful quantities for computing $|\phi|^2$, although up to now they seem to be effective only in the axially symmetric case. As I remarked earlier $|\phi|^2$ contains virtually every thing we want to know about the solution. $\Delta|\phi|^2$ gives the energy density, etc. $|\phi|^2$ is easily expressed in terms of the g matrix as

$$|\phi|^2 = \frac{1}{2} \mathrm{Tr}\ (g'_z\ g^{-1})^2 \qquad (47)$$

In the case of axial symmetry we can define a function τ in the following way :

$$\tau_{,\rho} = \frac{\rho}{4} \mathrm{Tr}\ [(g'_\rho\ g^{-1})^2 - (g_{,z} g^{-1})^2]$$
$$\tau'_z = \frac{\rho}{2} \mathrm{Tr}\ |(g'_\rho\ g^{-1})(g_{,z} g^{-1})| \qquad (48)$$

The integrability conditions of (48) are satisfied if g (ρ, z) solves (22). Then as one can verify

$$-\Delta\tau = \mathrm{Tr}\ (g'_z\ g^{-1})^2 \qquad (49)$$

where $\Delta = \partial_\rho^2 + \partial_z^2 + \frac{1}{\rho}\ \partial\rho$

It seems to be rather remarkable that one could find a completely explicit formula for τ in the case of SU(N) gauge group. I don't want to write out here (see [15] for details) the general expression for τ, what I think is important, that $\tau = \tau_0 + \ln \det\ + \ln F$, where τ_0 is potential for g_0, F is a complicated expression depending only on μ, $\bar{\mu}$, ρ . For the one monopole we get

$$\tau\ (1) = -\frac{1}{2}\rho^2 + \ln\{\sqrt{\rho}\,\frac{1}{R}\ \ \sinh R\}$$

which is a remarkable simple formula. It is perhaps worth quoting $\tau^{(2)}$ for the two-monopole as well :

$$\tau^{(2)} = \frac{\pi^2}{8}(1-\xi^2)(1+\eta^2) + \ln\left\{\frac{(1+\eta^2)(1-\xi^2)}{\xi^2+\eta^2}\left[\frac{\cos^2(\frac{\pi}{2}\xi)}{1-\xi^2} - \frac{\cosh^2(\frac{\pi}{2}\eta)}{1+\eta^2}\right]\right\}$$

where η ξ are oblate spheroidal coordinates :

$$\rho + iz = w\sqrt{(1-\xi^2)(1+\eta^2)} \qquad 0 \leq \eta \leq \infty$$
$$-1 \leq \xi \leq 1$$

where the value of $w = \frac{\pi}{2}$ was fixed to ensure finite energy.

The main advantage of the superpotential is that it is much simpler than the matrix g and for SU(N) N > 2 gauge groups this is essential in practical applications.

In the SU(2) case we found the axially symmetric n monopole solutions using Bäcklund transformations in a completely explicit form [22]. The BT's were generalized(still within axial symmetry) to arbitrary compact simple groups by Bais and Sasaki [23]. What is still lacking is their generalization for the SDE in 4 dimensions. These generalized BT's should of course reduce to the known ones in the axially symmetric case and they would hopefully facilitate to a certain extent to compute the monopole solutions (and perhaps also the instantons) in a more tranparent and direct way.

5 - The Riemann-Hilbert problem and its significance for gauge theories.

First we define the regular matrix Riemann-Hilbert problem (RHP). Consider a closed curve Γ in the complex λ plane. Define on Γ a matrix $G(\lambda)$ analytic in an annular neighbourhood of Γ and the problem is to find $\chi_1(\lambda)$ and $\chi_2(\lambda)$ analytic inside (resp. outside) Γ satisfying :

$$\chi_1^{-1}(\lambda)\,\chi_2(\lambda) = G(\lambda), \qquad \lambda \in \Gamma \tag{50}$$

We assume that $\det\chi_i \neq 0$ in their domains of analyticity. A simple way to connect our linear system (9) with the RHP is the following : for any two solutions ψ_1, ψ_2, of (9) :

$$(\lambda\partial_{\bar z}+\partial_y)(\psi_1^{-1}\psi_2) = 0$$
$$(-\lambda\,\partial_{\bar y}+\partial_z)(\psi_1^{-1}\psi_2) = 0$$

that is

$$\psi_2 = \psi_1\, G(\lambda, -\lambda y+\bar z, \lambda z+\bar y) \tag{51}$$

Noting that if $\psi(\lambda)$ solves (9) then $g\psi^{+-1}(-\frac{1}{\bar\lambda})$ also solves (9) we have an involution acting in the solution space of (9) / we used the fact that $g = g^+$/, thus

$$g\psi^{+-1}(-\frac{1}{\bar\lambda}) = \psi(\lambda)\, G(\lambda, -\lambda y+\bar z, \lambda z+\bar y) \tag{52}$$

If ψ is analytic around $\lambda=0$ then $g\psi^{+-1}(-\frac{1}{\bar\lambda})$ is analytic near $\lambda=\infty$, therefore G is analytic in the overlapping region of the two domains. Let us now suppose that there is given a nonsingular matrix G / i.e. $\det G \neq 0$ / satisfying

$$(\lambda\partial_{\bar z} + \partial_y)\,G = (-\lambda\,\partial_{\bar y}+\partial_z)\,G = 0$$
$$G(\lambda) = G^+(-\frac{1}{\bar\lambda})$$

and analytic in an annular neighbourhood of the unit circle. Split G according to (50) into the product of two matrices with nonvanishing determinants analytic inside (resp. outside) the circle with the additional property that $\chi_2(\lambda)$ is normalized to 1 at $\lambda=\infty$. This normalization is needed to ensure uniqueness of the solution. Acting now with ∇_i ($\nabla_1 = \lambda\,\partial_{\bar z}+\partial_y$, $\nabla_2 = -\lambda\partial_{\bar y}+\partial_z$) on eq. (50) we get

$$\nabla_i\chi_1\cdot\chi_1^{-1} = \nabla_i\,\chi_2\cdot\chi_2^{-1} \qquad \lambda\in\Gamma \tag{53}$$

Eq. (53) defines the analytic continuation of both sides from Γ to be the whole complex plane. As χ_2 is normalized to 1 at infinity, by Liouville's theorem both sides are independent of λ. Therefore

$$\nabla_i\,\chi_1\cdot\chi_1^{-1} = \nabla_i\,\chi_2\cdot\chi_2^{-1} = B_i \tag{54}$$

where the matrices B_i do not depend on λ. The compatibility of (54) implies

$$B_y = g_{,y}\, g^{-1} \quad , \quad B_z = g_{,z}\, g^{-1}$$

with g satisfying (8). To get all solutions of the SDE one should consider a RHP with zeroes, that is allow $\det\chi_i = 0$ in their domains of analyticity.

The regular RHP is equivalent to the following linear singular integral equation :

$$\frac{1}{i\pi}\left[P\int_{\Gamma} \frac{\sigma(\zeta)}{\zeta-\lambda}\, d\zeta + 1\right] T(\lambda) + \sigma(\lambda) = 0 \qquad \lambda \in \Gamma \qquad (55)$$

where $T(\lambda) = [G(\lambda) - 1][G(\lambda)+1]$, and we used the following integral representations for $\chi_i(\lambda)$'s :

$$\chi_1(\lambda) = 1 + \int_{\Gamma} \frac{\sigma(\zeta)}{\zeta-\lambda}\, d\zeta \qquad \text{inside } \Gamma$$

$$\chi_2(\lambda) = 1 + \int_{\Gamma} \frac{\sigma(\zeta)}{\zeta-\lambda}\, d\zeta \qquad \text{outside } \Gamma \qquad (56)$$

The RHP with zeroes is equivalent to a system of linear integral equations coupled to algebraic equations determining the residua of $\chi_i(\lambda)$'s at the pole sites [3].

It we take in (52) $G(\lambda) = \psi_0\, G_0\, \psi_0^{-1}$ where ψ_0 is a solution of (9) and $G_0 = G_0(\lambda, -\lambda y+\bar{z}, \lambda z+\bar{y})$, $G_0^{+}(-\frac{1}{\bar{\lambda}}) = G_0(\lambda)$, $\det G_0 = 1$, the solution of this RHP generates a new solution of the linear system (9)in the form

$$\psi = \chi_1(\lambda)\, \psi_0(\lambda).$$

The construction of self-dual solutions described in the previous sections corresponds to solving the RHP with zeroes only. This means that we take $G_0 = 1$, and construct a meromorphic $\chi(\lambda)$ with simple poles only. These solutions are referred to as "purely solitonic" in analogy with the known two dimensional models.

It is not known how to solve the RHP explicitly for an arbitrary matrix G. However, in some special cases when G is diagonal or triangular the splitting can be done explicitly. In fact the Atiyah-Ward Ansätze [24] correspond to splitting a matrix of the form

$$G_k = \begin{pmatrix} \lambda & \rho \\ 0 & \lambda^{-k} \end{pmatrix}$$

The splitting of G_k, $\alpha(\lambda)\,\beta(\lambda) = G_k$, was carried out in detail by Corrigan et al. |20|. As it is clear form the above the RHP is extremely powerful to generate solutions of the SDE. Also from the RHP an infinite dimensional invariance group (of the Kac-Moody type) can be constructed [11,25], with infinitely many conserved quantities. It is an extremely important problem to try to extend these ideas for the nonselfdual sector. Isenberg et al. [26] and Witten [27] gave a twistorial interpretation of the second order (sourceless) gauge field equations. We were able to derive RHP's for the nonself-dual sector using Witten's ideas [28].

In an eight dimensional space with coordinates a^i, b^i $i = 1, \dots 4$ we consider the following system of equations :

$$F_{a_i a_j} = {}^*F_{a_i a_j} \; ; \; F_{b_i b_j} = -{}^*F_{b_i b_j} \; ; \; F_{a_i b_j} = 0 \qquad (57)$$

It can be verified that solutions of (57) satisfy on the $x^i=\frac{1}{2}(a^i+b^i)$ diagonal subspace the four dimensional source free Yang-Mills equations

$$D_\mu F^{\mu\nu} = 0$$

First we were able to linearize (57) :

$$\begin{aligned} D_1\psi &= (-\lambda_1 \partial_{\bar{y}2} + \partial_{y1})\psi = g'_{y1} g^{-1} \psi \\ D_2\psi &= (\lambda_1 \partial_{\bar{y}1} + \partial_{y2})\psi = g'_{y2} g^{-1} \psi \\ D_3\psi &= (-\lambda_2 \partial_{\bar{z}2} + \partial_{z1})\psi = g'_{z1} g^{-1} \psi \\ D_4\psi &= (\lambda_2 \partial_{\bar{z}1} + \partial_{z2})\psi = g'_{z2} g^{-1} \psi \end{aligned} \qquad (58)$$

where $y_1 = b_1 + ib_2$; $z_1 = a_1 + ia_2$, etc,
$g = D^+D$, $D = D(y_i, z_i) \in SL(N,C)$ and the gauge fields,
$A_{u_i} = - D_{,ui} D^{-1}$ $u = y,z$, $A_{\bar{u}i} = D^{+-1} D^+_{,ui}$. and (57) becomes

$$(g_{,u} g^{-1})_{,\bar{u}_1} + (g_{,u2} g^{-1})_{\bar{u}2} = 0 \qquad u = y,z \qquad (59.a)$$

$$(g_{,yi} g^{-1})_{,\bar{z}j} = (g_{,zi} g^{-1})_{yj} = 0 \qquad i,j = 1;2 \qquad (59.b)$$

Now one can derive a RHP for eqs. (58), however since there are two spectral parameters here, (λ_1, λ_2) the resulting RHP with two variables is very difficult to solve. / see [28] for details /.
We were able to introduce a RHP with one variable only, it is possible to take in (58)

$$\lambda_2 = \lambda_1^3$$

Then we can write down the following RHP :

$$g\,\chi^{+-1}(-\tfrac{1}{\bar{\lambda}})g_o^{-1} = \chi(\lambda)\,\psi_o(\lambda)G_o(\lambda, \lambda y_1+\bar{y}_2, \lambda y_2-\bar{y}_1, \lambda^3 z_1+\bar{z}_2, \lambda^3 z_2-\bar{z}_1)\,\psi_o^{-1}(\lambda) \qquad (60)$$

Although we were not able to generate new solutions of the Yang-Mills eqs. with (58) or (60) this approach should be further studied, in particular to derive a Kac-Moody-type algebra for the Yang-Mills eqs. Also it would be important to connect our approach to the ideas of Volovich [29].

6 - Conclusions and outlook

I tried to describe here the soliton-theoretic framework we developed to produce explicit solutions of the SDE. We were mostly interested in multimonopole solutions, describing static finite energy monopoles of like charge. These solutions exist due to the fact that the net force

cancels between these monopoles.
I showed how to construct the most general charge n monopole configuration depending on 4n-1 parameter.
The solution is extremely complicated. Also to prove that it is regular everywhere is very hard and for the simples nontrivial case (the separated two monopole) we proved it numerically. Presumably Nahm's method is the best to prove that these solutions are indeed regular [30].
If we impose axial symmetry the solutions are completely explicit, although then it is much better to generate them by Bäcklund transformations
In the SU(2) case on can exploit the equivalence between the Bogomolny eqs. and the Ernst eq. of general relativity to use the BT's for the Ernst eq.
Further connections exist among the Einstein eqs.,the SDE and non-linear sigma models see for example [31].
I did not give an exhaustive review here, as that would easily fill a book, and I did not talk about other methods for solving the SDE, but as Nahm also gave a talk I think it is not necessary.
Clearly there are many things to be done. It would be very important to find time dependent solutions for example. In 2.1 dim $\mathbb{C}P^n$ models a step was made in this direction. [32].
Even more important is to clarify the structure of the full Yang-Mills equations.

Acknowledgement :

I am grateful for being invited to the Meudon seminars on "Non linear methods in field theories" to the organizers.

References :

| 1 | E.B. Bogomolny, Sov.J.Nucl.Phys. 24, 861 (1976).
| 2 | C.H. Taubes, Commun.Math.Phys. 86, 257 (1982).
C.H. Taubes, Commun.Math.Phys. 86, 299 (1982).
| 3 | V.E. Zakharov, S.V. Manakov, S.P. Novikov and S.P. Pitaevski, Theory of Solitons /Nauka, Moscow, 1980/ (in russian).
| 4 | M.K. Prasad, C.M. Sommerfield, Phys.Rev.Lett. 35, 760 (1975).
| 5 | N.S. Manton, Nucl.Phys. B135, 319 (1978).
| 6 | P. Rossi, Phys.Rep. 86, 317 (1982).
| 7 | A. Chakrabarti, Phys.Rev. D25, 3282 (1982).

|7| A. Chakrabarti, F. Koukiou, Phys. Rev. D26, 1425 (1982).
A. Chakrabarti, Phys.Rev. D28, 989 (1983).
A. Chakrabarti, Ecole Polytechnique preprint, A 594.0284 (1984).

|8| H. Boutaleb-Joutei, A. Chakrabarti, A. Comtet, Phys.rev. D23, 1781 (1981), ibid D24, 3146 (1981).

|9| C.N. Yang, Phys.Rev.Lett. 38, 1377 (1977).

|10| Y. Brihaye, D.B. Fairlie, J. Nuyts, R.G. Yates, P. Goddard, J.Math. Phys. 19, 2528 (1978).

|11| P. Forgács, Z. Horváth, L. Palla, Phys.Rev. D23, 1876 (1981).

|12| P. Forgács, Z. Horváth, L. Palla, Nucl.Phys. B229, 77 (1983).

|13| F.J. Ernst, Phys.Rev. 167, 1175 (1968).

|14| D. Maison, Phys.Rev.Lett. 41, 521 (1978).
J.Math.Phys. 20, 871 (1979).
V.A. Belinski, V.E. Zakharov, Sov.Phys. JETP 50, 1 (1979).

|15| P. Forgács, Z. Horváth, L. Palla, Nucl.Phys. B221, 235 (1983).

|16| P. Houston, L. O'Raifeartaigh, Phys.Lett. 93B, 151 (1980).

|17| W. Nahm, Phys.Lett. 85B, 373 (1979).

|18| E. Weinberg, Phys.Rev. D20, 936 (1979).

|19| P. Forgács, Z. Horváth, L. Palla, Phys.Lett. 109B, 200 (1982).

|20| E. Corrigan, P. Goddard, Comm.Math.Phys. 80, 151 (1981).

|21| M. Abramowitz, J.A. Stegun, Handbook of Mathematical Functions, Dover Publ. /N.Y./ (1972).

|22| P. Forgács, Z. Horváth, L. Palla, Nucl.Phys. 192B, 141 (1981) and Ann.Phys. /N.Y./ 136, 371 (1981).

|23| F.A. Bais, R. Sasaki, Phys.Lett. 113B, 35 (1982), Nucl.Phys. 195B, 522 (1982).

|24| M.F. Atiyah, R.S. Ward, Comm.Math.Phys. 25, 117 (1977).
R.S. Ward, Comm.Math.Phys. 79, 317 (1981).

|25| Y.S. Wu, Comm.Math.Phys. 90, 461 (1983).
K Ueno, Y. Nakamura, Phys.Lett. 109B, 273 (1982).

|26| J. Isenberg, P.B. Yasskin, P.S. Green, Phys.Lett. 78B, 462 (1978).

|27| E. Witten, Phys.Lett. 77B, 394 (1978).

|28| P. Forgács, Z. Horváth, L. Palla, Phys.Lett. 115B, 463 (1982).

|29| I.V. Volovich, Phys.Lett. 123B, 329 (1983).

|30| W. Nahm, CERN preprint TH 3172 (1981).

|31| N. Sánchez, Phys.Rev. D26, 2589 (1982), " Einstein Equations, Self-Dual Yang-Mills fields and non-linear sigma models" (Lect. Notes in Math., Springer-Verlag, to appear).

|32| P. Forgács, W.J. Zakrzewski, Z. Horváth, CERN preprint TH 3863 (1984)

CLASSICAL SOLUTIONS OF YANG-MILLS FIELDS (Selected Topics).

A. CHAKRABARTI
Centre de Physique Théorique de l'Ecole Polytechnique
Plateau de Palaiseau - 91128 Palaiseau-Cedex - France.

"Groupe de Recherche du C.N.R.S. n° 48"

PART 1 : Topological aspects of Yang-Mills fields in curved spaces. (Exact solutions).

PART 2 : Flat space instanton chains with monopole limits.

As will be seen, the static, spherically symmetric, conformally flat de Sitter space provides the link between the parts 1 and 2. But Part 2 can very well be considered entirely separately. To avoid making this report longer I have assumed certain results as known. In Part 1 this includes standard ones in general relativity as well as various aspects of gravitational instantons. In Part 2 familiarity with standard constructions of instantons and of selfdual monopoles, which emerge as limits in our formalism, has been assumed. I present exlusively our results and not comprehensive reviews of these domains.

Part 1 : To explore in its full richness the topological possibilities of gauge fields one should allow for simultaneous presence of gravitational and Yang-Mills ones. Thus if the integral topological indices of the Yang-Mills field for a flat Euclidean base space is associated with the structure of the vacuum, one may ask among other questions of interest, how this spectrum might be modified when the base space itself has non trivial indices. Exact solutions of SU(2) Yang-Mills fields are presented for metrics corresponding to well-known gravitational instantons. Such selfdual solutions, with vanishing energy momentum tensor $T_{\mu\nu}$ for Euclidean signature of the base space, do not perturb the metric. Thus they provide solutions of the combined gravitational Y.M. system. New topological possibilities, such as finite action SU(2) fields with fractional indices for many centre metrics are displayed explicitly. As another type of possibility non selfdual, finite action solutions are constructed explicitly on Schwarzschild and de Sitter metrics, the solution being real in the first and complex in second case respectively. It is also shown how various meron type solutions in flat space can be derived systematically from a very simple static solution in de Sitter.

Part 2 : Monopoles in the Bogomolny-Prasad-Sommerfield limit have been constructed, often using instanton like techniques, but separately starting all over again. It is possible to do better. The well known fact that (replacing the Higgs scalar Φ by A_t of the Euclidean gauge potentials) such solutions can be considered as infinite action limits of self-dual Yang-Mills field ones can be exploited more deeply and fruitfully. Instead of constructing one monopole (or one instanton) one can, in a single stroke construct an infinite sequence of instantons such that a monopole solution emerges practically trivially in a particularly simple scaling limit. Typically one constructs a single solution involving a parameter whose admissible values give the entire spectrum of indices for members of this sequence (chain). Then the rescaling is done through this parameter to obtain a finite energy but infinite action selfdual monopole. Thus in a comparable number of steps one obtains not only the monopoles but much more. The chains of instantons thus displayed have remarkable properties well worth studying even if one is not directly interesed in monopoles. I start by discussing examples of such fascinating special properties for the simplest sequence (or 1-chain) leading in the limit to the PS monopole of unit charge. Then techniques are presented for building a hierarchy of such chains giving multicharged monopoles as limits. For the higher chains our constructions present remarkable fully explicit examples of instantons in the Atiyah-Ward classes greater than one. The basic idea here is to use the hyperbolic line element to construct "static" finite action solutions, which becomes possible due to a compactification of the "time" direction.

REFERENCES :

|1| H. BOUTALEB-JOUTEI, A. CHAKRABARTI and A. COMTET, Phys.Rev. D24, 3146 (1981).

|2| A. CHAKRABARTI and F. KOUKIOU, Phys.Rev. D26, 1425 (1982).

|3| A. CHAKRABARTI, Phys.Rev. D28, 989 (1983).

|4| A. CHAKRABARTI, Proc. of the Asia Pacific Physics Conference, World Sci. PubK Co Singapore (to appear).

Stationary Solutions of the Einstein-Maxwell Equations

D. MAISON

Max-Planck-Institut für Physik und Astrophysik
– Werner Heisenberg Institut für Physik –
P.O.Box 40 12 12, Munich (Fed. Rep. Germany)

1 INTRODUCTION

In this lecture I shall describe stationary solutions of the Einstein-Maxwell equations in the framework of 'dimensional reduction' with particular emphasis on the group-theoretical aspects. Although the origin of this approach lies in work on Einstein's theory [1–4] it has profitted much from more recent work on supergravity [5]. Providing a bridge to a well-developped field of mathematics it not only allows for a better understanding of many results on stationary solutions of Einstein's theory (explicit construction and uniqueness of 'vacuum' and blackhole solutions) but also for an easy extension of these results to a large family of theories related to the bosonic part of ten-dimensional supergravity [6]. The simplest extension of Einstein's theory contained is the Einstein-Maxwell system which is the bosonic part of $N = 2$ supergravity. Therefore I shall restrict myself here to this theory. But before I dive into the deep waters of mathematical formalism I would like to motivate my own interest in these solutions.

Yang-Mills theories and Einstein's gravity theory are presently considered to provide the most adequate description of all the elementary forces of nature. In this context it is extremely remarkable that both types of theories can be unified into a common framework following the ideas of Kałuza and Klein [7]. In this type of theories the world is some higher dimensional manifold of which only four dimensions — the usual space-time manifold — are visible to our eyes due to some symmetry property of the actual vacuum and its low-energy excitations. The factorized classical 'vacuum'-manifold serving as a background for the quantized theory is supposed to be distinguished by its stability. Elementary particles living on this background are usually considered to be of purely quantum-mechanical origin. However if one looks for the corresponding classical description of quantized particles one is lead to the concept of point-particles, a concept which has little meaning in a theory including General Relativity. In fact the adequate concept of a massive particle in General Relativity is that of a black-hole [8]. I personally favour the idea that massive elementary particles are quantized black-holes, i.e. have a classical black-hole background, in contrast to massless excitations (photons etc.) which are quantum excitations of the classical vacuum.

Black-holes are the solitons of General Relativity resp. its generalizations. This viewpoint is supported by several facts:

i) black-holes are localized, particle-like objects

ii) black-holes are distinguished by uniqueness properties that give them stability

iii) black-holes are stationary, axially-symmetric solutions and therefore constructable via the Inverse Scattering Method (resp. Bæcklund transformations similar to the solitons of the KdV- or Sine-Gordon equation.

The latter aspect will be shortly discussed in section 4. Section 2 is devoted to the mathematical structure of the equations for stationary solutions, whereas in sect. 3 I discuss special types of solutions.

2 Field Equations and Duality Transformations

The Einstein-Maxwell equations for the vierbein $e_\mu{}^a$ and the vector potential A_μ are

$$R_{\mu\nu} = F_{\mu\lambda}F^{\lambda}{}_{\nu} - \frac{1}{4}g_{\mu\nu}F_{\kappa\lambda}F^{\kappa\lambda} \tag{1a}$$

$$F_{\mu\nu}{}^{;\nu} = 0 \tag{1b}$$

$$ {}^*F_{\mu\nu}{}^{;\nu} \equiv \frac{1}{2}\epsilon_{\mu\nu\kappa\lambda}F^{\kappa\lambda;\nu} = 0 \tag{1c} $$

Eqs. $(1a, b)$, the equations of motion, are derived from the action

$$ S = \int \frac{e}{2}\left(R - \frac{1}{2}F_{\mu\nu}F^{\mu\nu}\right) d^4x \tag{2} $$

Eq. $(1c)$ is the Bianchi identity. Eq. $(1a)$ can also be written in the form

$$ R_{\mu\nu} = \frac{1}{2}(F_{\mu\lambda}F^{\lambda}{}_{\nu} + {}^*F_{\mu\lambda}{}^*F^{\lambda}{}_{\nu}) \tag{3} $$

which displays the explicit duality invariance

$$ \begin{aligned} F_{\mu\nu} &\longrightarrow {}^*F_{\mu\nu} \\ {}^*F_{\mu\nu} &\longrightarrow -F_{\mu\nu} \end{aligned} \tag{4} $$

of eqs. (1). Interpreting eq. $(1b)$ as the Bianchi-identity we may introduce a vector-potential C_μ for $G_{\mu\nu} \equiv -{}^*F_{\mu\nu}$ and put

$$ \mathcal{A}_\mu = \begin{pmatrix} A_\mu \\ C_\mu \end{pmatrix} \qquad \text{and} \qquad \mathcal{F}_{\mu\nu} = \begin{pmatrix} F_{\mu\nu} \\ G_{\mu\nu} \end{pmatrix} \tag{5} $$

then

$$ \mathcal{F}_{\mu\nu} = \begin{pmatrix} 0 & 1 \\ -1 & 0 \end{pmatrix} {}^*\mathcal{F}_{\mu\nu} = \Omega^*\mathcal{F}_{\mu\nu} \tag{6} $$

with $\Omega^2 = -1$. The equation $\mathcal{F}_{\mu\nu}{}^{;\nu} = 0$ is now completely equivalent to ${}^*\mathcal{F}_{\mu\nu}{}^{;\nu} = 0$ due to eq. (6)

A similar type of on-shell duality symmetry was observed by E. Cremmer and B. Julia [5] for $N = 8$ supergravity; another case related to ten-dimensional supergravity is discussed in [6]. The general structure found is the following: there are vector fields A^i_μ $(i = 1, \ldots, n)$ transforming under a representation of a group $\bar{G}$ with the 'pseudo-metric' Ω (with $\Omega^2 = -1$) and scalars V parametrizing a homogeneous space $\bar{G}/\bar{H}$. The generalization of eq. (6) is

$$ \mathcal{F}_{\mu\nu} = \Omega V^*\mathcal{F}_{\mu\nu} \qquad \text{with} \qquad (\Omega V)^2 = -1 \tag{7} $$

To give a few examples:

$N = 2$	supergravity	$\bar{G} = U(1)$	$\bar{H} = U(1)$
$N = 4$	supergravity	$\bar{G} = SO(6) \times SU(1,1)$	$\bar{H} = SO(6) \times U(1)$
$N = 8$	supergravity	$\bar{G} = E_7$	$\bar{H} = SU(8)$

'Jordan-Thiry' theories
(Einstein's theory in $d > 4$ dimensions) $\bar{G} = SL(d-4)$ $\bar{H} = SO(d-4)$

Since we are only interested in stationary solutions we can perform the dimensional reduction from four to three dimensions. We assume the existence of a time-like Killing vector field ξ^μ such that the Lie-derivative $\mathcal{L}_\xi F_{\mu\nu} = 0$. For the vierbein we make the usual decomposition

$$e_\mu{}^a = \begin{pmatrix} \Delta^{-1/2}\hat{e}_m{}^a & \Delta^{1/2}B_m \\ 0 & \Delta^{1/2} \end{pmatrix} \tag{8}$$

The isometry-group G_1 related to ξ^μ is supposed to act regularly on the 4-dimensional space-time manifold M_4 producing the 3-dimensional orbit space $\Sigma = M_4/G_1$. From $e_\mu{}^a$ and A_μ we obtain the following fields living on Σ

$$\begin{array}{ll} \hat{e}_m{}^a & \text{dreibein} \\ B_m, \quad \hat{A}_m \equiv A_m - \varphi B_m & \text{vector potentials} \\ \Delta, \quad \varphi \equiv \Delta^{1/2} A_0 & \text{scalars} \end{array} \tag{9}$$

The 3-dimensional action obtained from eq. (2) is

$$\begin{aligned} \hat{S} = \frac{1}{2}\int_\Sigma \hat{e}[-\hat{R} + \frac{1}{2}\Delta^{-2}(\partial_m\Delta)^2 + \frac{1}{4}\Delta^2 B_{mn}^2 \\ \mp \frac{1}{2}\Delta(\hat{F}_{mn} + \varphi B_{mn})^2 \mp \Delta^{-1}(\partial_m\varphi)^2]\, d^3x \end{aligned} \tag{10}$$

with $B_{mn} = \partial_m B_n - \partial_n B_m$. The lower sign refers to the case of a space-like Killing vector which we include for later convenience. Remarkably it is possible to extend the duality-symmetry of the 4-dimensional Einstein-Maxwell equations corresponding to the group $U(1)$ to the much larger group $SU(2,1)$ acting on the 3-dimensional fields of eq. (9). This comes about as follows. The field equations for the vectors $\hat{A}_m$ and B_m

$$\begin{aligned} (\Delta\hat{F}_{mn} + \varphi B_{mn})^{;n} = 0 \\ \left(\Delta^2 B_{mn} - 2\Delta\varphi(\hat{F}_{mn} + \varphi B_{mn})\right)^{;n} = 0 \end{aligned} \tag{11}$$

are considered as 'Bianchi-identities' for scalar fields ψ and ω

$$\begin{aligned}(\Delta\hat{F}_{mn} + \varphi B_{mn}) &= \epsilon_{mn}{}^{l}\partial_l\psi \\ \Delta^2 B_{mn} &= \epsilon_{mn}{}^{l}(\partial_l\omega + 2\psi\partial_l\varphi - 2\varphi\partial_l\psi)\end{aligned} \tag{12}$$

The scalars ψ and ω replace the vectors $\hat{A}_m$ and B_m . This can also be achieved by adding the Bianchi-identities for B_{mn} and $\hat{F}_{mn}$ with Lagrangian multipliers to $\hat{S}$:

$$\hat{S} \rightarrow \hat{S} - \int_{\Sigma} \frac{\hat{e}}{2}\epsilon^{lmn}[\psi\partial_l\hat{F}_{mn} + \frac{1}{2}(\omega + \psi\varphi)\partial_l B_{mn}]\, d^3x \tag{13}$$

'Integrating out' $\hat{F}_{mn}$ and B_{mn} one obtains

$$\hat{S} = \frac{1}{2}\int_{\Sigma} \hat{e}[-\hat{R} + \frac{1}{2}\Delta^{-2}\left((\partial_m\Delta)^2 + \omega_m^2\right) \mp \Delta^{-1}\left((\partial_m\varphi)^2 + (\partial_m\psi)^2\right)]\, d^3x \tag{14}$$

with $\omega_m \equiv \partial_m\omega + 2\psi\partial_m\varphi - 2\varphi\partial_m\psi$.

The line-element

$$ds^2 = \frac{1}{2}\Delta^{-2}\left((d\Delta)^2 + (d\omega + 2\psi d\varphi - 2\varphi d\psi)^2\right) \mp \Delta^{-1}\left((d\varphi)^2 + (d\psi)^2\right) \tag{15}$$

turns out to describe the invariant metric of the Riemannian symmetric space $S = SU(2,1)/SU(1,1) \times U(1)$ for the upper sign and $S = SU(2,1)/SU(2) \times U(1)$ for the lower sign (corresponding to a space-like Killing vector) in some particular coordinates. This is revealed by the following parametrization of S

$$\begin{aligned}V &\equiv \exp\begin{pmatrix}\Delta/2 & 0 & 0 \\ 0 & 0 & 0 \\ 0 & 0 & -\Delta/2\end{pmatrix} \exp\begin{pmatrix}0 & 0 & 0 \\ i\sqrt{2}\phi & 0 & 0 \\ \omega & \sqrt{2}\bar{\phi} & 0\end{pmatrix} \\ &= \begin{pmatrix}\sqrt{\Delta} & 0 & 0 \\ i\sqrt{2}\phi & 1 & 0 \\ \frac{\omega + i|\phi|^2}{\sqrt{\Delta}} & \frac{\sqrt{2}\bar{\phi}}{\sqrt{\Delta}} & \frac{1}{\sqrt{\Delta}}\end{pmatrix}\end{aligned} \tag{16}$$

with $\phi = \varphi + i\psi$.

The action of $g \in SU(2,1)$ on V is

$$V \longrightarrow h(V,g)Vg^{-1} \tag{17}$$

where $h(V,g) \in SU(1,1) \times U(1)$ resp. $SU(2) \times U(1)$ refurnishes the triangular gauge of V. The difference between the two sign choices will turn out to be of some significance later. $SU(2) \times U(1)$ is the maximal compact subgroup of $SU(2,1)$ and hence $SU(2,1)/SU(2) \times U(1)$ is a Riemannian symmetric space, a non-compact form of CP_2. In contrast $SU(2,1)/SU(1,1) \times U(1)$ has signature $(++--)$, it is pseudo-Riemannian. The matrix V satisfies

$$V^+ \eta V = \eta \quad \text{with} \quad \eta = \begin{pmatrix} 0 & 0 & i \\ 0 & 1 & 0 \\ -i & 0 & 0 \end{pmatrix} \tag{18}$$

i.e. is unitary with respect to the metric η of $SU(2,1)$. An alternative parametrization [3] of $\mathcal{S}$ is obtained diagonalizing $\eta \to A\eta A^+$ and taking one column of the matrix AVA^+ similar to the usual parametrization of CP_2.

From eq. (16) we find

$$\partial V V^{-1} = \begin{pmatrix} \frac{1}{2}\Delta^{-1}\partial\Delta & 0 & 0 \\ i\sqrt{\frac{2}{\Delta}}\partial\phi & 0 & 0 \\ \Delta^{-1}\left(\partial\omega + i(\bar\phi\partial\phi - \phi\partial\bar\phi)\right) & \sqrt{\frac{2}{\Delta}}\partial\bar\phi & -\frac{1}{2}\Delta^{-1}\partial\Delta \end{pmatrix}$$

$$= \left.\begin{pmatrix} \frac{1}{2}\Delta^{-1}\partial\Delta & \mp\frac{i\partial\bar\phi}{\sqrt{2\Delta}} & \frac{1}{2}\Delta^{-1}(\partial\omega + \ldots) \\ \frac{i\partial\phi}{\sqrt{2\Delta}} & 0 & \frac{\pm\partial\phi}{\sqrt{2\Delta}} \\ \frac{1}{2}\Delta^{-1}(\partial\omega + \ldots)) & \frac{\partial\bar\phi}{\sqrt{2\Delta}} & -\frac{1}{2}\Delta^{-1}\partial\Delta \end{pmatrix}\right\} \mathcal{K} \tag{19}$$

$$+ \left.\begin{pmatrix} 0 & \pm\frac{i\partial\bar\phi}{\sqrt{2\Delta}} & -\frac{1}{2}\Delta^{-1}(\partial\omega + \ldots) \\ \frac{i\partial\phi}{\sqrt{2\Delta}} & 0 & \mp\frac{i\partial\phi}{\sqrt{2\Delta}} \\ \frac{1}{2}\Delta^{-1}(\partial\omega + \ldots) & \frac{\partial\bar\phi}{\sqrt{2\Delta}} & 0 \end{pmatrix}\right\} \mathcal{A}$$

$\mathcal{A}$ is an element of the Lie-algebra of $SU(1,1) \times U(1)$ resp. $SU(2) \times U(1)$ whereas $\mathcal{K}$ belongs to its orthogonal complement.

$\mathcal{A}$ plays the role of a gauge-potential and does not contribute to the action $\hat{S}$ which can be written

$$\hat{S} = \frac{1}{2}\int_{\Sigma} \hat{e}(-\hat{R} + TrK^2)\, d^3x \tag{20}$$

describing a generally covariant non-linear σ-model [9, 10]. It is also possible to use a 'gauge-invariant' parametrization [9] of S with the help of the isometric involution σ defining this symmetric space:

$$\sigma(V) = \xi(V^+)^{-1}\xi \quad \text{with} \quad \xi = \begin{pmatrix} 1 & 0 & 0 \\ 0 & \pm 1 & 0 \\ 0 & 0 & 1 \end{pmatrix} \tag{21}$$

Instead of V we introduce $\chi \equiv \sigma(V^{-1})V = \xi V^+ \xi V$ and find $\chi \to g\chi g^{-1}$ for $g \in SU(2,1)$. in contrast to eq. (17).

In terms of χ the action takes the simple form

$$\hat{S} = \frac{1}{2}\int_{\Sigma} \hat{e}[-\hat{R} + Tr(\chi^{-1}\partial\chi)^2]\, d^3x \tag{22}$$

and the equations of motion are

$$\hat{R}_{mn} = \frac{1}{2} Tr(\chi^{-1}\partial_m\chi\chi^{-1}\partial_n\chi) \tag{23a}$$

$$(\chi^{-1}\partial_m\chi)^{;m} = 0 \tag{23b}$$

Mathematicians would say that such χ's describe harmonic maps $\Sigma \longrightarrow S$ [11].

The hidden symmetry group $SU(2,1)$ can be used to produce a whole family of stationary solutions of the Einstein-Maxwell equations from any given one. The various elements of the Lie-algebra of $SU(2,1)$ correspond to the following infinitesimal transformations of the solution:

$$\begin{pmatrix} 0 & 0 & 0 \\ \alpha & 0 & 0 \\ \beta & \gamma & 0 \end{pmatrix}$$ are 'gauge' transformations of ω and ϕ which do not change the solution,

$$\begin{pmatrix} 0 & 0 & \alpha \\ 0 & 0 & 0 \\ 0 & 0 & 0 \end{pmatrix}$$ is a Ehlers transformation [2] changing $\Delta \to \omega$,

$$\begin{pmatrix} i\alpha & & \\ & -2i\alpha & \\ & & i\alpha \end{pmatrix}$$ is an e.m. duality transformation,

$$\begin{pmatrix} \alpha & & \\ & 0 & \\ & & -\alpha \end{pmatrix}$$ is a scale transformation, $\Delta, \omega, \phi \to e^{\alpha}\Delta, e^{\alpha}\omega, e^{\alpha/2}\phi$

$$\begin{pmatrix} 0 & i\alpha & 0 \\ 0 & 0 & \bar{\alpha} \\ 0 & 0 & 0 \end{pmatrix}$$ is a Harrison transformation [12] changing $\Delta \to \phi$.

3 SOLUTIONS

We shall now discuss what is known about various types of solutions of eqs. (23) describing stationary solutions of the Einstein-Maxwell system. For physical reasons we shall only be interested in asymptotically flat (Minkowskian) solutions.

3.1 REGULAR SOLUTIONS

From $(\chi^{-1}\partial_m\chi)^{;m} = 0$ we find

$$\int_{\Sigma} (\chi^{-1}\partial_m\chi)^{;m}\hat{e}\, d^3x = \int_{\partial\Sigma_\infty} (\chi^{-1}\partial_m\chi)\, d\sigma^m = 0 \tag{24}$$

where $\partial\Sigma_\infty$ is the surface $r = \infty$. From the field eqs. (23) it follows [13] that regular, asymptotically flat solutions have an expansion
$\chi = \chi_0 + \frac{1}{r}\chi_1 + \cdots$ for $r \to \infty$, where χ_o and χ_1 are constants. Hence we get $\chi_1 = 0$. This means that there is no mass and no charge. The 'Positive Mass

Theorem' [14] tells us that therefore $\chi = \chi_0$ and the solution is Minkowski space with vanishing e.m. field.

3.2 SPHERICALLY SYMMETRIC SOLUTIONS

Using polar coordinates the metric on Σ can be parametrized as

$$\hat{h}_{mn} \equiv -\hat{e}_m{}^a \hat{e}_n{}^b \delta_{ab} = \begin{pmatrix} 1 & & \\ & f^2 & \\ & & f^2 \sin^2\theta \end{pmatrix} \tag{25}$$

where f and χ depend only on r and eqs. (23) yield

$$\hat{R}_{rr} \equiv -\frac{2f''}{f} = \frac{1}{2} Tr(\chi^{-1}\chi')^2 \tag{26a}$$

$$\hat{R}_{\theta\theta} \equiv \frac{f''}{f} + \left(\frac{f'}{f}\right)^2 - \frac{1}{f^2} = 0 \tag{26b}$$

$$(f^2\chi^{-1}\chi')' = 0 \tag{26c}$$

The last eq. gives $(f^2\chi^{-1}\chi')' = \mu = const \in SU(2,1)$ which can be integrated to

$$\chi = \chi_0 \exp t(r)\mu \qquad \text{with} \quad t(r) = \int^r \frac{dr}{f^2} \tag{27}$$

Eqs. (26) yield $f^2 = (r - r_0)^2 - \frac{a^2}{4}$ with $a^2 = Tr\mu^2$. Since $t(r) \to 0$ for $r \to \infty$ we get $\chi_0 = \chi(\infty)$.

For vanishing NUT-parameter this is the electrically and magnetically charged Reissner-Nordstrøm solution. There is an interesting special case

$$a^2 = o \quad , \quad \hat{R}_{mn} = 0 \quad , \quad t(r) = -\frac{1}{r} \tag{28}$$

which is the extremal Reissner-Nordstrøm solution. The possibility to have a nontrivial μ for $Tr\mu^2 = 0$ is due to the signature $(++--)$ of $SU(2,1)/SU(1,1)\times U(1)$. For pure gravity, where $S = SL(2,R)/SO(2)$ with signature $(++)$ no such solution exists. The solution eq. (27) can be viewed from a different aspect, it defines maps

$$\Sigma \xrightarrow{t} R^1 \xrightarrow{\chi} S$$

$t(r)$ is a solution of $(f^2t')' = 0$ which means it is a harmonic function on Σ relative to the metric $\hat{h}$. χ solves $\frac{d}{dt}(\chi^{-1}\dot{\chi}) = 0$, which is the equation for a geodesic of S. The fact that the combined map $\chi \circ t$ is harmonic is a special case of a

Theorem [11]: *If t is harmonic and χ is totally geodesic than $\chi \circ t$ is harmonic.*

3.3 Static, single black-hole solutions

We know already from 3.1 that in order to have charge there must be sources (= singularities). For static black-holes the following theorem holds.

Theorem 1: *If a static black-hole is uncharged, then A_μ vanishes pointwise.*

This is a special case of a theorem for the larger class of theories mentioned in the introduction. A proof will be given in [15].

There is a well-known uniqueness theorem:

Theorem 2 (Israel [16]): *The only static black-hole of Einstein's theory is the Schwarzschild solution.*

Since by a Harrison transformation we can transform away the charge of a black-hole, i.e. transform a charged static black-hole into the Schwarzschild solution, we conclude conversely that any static, charged black-hole can be obtained by a Harrison transformation from the Schwarzschild solution. This yields just the Reissner-Nordstrøm solution. Hence we have obtained a uniqueness theorem for the latter solution without using axial symmetry.

3.4 Stationary, single black-holes

From Hawking's rigidity theorem [16] we may deduce that single black-holes are axially-symmetric. Our knowledge on such axially-symmetric single black-holes is again quite complete due to the following theorems

Theorem 3 (Robinson [17]): *The Kerr solution is the only axially-symmetric, single, uncharged black-hole for a given mass and angular-momentum.*

Theorem 4 (Mazur [18]): *The Kerr-Newman solution is the only axially-symmetric, single black-hole for a given mass, angular-momentum, electric and magnetic charge.*

A similar theorem holds again for all the mentioned generalizations of Einstein's theory [18, 15]. Without the use of axial symmetry things are not so clear as for static solutions since the proof for the analogue of Theorem 1 given in [16] unfortunately contains a serious flaw [15].

3.5 Conform-stationary Solutions

It is highly remarkable that it is possible to superpose certain black-holes in a stationary fashion. The simplest example are extremal Reissner-Nordstrøm solutions with $e_i = m_i$ mentioned above. They are characterized by $\hat{R}_{mn} = 0$ (i.e. $\hat{h}_{mn} = \delta_{mn}$) and $\varphi(x) = -\sum \frac{m_i}{|x-x_i|}$. There is a more general family of solutions of this type — those of Israel-Wilson and Perjés [19] — given in terms of two harmonic functions, but the non-static ones have naked singularities [20]. The existence of these Bogomolny- type solutions is due to the fact — already mentioned above — that the signature of $SU(2,1)/SU(1,1) \times U(1)$ is $(++--)$. This permits to find nontrivial χ's with light-like $\chi^{-1}\partial\chi$. Analogous solutions exist again for more general theories [21, 22]. There is an interesting connection of these solutions with supersymmetry [23, 24]. Let us consider $N = 2$ supergravity whose bosonic part is the Einstein-Maxwell theory. A supersymmetric bosonic solution of this theory is characterized by the vanishing of all supercharges Q_i which is equivalent to the existence of a full set of super-covariantly constant spinors (Killing spinors). A Witten-type argument shows that this situation leads to Minkowski space as the unique solution. On the other hand it is possible to have only half of the Q_i's vanish, corresponding to a half-full set of Killing spinors. All these solutions are characterized by the

Theorem 5 (Tod [25]): *The existence of one supercovariantly constant Dirac spinor (= half-full set) leads to the Israel-Wilson-Perjés class of solutions.*

4 Stationary, axially-symmetric solutions

In order to describe stationary, axially-symmetric solutions we perform a further 'dimensional reduction' from 3 to 2 dimensions. This concerns only the dreibein $\hat{e}_m{}^a$ since all the other fields are already scalars in 3 dimensions. One has however the choice to use either the time-like or the space-like Killing vector to step down from 4 to 3 dimensions. As we have seen already this leads to a slightly different structure of the symmetric space of scalars, a pseudo-Riemannian versus a

Riemannian space. Lastly it turns out, however, that it is possible to incorporate both possibilities into a single representation leading to an infinite dimensional group of transformations acting on these solutions. Introducing isothermal coordinates the dreibein takes the form

$$\hat{e}_m{}^a = \begin{pmatrix} h & & \\ & h & \\ & & \rho \end{pmatrix} \tag{29}$$

where ρ is a harmonic function on the remaining 2-dimensional space, which we can choose as a coordinate as long as $\partial_m \rho \neq 0$. There is a conjugate harmonic function z leading to a natural system of cylindrical coordinates (ρ, z). The equation for the 'matter' fields represented by the scalars taking values in $SU(2,1)/SU(1,1) \times U(1)$ resp. $SU(2,1)/SU(2) \times U(1)$ becomes

$$(\chi^{-1}\partial_m \chi)^{;m} = 0 \longrightarrow \partial^m (\rho \chi^{-1} \partial_m \chi) = 0 \tag{30}$$

independent of h. The equation (23a) can be integrated to yield h as a function of $\chi(\rho, z)$ once the latter has been determined from eq. (30). The eq. (30) can be considered as integrability condition for the existence of a scalar potential ω

$$\rho \chi^{-1} \partial_m \chi = \epsilon_m{}^n \partial_n \omega \tag{31}$$

The generalized 'duality' transformation (remember that σ is the automorphism discussed above)

$$\delta \chi = \sigma(\omega) \chi - \chi \omega \tag{32}$$

turns out to be a symmetry of eq. (30) because ($\jmath_m \equiv \rho \chi^{-1} \partial_m \chi$)

$$\delta \jmath_m = -2\rho^* \jmath_m + [\omega, \jmath_m] \tag{33}$$

and hence

$$\partial^m \delta \jmath_m = -2\partial^m \rho^* \jmath_m + [{}^*\jmath^m, \jmath_m] = 0 \tag{34}$$

since

$$\partial_m \jmath_n - \partial_n \jmath_m + \rho^{-1}[\jmath_m, \jmath_n] - \rho^{-1}\partial_m \rho \jmath_n + \rho^{-1} \partial_n \rho \jmath_m = 0$$

The infinitesimal transformation eq. (32) can in fact be exponentiated with the help of some $SU(2,1)$-matrix $U(s,\chi) = 1 - s\omega + o(s)$

$$\chi(s) = \sigma\left(U(s,\chi)^{-1}\right)\chi(0)U(s,\chi) \tag{35}$$

U has to fulfil the linear system of differential eqs. [26]:

$$\partial_m U = \frac{t}{1+t^2}({}^*j_m - tj_m)U \tag{36}$$

where $t(s,\rho,z) = \frac{1}{\rho}\left(\frac{1}{2s} - z - \sqrt{(\frac{1}{2s} - z)^2 + \rho^2}\right)$.

Apart from the fact that it is possible to create a 1-parameter family of solutions from any given one, there is another, even more important aspect of these equations. The integrability conditions for the linear system eq. (36) are $\partial^m j_m = 0$ and the integrability condition for $j_m = \rho\chi^{-1}\partial_m\chi$. Yet, $\partial^m j_m = 0$ is the equation of motion (30). This is reminiscent of the situation for completely integrable dynamical systems as e.g. the KdV- or the Sine-Gordon-equation [27]. In fact it turns out that the methods developped for the solution of these equations are also applicable in the present case. This has lead to a number of very interesting papers and to the construction of a large number of new exact solutions of Einstein's resp. of the Einstein-Maxwell equations. There are in particular the analogues of the multi-soliton solutions, the so-called multi-Kerr-solutions [28]. One might hope that among these there are stationary solutions describing charged, rotating black-holes in equilibrium, but this hope could not be materialized up to now.

The most general approach to the integration of eq. (30) is the Riemann-Hilbert-Method [29] derived from the Inverse-Scattering-Method so successful for the KdV- and Sine-Gordon equation [27]. The essential point is the presence of the 'spectral'-parameter t in the linear system eq. (36). As a consequence the differential $\partial_m U(s)U(s)^{-1}$ belongs to the so-called affine Lie-algebra $su(2,1)^{(1)}$ [30]. The latter can be represented by formal a Laurent series in t

$$X(t) = \sum_{n=-\infty}^{+\infty} t^n X_n \quad \text{with} \quad X_n \in su(2,1) \tag{37}$$

In order to exponentiate the Lie-algebra elements in eq. (37) one has to introduce some notion of convergence for formal power series. In this case the most appropriate way is to require analyticity in a domain containing $t = 0$ [31]. The corresponding transformation group is that of the so-called Riemann-Hilbert-Transformations

[32]. A detailed description of this powerful tool will be given in another publication [33] initiated with the desire to clarify the connection between the various methods developped to construct all stationary axially-symmetric solutions of Einstein's equations.

ACKNOWLEDGEMENTS

I profitted much from a collaboration with P. Breitenlohner and G.W. Gibbons. Some of the results presented in this lecture are due to this collaboration and will appear in separate publications.

REFERENCES

[1] D. Kramer and G. Neugebauer, Ann.Phys.(Leipzig) 24 (1969) 62.

[2] R. Geroch, J.Math.Phys. 12 (1971) 918.

[3] W. Kinnersley, J.Math.Phys. 14 (1973) 651.

[4] D. Maison, Gen.Rel.Grav.10 (1979) 717.

[5] E. Cremmer and B. Julia, Nucl.Phys. B159 (1979) 141.

[6] P. Breitenlohner and D. Maison, in *International Seminar on Exact Solutions of Einstein's Equations*, W. Dietz and C. Hoenselaers eds.

[7] T. Kałuza, Sitzungsber.Preuss.Akad.Wiss. 1921, 966.
O. Klein, Z.Phys. 37 (1925) 895.
Y.M. Cho, J.Math.Phys. 16 (1975) 2029.

[8] T. Damour, in *Gravitational Radiation*, N. Deruelle and T. Piran eds.

[9] H. Eichenherr and M. Forger, Nucl.Phys. B155 (1979) 381.

[10] D. Maison, in *Developments in the Theory of Fundamental Interactions*, L. Turko and A. Pekalski , eds.

[11] J. Eells, Jr. and J.H. Sampson, Amer.J.Math. 86 (1964) 109.

[12] B.K. Harrison, J.Math.Phys. 9 (1968) 1744.

[13] W. Simon and R. Beig, J.Math.Phys. 24 (1983) 1163.

[14] E. Witten, Commun.Math.Phys. 80 (1981) 381.

[15] P. Breitenlohner, G. Gibbons and D. Maison, in preparation.

[16] B. Carter, in *Black Holes*, C. de Wit and B.S. de Wit eds.

[17] D.C. Robinson, P.R. Lett. 34 (1975) 905.

[18] P.O. Mazur, J.Phys.A. Math.Gen. 15 (1982) 3173.

[19] W. Israel, G.A. Wilson; J.Math.Phys. 13 (1972) 865.
Z. Perjés, P.R. Lett. 27 (1971) 1668.

[20] J.B. Hartle and S.W.Hawking, Commun.Math.Phys. 26 (1972) 87.

[21] P. Dobiasch, D. Maison. Gen.Rel.Grav. 14 (1982) 231.

[22] G.W. Gibbons, Nucl.Phys.B207 (1982)337.

[23] G.W. Gibbons, in *Unified Theories of Elementary Particles*, P. Breitenlohner and H.P. Dürr eds.
G.W. Gibbons and C.M. Hull, Phys.Lett. 109B (1982) 190.

[24] D. Maison, in *Supersymmetry and Supergravity 1983*, B. Milewski ed.

[25] K.P. Tod, Phys.Lett. 121B (1983) 241.

[26] D. Maison, P.R.Lett. 41 (1978) 521.

[27] A.C. Scott, E.Y.F. Chu and D.W. McLaughlin, Proc. IEEE 61, (1973) 1443.

[28] D Kramer and G. Neugebauer, Phys.Lett. 75A (1980) 259.
A. Tomimatsu and M. Kihara, Prog.Theor.Phys. 67 91982) 1406.

[29] V.A. Belinskii and V.E. Zakharov, Sov.Phys. JETP 48 (1978) 985.

[30] V. Kac, Funct.Anal.Appl. 1 (1967) 82.
R. Moody, Bull.Am.Math.Soc. 73 (1967) 217.
B. Julia, in *Proceedings of the John Hopkins Workshop on Particle Theory*, 1981.

[31] I. Hauser and F.J. Ernst, J.Math.Phys. 21 (1980) 1418.

[32] K. Ueno and Y. Nakamura, Phys.Lett. 117B (1982) 208.

[33] P. Breitenlohner and D. Maison, in preparation

NON LINEAR SIGMA MODELS:

A GEOMETRICAL APPROACH IN QUANTUM FIELD THEORY

Elcio Abdalla
The Niels Bohr Institute, University of Copenhagen
DK-2100 Copenhagen Ø, Denmark

Contents

1. INTRODUCTION

Non linear sigma models have been introduced in the 60'ties to describe low energy phenomena of field theory[1]. However, a four dimensional model could not be formulated due to non renormalizability[2].

The major discovery which promoted the models to an outstanding position in quantum field theory is the similarity between two dimensional non linear sigma models, and four dimensional non abelian gauge theories, both classically as well as quantum mechanically. At the classical level we have the actions

$$S_G = -\frac{1}{4}\int \langle F_{\mu\nu} | F_{\rho\sigma} \rangle g^{\mu\rho} g^{\nu\sigma} d^4x \tag{1.1}$$

$$S_{SM} = \frac{1}{2}\int g^{\mu\nu} \langle \partial_\mu q | \partial_\nu q \rangle d^2x \tag{1.2}$$

or

$$S_{SM} = \frac{1}{2}\int g^{\mu\nu} \langle D_\mu g | D_\nu g \rangle d^2x \tag{1.3}$$

presenting, among various similar mathematical properties in the context of differential geometry[3], conformal invariance, non trivial topology if $\pi_2(G/H) \neq 0$ for non linear sigma models, and if $\pi_3(G) \neq 0$ for gauge theories. Quantum mechanically both class of models are asymptotically free. One expects that gauge theories confine, a property also believed to be true for sigma models whose gauge group H is not simple (has non trivial ideals), since this is the non factorizable case; this was shown for CP^{N-1} [4] and Grassmannian models[5].

Interaction with fermions can be implemented[5,6]. In general, for confining pure models, fermions are confined as well[6]. Nevertheless, if the interaction is defined in a geometrical way, in a sense to be defined later, the model has exact factorization properties, so that in general a factorizable S-matrix can be found[7]. In the CP^{N-1} and Grassmannian case this means that the gauge field no longer has a pole, and the long range confining force disappears, a conclusion that is probably true in general. The supersymmetric model is a particular case. It presents a strong analogy with supersymmetric gauge theories in four dimensions. Those models are probably no longer confining, what is true at least in the case of N=4 supersymmetric Yang-Mills theory where the β function vanishes[8]. In N=2, the β function has no radiative corrections[9], the same being true for supersymmetric sigma models[10].

Also in four dimensions, sigma models turn out to be important. They appear in a natural way in extended supergravity[11], describing the spin zero sector of the theory. However, those sigma models are of the non compact type[12]. Instead of being a disease, this could cure the non renormalizability problem[13], quantizing the negative metric field in a suitable way. This quantization procedure leads to a break of the non compact group into its maximal compact subgroup[14].

2. PURE TWO DIMENSIONAL NON LINEAR SIGMA MODELS

2.1. Classical models and integrability

Classical sigma models are defined in terms of fields $q(x)$ mapping two dimensional Minkowski (or Euclidean) space time into a manifold M. with a free field Lagrangian. The necessary and sufficient condition for the sigma models to be integrable is that M is a pseudo-Riemannian symmetric space[15]. This means that M is a coset space, namely $M = G/H$, where G is a connected Lie group with Lie algebra $\mathfrak{g}$, and H a closed subgroup with Lie subalgebra $\mathfrak{h}$, such that the decomposition $\mathfrak{g} = \mathfrak{h} + \mathfrak{m}$ holds, and the following commutation relations are true:

$$[\mathfrak{h}, \mathfrak{h}] \subset \mathfrak{h} \quad , \quad [\mathfrak{h}, \mathfrak{m}] \subset \mathfrak{m}$$

$$\text{and} \quad [\mathfrak{m}, \mathfrak{m}] \subset \mathfrak{h} \tag{2.1}$$

Symmetric spaces have a complete classification, analogous to the Kartan list[16].

The field configuration $q(x)$ can be lifted to a field $g(x)$ taking values in G, subjected to the natural gauge equivalence

$$g_2(x) \sim g_1(x) \qquad \text{iff} \qquad q_2(x) = q_1(x) \tag{2.2}$$

and there is a field $h(x)$ taking values in H, such that

$$g_2(x) = g_1(x)\, h(x) \tag{2.3}$$

In terms of $g(x)$ the lagrangian looks like the r.h.s. of (1.3).

The Lie algebra $\mathfrak{h}$ can be decomposed in simple ideals

$$\mathfrak{h} = \mathfrak{h}^0 \oplus \mathfrak{h}^1 \oplus \dots \oplus \mathfrak{h}^r \tag{2.4}$$

so that

$$\mathfrak{g} = \mathfrak{m} \oplus \mathfrak{h}^0 \oplus \dots \oplus \mathfrak{h}^r \tag{2.5}$$

The left translated derivative $g^{-1}\partial_\mu g$ can be split into a gauge potential A_μ, lying in $\mathfrak{h}$, and a gauge covariant object K_μ, lying in $\mathfrak{m}$:

$$A_\mu = (g^{-1}\partial_\mu g)_{\mathfrak{h}} \tag{2.6}$$

$$K_\mu = g^{-1} D_\mu g = (g^{-1}\partial_\mu g)_{\mathfrak{m}} \tag{2.7}$$

The gauge potential can be split into components along the ideals $\mathfrak{h}_i$

$$A_\mu = A_\mu^0 + \dots + A_\mu^r \tag{2.8}$$

and analogous formula holds for the corresponding field strengths:

$$F_{\mu\nu} = \partial_\mu A_\nu - \partial_\nu A_\mu + [A_\mu, A_\nu] = \sum_{i=0}^{r} F_{\mu\nu}^{(i)} \tag{2.9a}$$

$$F_{\mu\nu}^{i} = \partial_\mu A_\nu^i - \partial_\nu A_\mu^i + [A_\mu^i, A_\nu^i] \tag{2.9b}$$

We can also define the gauge invariant quantities

$$G_{\mu\nu} = g F_{\mu\nu} g^{-1} \tag{2.10a}$$

(resp. $G_{\mu\nu}^{(i)}$)

$$J_{\mu\nu} = g D_\mu K_\nu g^{-1} \tag{2.10b}$$

and

$$j_\mu = -g K_\mu g^{-1} = -D_\mu g \, g^{-1} \tag{2.10c}$$

The last quantity corresponds to the Noether current of the models. This is explicited writing out the action in terms of the g fields, eq. (1.3), with the usual definition of the scalar product.

Due to the symmetric space structure of the manifold M, if we take the horizontal and vertical parts of the identity

$$\partial_\mu (g^{-1} \partial_\nu g) - g^{-1} \partial_\mu g \, g^{-1} \partial_\nu g = \partial_\nu (g^{-1} \partial_\mu g) - g^{-1} \partial_\nu g \, g^{-1} \partial_\mu g \tag{2.11}$$

we have, respectively

$$D_\mu K_\nu = D_\nu K_\mu \tag{2.12a}$$

$$F_{\mu\nu} = -[K_\mu, K_\nu] \tag{2.12b}$$

Also the following identities hold

$$\partial_\mu j_\nu = J_{\mu\nu} + G_{\mu\nu} \tag{2.13a}$$

and

$$G_{\mu\nu} = -[j_\mu, j_\nu] \tag{2.13b}$$

As a consequence

$$\partial_\mu j_\nu - \partial_\nu j_\mu + 2[j_\mu, j_\nu] = 0 \tag{2.14}$$

This is the main result. It implies the integrability of the system of first order linear differential equations with a real parameter λ

$$\partial_\mu U^{(\lambda)} = U^{(\lambda)}\left\{ j_\mu (1 - ch\,\lambda) - \varepsilon_{\mu\nu} j^\nu sh\,\lambda \right\} \tag{2.15}$$

An infinite number of local[17] and non local[15] conserved charges can be obtained. In particular, (2.14) immediately implies conservation of the first non local charge

$$Q^{(1)}(t) = \int dy_1\, dy_2\, \varepsilon(y_1 - y_2)\, j_0(t, y_1)\, j_0(t, y_2) - \int dt\, j_1(t, y) \tag{2.16}$$

2.2. Quantization and anomalies

In order to have information about the dynamics of the model in the quantum theory, we shall make use of the first non local charge. For this purpose we shall now work in a faithful finite dimensional unitary representation of G[18]. This means that we are excluding non compact models from the discussion. We shall discuss those models later on as a special case. Thus we can substitute g^{-1} (which depends non linearly on g) by g^+.

In order to define products of field operators at the same point, we are led to introduce a normal product prescription, which we denote by $N[\mathcal{O}(x)]$, where $\mathcal{O}(x)$ is a formal product of fields[19]. Inside normal products we have a constraint on the fields

$$N[g^+ g\, \mathcal{O}] = c\, N[\mathcal{O}] = N[\mathcal{O}\, g\, g^+] \tag{2.17}$$

where c is a renormalization dependent constant[20]. Other constraints defining G as a subgroup of U(N) should be handled similarly. We also require that internal symmetry (namely global G transformation) and local H-gauge transformations are preserved by the normal product prescription.

Aiming at a correct definition of the first quantum non local charge, we examine the Wilson expansion of the commutator between two currents:

$$\left[j_\mu(x+\varepsilon), j_\nu(x) \right] = \sum_k C^{(k)}_{\mu\nu}(\varepsilon)\, N\left[\mathcal{O}_k(x) \right] \quad ; \ (\varepsilon^2 < 0) \tag{2.18}$$

The correct (finite) definition of the quantum non local charge, requires a classification of the operators appearing in the right hand side of (2.18). Because of asymptotic freedom[21], they are classified by their naive dimension. The operators are constructed out of the field g (dimension zero) and the covariant derivative D_μ (dimension 1). Because of (2.17), the gauge invariant operators $\mathcal{O}_k$ are[18]:

dimension 1: j_μ

dimension 2: $J_{\mu\nu}$; $\partial_\mu j_\nu - \partial_\nu j_\mu$; $G^{(i)}_{\mu\nu}$

The Wilson expansion now reads:

$$\left[j_\mu(x+\varepsilon), j_\nu(x-\varepsilon) \right] = C^{\rho}_{\mu\nu}(\varepsilon)\, j_\rho(x) + D^{\sigma\rho}_{\mu\nu}(\varepsilon)\, \partial_\rho j_\sigma(x) + \sum_{i=0}^{r-1} D^{\sigma\rho\,(i)}_{\mu\nu}(\varepsilon)\, G_{\sigma\rho}(x) \tag{2.19}$$

Since the only operator of dimension 1 is the current j_μ, the finite quantum non local charge can be defined by means of a renormalization of the second term of the charge:

$$Q^{(1)} = \lim_{\delta\to 0} Q^{(1)}_\delta = \lim_{\delta\to 0} \int_{|y_1-y_2|\geq\delta} dy_1\, dy_2\, \varepsilon(y_1-y_2)\, j_0(t,y_1)\, j_0(t,y_2) - Z_\delta \int dy\, j_1(t,y) \tag{2.20}$$

with Z_δ = const. $\ln \mu\delta$

where the constant is model dependent and is chosen to make $Q^{(1)}$ finite. The time derivative of the charge can be readily calculated. As a result of general principles (Lorentz covariance, current conservation locality and C, P, T conservation) the coefficients C, D can be determined[22,23,24] and D^i can be shown to be proportional to $\varepsilon_{\mu\nu}\,\varepsilon^{\rho\sigma}$. As a result

$$\frac{dQ}{dt} = \sum_{i \text{ indep.}} a_i \int dy\, \varepsilon^{\mu\nu}\, G^{(i)}_{\mu\nu}(t,y) \tag{2.21}$$

This means that we have a very simple criterion for the existence of anomalies in pure non linear sigma models namely[18]:

1) $\mathfrak{h}$ is simple. There is no anomaly, and the charge is conserved. The quantum in and out charges are equal. Equality of their matrix elements determine the structure of the S-matrix. With a suitable ansatz for the bound states, we can define the exact S-matrix. This is done for the O(N)/SO(N-1) model[22] (non linear σ model on the N-sphere and for the principal chiral model[25] (SU(N) x SU(N)/SU(N)). Results are discussed in the next section.

2) $\mathfrak{h}$ has non trivial ideals. The non local charge is not conserved. Anomaly coefficients can be calculated by a 1/N expansion. In the known cases the anomaly obeys an Adler-Bardeen type theorem. This was done in CP^{N-1} [20,24] and Grassmann[5] models.

2.3. Some exact S-matrices

In the cases of models on a sphere $S^{N-1} = O(N)/SO(N-1)$ and principal chiral model $SU(N) \times SU(N)/SU(N)$, the non local charges are conserved and can be explicitly calculated in terms of in and out fields. For the principal model, due to the symmetry $g \longleftrightarrow g^+$, we have in fact 2 Noether currents and 2 non local charges, both conserved. As a consequence of the conservation of the non local charge, the S-matrix is of factorizable type, not allowing pair production[26]. All S-matrix elements can be written as a product of $2 \longrightarrow 2$ elements. We can use the equality of in and out charges to have a system of linear equations, since:

$$Q^{ij}_{in} \, |m\theta_1, n\theta_2\rangle = (M^{ij}_{in})_{mnm'n'} \, |m'\theta_1 \, n'\theta_2\rangle \tag{2.22a}$$

$$\langle m\theta_1 \, n\theta_2| \, Q^{ij}_{OUT} = \langle m'\theta_1 \, n'\theta_2| \, (M^{ij}_{OUT})_{mnm'n'} \tag{2.22b}$$

We write the most general ansatz for the S-matrix and obtain the factorization equations. For the model S^{N-1} we have

$$\langle \theta_1' k, \theta_2' l \,|\, \theta_1 i \; \theta_2 j \rangle = \delta(\theta_1 - \theta_1')\, \delta(\theta_2 - \theta_2') \Big\{ \sigma_1(\theta)\, \delta^{kl} \delta^{ij} + \sigma_2(\theta)\, \delta^{ik} \delta^{jl} + \sigma_3(\theta)\, \delta^{kj} \delta^{il} \Big\} + (\theta_1 \leftrightarrow \theta_2) \tag{2.23}$$

and as a result of the linear system implied by the charge conservation

$$\sigma_3 = \frac{-2i\pi}{(N-2)\theta}\, \sigma_2 \tag{2.24a}$$

$$\sigma_1 = \frac{-2i\pi}{(N-2)(i\pi - \theta)}\, \sigma_2 \tag{2.24b}$$

where $\theta = \theta_1 - \theta_2$

Using general properties of the S-matrix, and absence of bound states, we obtain[26]

$$\sigma_2(\theta) = \frac{\Gamma\left(\frac{1}{N-2}+\frac{\theta}{2\pi i}\right)\Gamma\left(\frac{1}{2}+\frac{\theta}{2\pi i}\right)\Gamma\left(\frac{1}{2}+\frac{1}{N-2}-\frac{\theta}{2\pi i}\right)\Gamma\left(1-\frac{\theta}{2\pi i}\right)}{\Gamma\left(\frac{\theta}{2\pi i}\right)\Gamma\left(\frac{1}{2}+\frac{1}{N-2}+\frac{\theta}{2\pi i}\right)\Gamma\left(\frac{1}{2}-\frac{\theta}{2\pi i}\right)\Gamma\left(1+\frac{1}{N-2}-\frac{\theta}{2\pi i}\right)} \tag{2.25}$$

which can be verified up to 2'nd order in a 1/N expansion[27].

For the principal model we have the more complicated SU(N) x SU(N) invariant ansatz:

$$\begin{aligned}\langle \theta_1' c_1' \alpha_1', \theta_2' c_2' \alpha_2' | \theta_1 c_1 \alpha_1, \theta_2 c_2 \alpha_2 \rangle &= \delta(\theta_1-\theta_1')\,\delta(\theta_2-\theta_2')\Big\{\Big[u_1(\theta)\,\delta^{c_1c_1'}\delta^{c_2c_2'} + \\ &+ u_2(\theta)\,\delta^{c_1c_2'}\delta^{c_2c_1'}\Big]\delta^{\alpha_1\alpha_1'}\delta^{\alpha_2\alpha_2'} + \Big[u_3(\theta)\,\delta^{c_1c_1'}\delta^{c_2c_2'} + u_4(\theta)\,\delta^{c_1c_2'}\delta^{c_2c_1'}\Big]\delta^{\alpha_1\alpha_2'}\delta^{\alpha_2\alpha_1'}\Big\} \\ &+ \delta(\theta_1-\theta_2')\,\delta(\theta_2-\theta_1')\Big\{\Big[u_1(\theta)\,\delta^{c_1c_2'}\delta^{c_2c_1'} + u_2(\theta)\,\delta^{c_1c_1'}\delta^{c_2c_2'}\Big]\delta^{\alpha_1\alpha_2'}\delta^{\alpha_2\alpha_1'} \\ &+ \Big[u_3(\theta)\,\delta^{c_1c_2'}\delta^{c_2c_1'} + u_4(\theta)\,\delta^{c_1c_1'}\delta^{c_2c_2'}\Big]\delta^{\alpha_1\alpha_1'}\delta^{\alpha_2\alpha_2'}\Big\}\end{aligned} \tag{2.26}$$

$$\begin{aligned}\langle \bar\theta_1' \bar c_1' \bar\alpha_1', \theta_2 c_2 \alpha_2 | \bar\theta_1 \bar c_1 \bar\alpha_1, \theta_2 c_2 \alpha_2 \rangle &= \delta(\theta_1-\theta_1')\,\delta(\theta_2-\theta_2')\Big\{\Big[t_1(\theta)\,\delta^{c_1c_1'}\delta^{c_2c_2'} + \\ &+ t_2(\theta)\,\delta^{c_1c_2}\delta^{c_1'c_2'}\Big]\delta^{\alpha_1\alpha_1'}\delta^{\alpha_2\alpha_2'} + \Big[t_3(\theta)\,\delta^{c_1c_1'}\delta^{c_2c_2'} + t_4(\theta)\,\delta^{c_1c_2}\delta^{c_1'c_2'}\Big]\delta^{\alpha_1\alpha_2}\delta^{\alpha_1'\alpha_2'}\Big\} \\ &+ \delta(\theta_1-\theta_2')\,\delta(\theta_2-\theta_1')\Big\{\Big[r_1(\theta)\,\delta^{c_1c_1'}\delta^{c_2c_2'} + r_2(\theta)\,\delta^{c_1c_2}\delta^{c_1'c_2'}\Big]\delta^{\alpha_1\alpha_1'}\delta^{\alpha_2\alpha_2'} \\ &+ \Big[r_3(\theta)\,\delta^{c_1c_1'}\delta^{c_2c_2'} + r_4(\theta)\,\delta^{c_1c_2}\delta^{c_1'c_2'}\Big]\delta^{\alpha_1\alpha_2}\delta^{\alpha_1'\alpha_2'}\end{aligned} \tag{2.27}$$

where $\theta = \theta_1 - \theta_2$.

From the non local charge conservation, and the symmetry $g \longleftrightarrow g^{-1}$ (implying the existence of a second non local charge) we obtain:

$$u_2(\theta) = -(2\pi i/N\theta)\, u_1(\theta) \;,\quad u_4(\theta) = -(2\pi i/N\theta)\, u_3(\theta)$$
$$t_2(\theta) = -\left[2\pi i/N(i\pi-\theta)\right] t_1(\theta) \;,\quad t_4(\theta) = -\left[2\pi i/N(i\pi-\theta)\right] t_3(\theta)$$
$$r_2(\theta) = r_4(\theta) = 0$$
$$u_2(\theta) = u_3(\theta) \,,\quad t_2(\theta) = t_3(\theta) \,,\quad r_1(\theta) = r_3(\theta) = 0 \tag{2.28}$$

Using an argument similar to those followed to obtain Z_n [28] and chiral Gross-Neveu[29] S-matrices, we use the unity determinant condition

$$\det g = \varepsilon^{i_1 \dots i_N} g_{1 i_1} \cdots g_{N i_N} \text{ "="} \; 1 \tag{2.29}$$

interpreting it as the fact that an antiparticle is the bound state of N-1 particles. This implies, by the fusion method[30], the bound state spectrum

$$m_n = m \frac{\sin n\pi/N}{\sin \pi/N} \tag{2.30}$$

and

$$u_1(\theta) = \frac{\operatorname{sh}\frac{1}{2}\left(\theta + \frac{2\pi i}{N}\right)}{\operatorname{sh}\frac{1}{2}\left(\theta - \frac{2\pi i}{N}\right)} \left\{ \frac{\Gamma\left(\frac{1}{N} + \frac{\theta}{2\pi i}\right)\Gamma\left(1 - \frac{\theta}{2\pi i}\right)}{\Gamma\left(\frac{\theta}{2\pi i}\right)\Gamma\left(1 + \frac{1}{N} - \frac{\theta}{2\pi i}\right)} \right\}^2 \tag{2.31}$$

Note that for N=2 this is in accordance with the O(4) result above. Also the result can be verified using the Bethe ansatz[31].

3. INTERACTION WITH FERMIONS

3.1. Classical model

The fermionic sector of the non linear sigma models can be specified by choosing a definite representation of the gauge group H . According to each case we can derive explicit expressions for the composite fields B_μ and j_μ^M, defined by

$$(B_\mu, X) = -\frac{i}{2}\bar{\Psi}\gamma_\mu X \Psi \,,\qquad X \in \mathfrak{h} \tag{3.1a}$$

$$j_\mu^M = g B_\mu g^{-1} \tag{3.1b}$$

For the Grassmannian models, as a very general example (notation as in (5))

$$j_\mu^M = z\, B_\mu^z\, z^+ + y\, B_\mu^y\, y^+ \tag{3.2}$$

The Lagrangian is defined as a sum of the pure non linear sigma model Lagrangian, the (gauge-covariantized) free fermion Lagrangian and a fermion self interaction $-1/4(B_\mu, B^\mu)$.

The representation of H we choose, determines further physical properties of the model. We have two main interesting cases, namely the fundamental and the adjoint representations.

If the fermion Ψ transforms according to the fundamental representation of H

$$\Psi \to h^{-1}\Psi$$

with covariant derivatives

$$D_\mu \Psi = \partial_\mu \Psi + A_\mu \Psi \tag{3.3}$$

the Lagrangian reads

$$\mathcal{L} = g^{\mu\nu}\left(D_\mu g, D_\nu g\right) - \frac{i}{2}\overline{\Psi}\overleftrightarrow{\not{D}}\Psi - \frac{1}{4}g^{\mu\nu}\left(B_\mu, B_\nu\right) \tag{3.4}$$

The Noether current is given by

$$J_\mu = j_\mu + j_\mu^M \tag{3.5}$$

where $\quad j_\mu^M = i z \overline{\Psi}\gamma_\mu \Psi \bar{z}$

satisfying the condition

$$\partial_\mu\left(J_\nu + j_\nu^M\right) - \partial_\nu\left(J_\mu + j_\mu^M\right) + 2\left[J_\mu, J_\nu\right] = 0 \tag{3.6}$$

implying compatibility of the system[32]

$$\partial_\mu \mathcal{U}^{(\lambda)} = \mathcal{U}^{(\lambda)}\Big\{(1-\mathrm{ch}\,\lambda)j_\mu - \mathrm{sh}\,\lambda\,\varepsilon_{\mu\nu}j^\nu + \\ + \frac{1}{2}(1-\mathrm{ch}\,2\lambda)j_\mu^M - \frac{1}{2}\mathrm{sh}\,2\lambda\,\varepsilon_{\mu\nu}j^{\nu M}\Big\} \tag{3.7}$$

and conservation of the charge

$$Q^{(1)} = \int dy_1 dy_2\, \varepsilon(y_1-y_2)\, J_0(t,y_1) J_0(t,y_2) - \int dy\left(J_1(t,y) + j_1^M(t,y)\right) \tag{3.8}$$

In this case we can define a new gauge field, in the Grassmann model[5] (CP^{N-1} as a particulat case[33])

$$\tilde{A}_\mu = A_\mu + \frac{i}{2}\bar{\psi}^z \gamma_\mu \psi^z \tag{3.9a}$$

$$J_\mu = j_\mu (\tilde{A}_\mu) \tag{3.9b}$$

and covariant derivatives $\tilde{D}_\mu$ accordingly. The Lagrangian takes the extremely simple form

$$\mathcal{L} = \overline{\tilde{D}_\mu z}\, \tilde{D}^\mu z + i \bar{\psi} \tilde{\not{D}} \psi \tag{3.10}$$

The adjoint representation is not less interesting. In that case we can redefine the fermion field

$$\tilde{\psi} = g\psi \tag{3.11}$$

which transforms just like the g field itself under gauge as well as global symmetry transformation. Writing explicitly the Lagrangian density in the case of U(N) symmetry we obtain

$$\mathcal{L} = \overline{\tilde{D}_\mu z}\, \tilde{D}^\mu z + i \bar{\tilde{\psi}} \tilde{D} \tilde{\psi} + \frac{1}{4}\left[(\bar{\tilde{\psi}}\tilde{\psi})^2 - (\bar{\tilde{\psi}}\gamma_5 \psi)^2 \right] \tag{3.12}$$

with $z^+ \tilde{\psi} = 0 = \tilde{\psi}^+ z$

which is the Lagrangian and constraints of the supersymmetric model! Supersymmetry happens to appear naturally just choosing a representation for the fermions under the gauge group.

Again integrability follows, and the non local charge is conserved, where in case of U(N) symmetry

$$J_\mu = j_\mu + j_\mu^M \tag{3.13a}$$

$$j_\mu^{M\,ij} = -i z^{ia} \bar{\tilde{\psi}}_a^k \gamma_\mu \tilde{\psi}_b^k \bar{z}^{bj} - i \bar{\tilde{\psi}}_a^i \gamma_\mu \tilde{\psi}_a^j \tag{3.13b}$$

and (3.8) for Q.

3.2. Quantization and cancellation of anomalies

The most striking feature of non linear sigma models interacting with fermions in a minimal or supersymmetric way is the exact cancellation of the anomaly (2.21) with the Adler anomaly of axial current divergence [33,24]. In order to prove this statement we proceed along the following. We first write the Wilson expansion to be used in the definition of the quantum non local charge in the minimal (M) and supersymmetric (S) cases:

$$\text{(M)} \quad \left[J_\mu(x+\varepsilon), J_\nu(x)\right] = C^{\rho}_{\mu\nu}(\varepsilon)\, J_\rho(x) + D^{\sigma\rho}_{\mu\nu}(\varepsilon)\, \partial_\sigma J_\rho(x) + \frac{N}{\pi} G_{\mu\nu}(x) + N\left[J_\mu, J_\nu\right] \tag{3.14}$$

$$\text{(S)} \quad \left[J_\mu(x+\varepsilon), J_\nu(x)\right] = C^{\rho}_{\mu\nu}(\varepsilon)\, J_\rho(x) + D^{\sigma\rho}_{\mu\nu}(\varepsilon)\, \partial_\sigma J_\rho(x) + C'^{\rho}_{\mu\nu}(\varepsilon)\, i_\rho(x) + D'^{\sigma\rho}_{\mu\nu}(\varepsilon)\, \partial_\sigma i_\rho(x) + \frac{N}{\pi} G_{\mu\nu}(x) + N\left[J_\mu, J_\nu\right] \tag{3.15}$$

Note that presence of the finite part. In the pure case it is simply equal to $\bar{z}z(\partial_\mu J_\nu - \partial_\nu J_\mu)$, which can either be included in the coefficient $D^{\sigma\rho}_{\mu\nu}$ or be put zero by defining [20] $\bar{z}z=0$. In the case with fermions this is no longer true, since eq. (3.6) contains the divergence of the axial current, and must be replaced by

$$\partial_\mu\left(J_\nu + j^M_\nu\right) - \partial_\nu\left(J_\mu + j^M_\mu\right) + 2\left[J_\mu, J_\nu\right] = -\frac{N}{\pi} G_{\mu\nu} \tag{3.16}$$

and there is an exact cancellation in the Wilson expansion, so that the quantum non local charges

$$\text{(M)} \quad Q = \int dy_1\, dy_2\, \varepsilon(y_1-y_2)\, J_0(t,y_1)\, J_0(t,y_2) + Z\int dy\, J_1(t,y) + i\int dy\, z\bar{\Psi}\gamma_1\Psi\bar{z}(t,y) \tag{3.17}$$

$$\text{(S)} \quad Q = \int dy_1\, dy_2\, \varepsilon(y_1-y_2)\, J_0(t,y_1)\, J_0(t,y_2) + Z\int dy\, (J_1 + i_1)(t,y) + i\int dy\, z\bar{\Psi}\gamma_1\Psi\bar{z}(t,y) \tag{3.18}$$

are conserved.

3.3. Exact S-matrices

Using the fact that non local charges are conserved, we can proceed along the steps followed in non anomalous pure non linear σ models and calculate the S-matrix of each model.

For the CP^{N-1} models with minimal fermions we can construct the S-matrix for the bosonic sector. The asymptotic charge is[17]

$$Q^{ij}_{\substack{in\\(OUT)}} = \underset{(-)}{+}\frac{1}{N}\int d\mu(p_1)\, d\mu(p_2)\, \tilde{\varepsilon}(p_1-p_2) : \Big[b^{i+}(p_1)\, b^{k}(p_1) - a^{k+}(p_1)\, a^{i}(p_1)\Big]\Big[a^{j+}(p_2)\, a^{k}(p_2) - b^{k+}(p_2)\, b^{j}(p_2)\Big] : + \frac{1}{i\pi}\int d\mu(p)\, \ln\frac{p^0+p}{m}\Big[b^{i+}(p)\, b^{j}(p) - a^{j+}(p)\, a^{i}(p)\Big] \tag{3.19}$$

with $d\mu(p) = \frac{dp}{2\pi\, 2p^0}$, and b, b^+, a, a^+ are the creation and annihilation operators of the z field. Conservation leads to the factorizable S-matrix

$$\langle k\theta_1', \ell\theta_2' \,|\, i\theta_1, j\theta_2\rangle = \delta(\theta_1-\theta_1')\,\delta(\theta_2-\theta_2')\Big[u_1(\theta)\,\delta_{ik}\delta_{j\ell} + u_2(\theta)\,\delta_{i\ell}\delta_{jk}\Big] + (\theta_1 \leftrightarrow \theta_2\,,\; i \leftrightarrow j) \tag{3.20}$$

$$\langle k\theta_1', \bar{\ell}\bar{\theta}_2' \,|\, i\theta_1, \bar{j}\bar{\theta}_2\rangle = \delta(\theta_1-\theta_1')\,\delta(\theta_2-\theta_2')\Big[t_1(\theta)\,\delta_{ik}\delta_{j\ell} + t_2(\theta)\,\delta_{ij}\delta_{k\ell}\Big] \tag{3.21}$$

with

$$t_1(\theta) = \frac{\Gamma\left(\frac{1}{2}+\frac{\theta}{2\pi i}\right)\Gamma\left(\frac{1}{2}+\frac{1}{N}-\frac{\theta}{2\pi i}\right)}{\Gamma\left(\frac{1}{2}-\frac{\theta}{2\pi i}\right)\Gamma\left(\frac{1}{2}+\frac{1}{N}+\frac{\theta}{2\pi i}\right)} \tag{3.22}$$

a result that can be checked by the 1/N expansion[34]. Physically, the charges are set free due to the screening of the fermion condensate. Note that the above S-matrix does not have bound states. For N=2 this is in contrast to the pure CP^1 model, which is exactly soluble[35], and whose bound states have the S-matrix of the O(3) non linear σ model.

The supersymmetric model is more interesting. The asymptotic charge is

$$Q^{ij}_{\substack{in\\(OUT)}} = \overset{+}{(-)} \frac{1}{N} \int d\mu(p_1)\, d\mu(p_2)\, \tilde{\mathcal{E}}(p_1 - p_2) : \Big[b^{i+}(p_1)\, b^k(p_1) -$$

$$- a^{k+}(p_1)\, a^i(p_1) + d^{i+}(p_1)\, d^k(p_1) - c^{k+}(p_1)\, c^i(p_1) \Big] \Big[a^{j+}(p_2)\, a^k(p_2)$$

$$- b^{j+}(p_2)\, b^k(p_2) + c^{j+}(p_2)\, c^k(p_2) - d^{j+}(p_2)\, d^k(p_2) \Big] : +$$

$$+ \frac{1}{i\pi} \int d\mu(p)\, \ln \frac{p^0 + p}{m} \Big[b^{i+}(p)\, b^j(p) - a^{j+}(p)\, a^i(p) +$$

$$+ d^{i+}(p)\, d^j(p) - c^{j+}(p)\, c^i(p) \Big] \tag{3.23}$$

and a factorizable S-matrix for both fermions and bosons can be computed:

$$\Big\langle b_\beta(\theta_1')\, b_\delta(\theta_2') \Big| b_\alpha(\theta_1)\, b_\gamma(\theta_2) \Big\rangle = \delta(\theta_1 - \theta_1')\, \delta(\theta_2 - \theta_2')\; {}_{\alpha\gamma}V_{\beta\delta}(\theta)$$

$$+ \delta(\theta_1 - \theta_2')\, \delta(\theta_2 - \theta_1')\; {}_{\alpha\gamma}V_{\delta\beta}(\theta) \tag{3.24}$$

$$\Big\langle f_\beta(\theta_1')\, f_\delta(\theta_2') \Big| f_\alpha(\theta_1)\, f_\gamma(\theta_2) \Big\rangle = \delta(\theta_1 - \theta_1')\, \delta(\theta_2 - \theta_2')\; {}_{\alpha\gamma}U_{\beta\delta}(\theta)$$

$$- \delta(\theta_1 - \theta_2')\, \delta(\theta_2 - \theta_1')\; {}_{\alpha\gamma}U_{\delta\beta}(\theta) \tag{3.25}$$

$$\Big\langle f_\beta(\theta_1')\, b_\delta(\theta_2') \Big| f_\alpha(\theta_1)\, b_\gamma(\theta_2) \Big\rangle = \delta(\theta_1 - \theta_1')\, \delta(\theta_2 - \theta_2')\; {}_{\alpha\gamma}C_{\beta\delta}(\theta)$$

$$- \delta(\theta_1 - \theta_2')\, \delta(\theta_2 - \theta_1')\; {}_{\alpha\gamma}D_{\beta\delta}(\theta) \tag{3.26}$$

with V, U, C, and D given by

$${}_{\alpha\gamma}V_{\beta\delta}(\theta) = v_1(\theta)\, \delta_{\alpha\beta}\, \delta_{\gamma\delta} + v_2(\theta)\, \delta_{\alpha\delta}\, \delta_{\gamma\beta} \tag{3.27}$$

(analogously for U, u_1, u_2, C, c_1, c_2, D, d_1, d_2) and the backward particle-antiparticle scattering amplitudes vanish.

Other particle-antiparticle amplitudes are obtained from crossing.

Moreover

$$v_1(\theta) = \frac{\sin\left[(\pi/2)(\theta/\pi i - 2/N)\right]}{\sin(\theta/2i)} C_1(\theta) \tag{3.28}$$

$$u_1(\theta) = \frac{\sin(\pi/2)(\theta/\pi i + 2/N)}{\sin(\theta/2i)} C_1(\theta) \tag{3.29}$$

$$d_1(\theta) = -\frac{\sin\pi/N}{\sin\theta/2i} C_1(\theta) \tag{3.30}$$

$$C_1(\theta) = \prod_{\ell=0}^{\infty} \frac{\Gamma(\theta/2\pi i + 1/N + \ell)\Gamma(1-\theta/2\pi i + \ell)\Gamma(\theta/2\pi i - 1/N + \ell)\Gamma(1-\theta/2\pi i + \ell)}{\Gamma(1-\theta/2\pi i + 1/N + \ell)\Gamma(\theta/2\pi i + \ell)\Gamma(1-\theta/2\pi i - 1/N + \ell)\Gamma(\theta/2\pi i + \ell)} \tag{3.31}$$

The mass spectrum is

$$m_\ell = m\frac{\sin\ell\pi/N}{\sin\pi/N} \qquad \ell = 1,\dots,N-1 \tag{3.32}$$

This can also be verified in a 1/n expansion[36]. It has the interesting property that the antiboson is a bound state of n-1 fermions and the antifermion is a bound state of n-2 bosons and 1 fermion. In fact the S-matrix was constructed to obey this fact, in a way analogous to the solution of the chiral Gross-Neveu model[29]. This is related to the screening of the chirality for physical fermions, which are massive and unconfined, the confining force being screened by chirality[29].

We should point out that the above construction does not work for Grassmann models, because the non local charge does not provide enough information. Only information about color singlets is given[5].

The S-matrix for the supersymmetric O(N) σ model is also known, but will not be presented here[37].

4. WESS-ZUMINO INTERACTION AND NON ABELIAN BOSONIZATION

Among the non linear sigma models, the principal model seems to be one of big importance and interest. It presents a chiral symmetry

SU(N) x SU(N), and has a highly non trivial 1/N expansion, the lowest order being represented by the sum of planar graphs, just like four dimensional non abelian gauge theories[38]. In spite of this fact, the exact S-matrix is very simple[25,31], being expandable in an asymptotic series of 1/N. In four dimensions the chiral model summed with a Wess-Zumino term[39] is a phenomenological low energy action for baryons[40].

In two dimensions the β function of the model is vanishing, if we fix the coefficient of the chiral term in a suitable way[41] and the Kac-Moody algebra obeyed by the current is the same as that of a current of a free fermion multiplet.

The model is defined by the action

$$S = \frac{1}{4\lambda^2} \int d^2x \; \mathrm{tr}\; \partial_\mu g^{-1} \partial^\mu g \; + \; n\Gamma \tag{4.1a}$$

$$\Gamma = \frac{1}{24\pi} \int_B d^3y \; \varepsilon^{ijk} \; \mathrm{tr}\; g^{-1}\partial_i g\, g^{-1}\partial_j g\, g^{-1}\partial_k g \tag{4.1b}$$

B is a ball with two dimensional (compactified) space as boundary. Γ is the Wess-Zumino term. The scale invariant case corresponds to $\lambda^2 = 4\pi/n$.

For general values of λ, the exact S-matrix of the model has been computed using the Bethe ansatz[42].

For the scale invariant model, we can have further information by a new kind of bosonization procedure, in which the equivalence of the model to a free fermion theory is proved[43]. In an operator language this is done by the following steps.

We define a free fermion theory

$$\mathcal{L} = i\, \overline{\psi}_j\, \partial\!\!\!/\, \psi_j \tag{4.2}$$

with ψ, belonging to the fundamental representation of SU(N). We can define the currents

$$J^F_{\pm ij}(x) = \,:\psi^+_{\pm i}(x)\, \psi_{\pm j}(x): \tag{4.3}$$

We define also the composite operator

$$g_{ij}(x) = \mu^{-1} : \psi^+_{-i}(x)\, \psi_{+j}(x) : \tag{4.4}$$

We analyze the operator product expansion of

$$J^B_{+ij}(x;\varepsilon) = \frac{i}{2\pi}\, g^{-1}_{ik}(x+\varepsilon)\, \partial_+ g_{kj}(x) \tag{4.5}$$

and

$$J^B_{-ij}(x;\varepsilon) = -\frac{i}{2\pi}\, \partial_- g_{ik}(x+\varepsilon)\, g^{-1}_{kj}(x) \tag{4.6}$$

We define (as usual[44]) the currents $J^B_{\pm ij}(x)$ as being the most divergent term of the operator product expansion of $J^B_{\pm ij}(x;\varepsilon)$, disregarding a trivial c-number:

$$J^B_{\pm ij}(x) = \lim_{\varepsilon \to 0} (\mu \varepsilon_\pm) \left[J^B_{\pm ij}(x;\varepsilon) - \langle 0 | J^B_{\pm ij}(x;\varepsilon) | 0 \rangle \right] \tag{4.7}$$

We verify that

$$J^B_{\pm ij}(x) = J^F_{\pm ij}(x) \tag{4.8}$$

As a consequence, the Hamiltonian is also the same for both[45] (fermionic and bosonic) cases. The bosonic equations of motion correspond to those of the scale invariant model, and all Green functions can be explicitly computed. This bosonization prescription can also be carried out by other methods[46].

5. CONCLUSION AND OUTLOOK

Besides the striking similarity to non Abelian theories, one of the most attractive features of non linear σ models is their integrability. Most striking is the integrability of supersymmetric models (or generally for models with fermions), since this is the case where finiteness, or at least a better ultraviolet behavior is expected for general field theories. Previous results show how this is obtained. A procedure to extend them to the Green functions can presently not be envisaged.

None the less other questions, not less interesting, can be raised. Non compact models deserved up to now not very much attention, due to the difficulty of quantizing fields with a non positive definite Euclidean action. It has been proposed that in general these models can display 2 phases: either the non compact symmetry is maintained in the quantization procedure, or it is broken in its maximal compact subgroup. Breaking a symmetry in two dimensions raises the serious question of treating massless bosons, so that it is very likely that at least in two dimensions the non compact symmetry remains unbroken, and the S-matrix should contain passive bystanders in order to maintain positive definiteness[47]. In higher dimensions the broken phase quantization procedure is very desirable due to cancellation of ultraviolet divergences in those cases.

Also the status of anomalous models should be better understood. It seems that confinement is behind the breaking of factorizability, although this has never been proved. This happens since those models presenting anomalies have a confining massless gauge field, and coupling fermions, the confinement disappears as long as the pole of the gauge field is driven away from zero due to the fermionic contribution to the gauge field propagator. In other words there is a cancellation between the gauge field pole (generated by the bosons), and the chirality carrying part of the fermions. Physical fermions are massive, and do not carry chirality.

Besides being a simplified theater of the extremely complicated non Abelian gauge field scenario, the models give explicit examples of non trivial S-matrices in quantum field theory, comparison with perturbative expansions, and explicit models where supersymmetry can be realized at all levels, namely Lagrangian (classical and quantum), physical states and S-matrices. Those simple facts are far from trivial in four dimensional quantum field theory, since non trivial models can have trivial scattering, perturbation expansion is not convergent, 1/N expansion had never been completely developed, and there are suspicions that supersymmetry cannot be realized quantum mechanically in a gauge theory.

Moreover, in integrable models other developments can been foreseen, towards the complete solution of the problem, by means of the study of the monodromy matrix[48].

We think that a complete understanding of these models is a very big progress in the development of ideas in quantum field theory.

REFERENCES

1 - B.W. Lee, "Chiral Dynamics". New York, Gordon & Breach 1972.
2 - J. Honerkamp, Nucl.Phys. B36 (1972) 130;
- G. Ecker and J. Honerkamp, Nucl.Phys. B35 (1971) 481.
3 - M. Forger, Ph.D. Thesis, Freie Univ. Berlin 1980 (unpublished);
- E. Abdalla, Modelos bidimensionais em teoria quântica de campos, Sâo Paulo Univ. 1982 (unpublished).
4 - A. D'Adda, P. Di Vecchia and M. Lüscher, Nucl.Phys. B146 (1978) 63.
5 - E. Abdalla, M. Forger and A. Lima-Santos, preprint NBI-HE-83-43.
6 - A. D'Adda, P. Di Vecchia and M. Lüscher, Nucl.Phys. B152 (1979) 125.
7 - E. Abdalla and A. Lima-Santos, Phys.Rev. D29 (1984) 1851.
8 - M. Grisaru, M. Rocek and W. Siegel, Nucl.Phys. B183 (1981) 141.
9 - M.T. Grisaru and W. Siegel, Nucl.Phys. B201 (1982) 292.
10 - L. Alvarez-Gaumé and D.Z. Friedman, Phys.Rev. D22 (1980) 846.
11 - E. Cremmer and B. Julia, Nucl.Phys. B159 (1979) 141.
12 - J. Ellis, C. Kouwnas and D.V. Nanopoulos, preprint CERN TH 3768;
- S. Ferrara and A. van Proyen, Phys.Lett. B138 (1984) 77.
13 - J.W. van Holten, Phys.Lett. B135 (1984) 427;
- N. Ohta, Phys.Lett. B134 (1984) 75.

14 - J.W. van Holten, preprint Wuppertal B-84-9;
- E. Abdalla, preprint NBI-HE-84-12, to be published in Phys.Lett.B.
15 - H. Eichenherr and M. Forger, Nucl.Phys. B155 (1979 381; B164 (1980) 528.
16 - S. Helgason, Differential Geometry and Symmetric Spaces, New York, Acad. Press 1962.
17 - H. Eichenherr and M. Forger, Commun.Math.Phys. 82 (1981) 227.
18 - E. Abdalla, M. Forger and M. Gomes, Nucl.Phys. B210 (1982) 181.
19 - W. Zimmermann, Lectures in Elementary Particles and Quantum Field Theory, Brandeis 1970.
20 - E. Abdalla, M.C.B. Abdalla and M. Gomes, Phys.Rev. D23 (1981) 1800.
21 - E. Brézin, S. Hikami and J. Zinn-Justin, Nucl.Phys. B165 (1980) 528.
22 - M. Lüscher, Nucl.Phys. B135 (1978) 1.
23 - E. Abdalla and A. Lima-Santos, Rev.Bras.Fis. 12 (1982) 293.
24 - E. Abdalla, M.C.B. Abdalla and M. Gomes, Phys.Rev. D27 (1983) 825.
25 - E. Abdalla, M.C.B. Abdalla and A. Lima-Santos, Phys.Lett. 140B (1984) 71.
26 - A.B. Zamodlodchikov and Al.B. Zamolodchikov, Ann.Phys. 120 (1979) 253.
27 - A.B. Zamolodchikov and Al.B. Zamolodchikov, Nucl.Phys. B133 (1978) 525.
28 - R. Köberle and J.A. Swieca, Phys.Lett. 86B (1979) 209.
29 - E. Abdalla, B. Berg and P. Weisz, Nucl.Phys. B157 (1979) 387;
- V. Kurak and J.A. Swieca, Phys.Lett. 82B (1979) 289;
- R. Köberle, V. Kurak and J.A. Swieca, Phys.Rev. D20 (1979) 897, Erratum D20 (1979) 2638.
30 - B. Schroer, T.T. Thruong and P. Weisz, Phys.Lett. 63B (1976) 422.
31 - P.B. Wiegmann, Phys.Lett. 141B (1984) 217.
32 - T.L. Curtright and C.K. Zachos, Phys.Rev. D21 (1980) 411.
33 - E. Abdalla, M.C.B. Abdalla and M. Gomes, Phys.Rev. D25 (1982) 452.
34 - R. Köberle and V. Kurak, IFUSP preprint 200 (unpublished).
35 - M.C.B. Abdalla and A. Lima-Santos, Acta Phys.Pol. Sept. 1984.
36 - R. Köberle and V. Kurak, IFUSP preprint 202 (unpublished).
37 - R. Shankar and E. Witten, Phys.Rev. D17 (1978) 2134.
38 - G. 't Hooft, Nucl.Phys. B72 (1974) 461.
39 - J. Wess and B. Zumino, Phys.Lett. 37B (1971) 95.
40 - E. Witten, Nucl.Phys. B223 (1983)422 and 433.
41 - E. Witten, Commun.Math.Phys. 92 (1984) 455.
42 - A.M. Polyakov and P.B. Wiegmann, Phys.Lett. 141B (1984) 223.
43 - E. Abdalla and M.C.B. Abdalla, preprint NBI-HE-84-11.
44 - J.A. Swieca, Fortschr. der Phys. 25 (1977) 303.
45 - H. Sugawara, Phys.Rev. 170 (1968) 1659;
- G.F. Dell'Antonio, Y. Frishman and D. Zwanziger, Phys.Rev. D6 (1973) 988.
46 - P. Di Vecchia, B. Durhuus and J.L. Petersen, preprint NBI-HE-84-02;
- P. Di Vecchia and P. Rossi, Phys.Lett. 140B (1984) 344.
47 - M. Gomes and Y.K. Ha, preprint Rockefeller Univ. RU84-B-78.
48 - H.J. de Vega, Proc. of the Johns Hopkins Workshop on current problems in particle theory 7, Bonn 1983.

AN APPROACH TOWARDS THE QUANTIZATION OF THE RELATIVISTIC CLOSED STRING BASED UPON SYMMETRIES

K. Pohlmeyer

Department of Physics
University of Freiburg
D-7800 Freiburg, W.Germany

In this talk I would like to convey the idea that it may by worthwhile to formulate and attack the problem of the quantization of the free relativistic bosonic string by algebraic means.

In general, strings and their quantum theory are of interest because the elementary excitations of QCD - if this is the theory of the hadrons - are string-like; quarks and/or antiquarks joined by strings of chromoelectric flux and possibly glue-balls as closed strings of chromoelectric flux.

I shall not even attempt to derive anything for these QCD-strings. Instead, I shall address myself to the Nambu-Goto-String [1] well known from the high-days of the dual resonance model - which - as Nambu suggested some four years ago [2] - should give us valuable information about the SU(N) Yang-Mills-Theory for $N \to \infty$. This Nambu-Goto-String theory, the generalization from random walks to random surface trajectories in Minkowski-space, has so far resisted all attempts to establish it as a quantum theory*. The unacceptable features of the quantization schemes were among others

i) lack of Lorentz invariance,

ii) tachyonic states,

iii) loss of reparametrization invariance.

The methods employed were

a) canonical quantization in some special (orthonormal) coordinate system [4],

b) functional integration in the so-called transverse gauge [5],

c) functional integration using an extended invariant measure [6].

Unfortunately, it is not yet clear whether method c), i.c. Polyakov's

---

* see, however, the remarkable WKB-treatment of the theory by Lüscher, Symanzik and Weisz [3]

treatment [6,7] cures the diseases listed above.

Whatever the situation may be, I am pursueing together with my collaborators C.Kimstedt and K.-H.Rehren in Freiburg the idea to treat the string theory in complete analogy to the quantum theory of the free massive relativistic particle as a representation problem for some (Lie)algebra of symmetry operations. For the free massive scalar particle the Klein-Gordon equation is interpreted as a representation condition for the Lorentz group. One approaches the problem by determining the Poisson-algebra of the conserved observable quantities $P_\mu, M_{\mu\nu}$ of the system whose dynamics is given by the reparametrization invariant action based upon the Lagrangian

$$\mathcal{L} = m\sqrt{\dot{x}^2}\ .$$

One replaces the Poisson-brackets by the commutator brackets and represents this (Poincaré) algebra irreducibly and unitarily such that the energy is positive. To this end one determines the Casimir operators P^2 and $W=-w_\lambda w^\lambda,\ w_\lambda=\frac{1}{2}\varepsilon_{\lambda\mu\nu\varsigma}M^{\mu\nu}P^\varsigma$ together with their spectrum and assigns to these operators one of the possible values, here: $P^2=m^2>0,\ W=m^2 s(s+1)=0$. Next one determines a Cartan-subalgebra, e.g. $\vec{P}$ and s_3^0 (= 3 component of the spin in the rest-frame) and constructs, starting from the eigenstates corresponding to $\vec{P}$ and s_3^0, by applying appropriate boosts the irreducible, unitary representation with positive energy of the Poincaré algebra and -group respectively.

This way of quantizing the free massive scalar particle is much more direct and gives - at least to me - much more insight than the calculation of the Green's function $K(x^{(f)},x^{(i)})$ by functional integration. Here $K(x^{(f)},x^{(i)})$ is the amplitude, which gives the probability that a particle in a volume element $d^3x^{(i)}$ near $\vec{x}^{(i)}$ at time $x^{(i)0}$ will be located in a volume element $d^3x^{(f)}$ near $\vec{x}^{(f)}$ at time $x^{(f)0}$.
I want to apply the algebraic procedure just outlined to the quantization of the free relativistic closed string. For that I have to set up a correspondance (analogy) of the so-called loop equations for the physical states $|\Phi\rangle$

$$R(\cdot,\sigma)|\Phi\rangle = 0 \quad , \quad L(\cdot,\sigma)|\Phi\rangle = 0$$

and the Klein-Gordon equation $(P^2-m^2)|\Phi\rangle=0$ or rather for the loop operators

$$R(\cdot,\sigma)\equiv x'(\cdot,\sigma)\cdot p(\cdot,\sigma) \quad , \quad L(\cdot,\sigma)\equiv\tfrac{1}{2}[m^4x'^2(\cdot,\sigma)+p^2(\cdot,\sigma)]$$

and the operator

$$P^2 - m^2 .$$

The two loop-operators are not on the same footing. The first one guarantees that one is dealing with geometrical objects independent of any parametrization, namely: curves in space - time together with certain tangential planes to these curves.
This equation has no correspondence for point particles. Hence it is the second loop operator which has to correspond to $P^2 - m^2$.
Then, what is the (Lie-)algebra for which $L(\cdot,\sigma)|\Phi\rangle = 0$ acts a representation condition? This algebra necessarily must be infinite-dimensional because infinitely many independent compatible conditions are imposed on its representation.
For the string theory two infinite dimensional Lie algebras have already been analyzed:

1) the Virasoro algebra of the infinitesimal generators of the conformal reparametrizations of the trajectory surfaces [8] ,
2) the Kac-Moody-Lie algebra generated with the help of the vertex-operator of the Koba-Nielsen formalism [9] .

However, since the elements of these algebras do not commute with $R(\cdot,\sigma)$ and $L(\cdot,\sigma)$, the algebras are not admissable. Thus we have to look for some new algebra $\mathfrak{g}$ on which the loop operators $R(\cdot,\sigma)$ and $L(\cdot,\sigma)$ act trivially such that they consist of Casimir operators of this algebra.
Before I turn to the construction of the desired algebra let me specify the theory by its classical action and let me introduce some notation.
The Nambu-Goto-action of the bosonic, free, relativistic string is given by

$$S = \frac{1}{2\pi\alpha'} \int d\tau\, d\sigma \sqrt{-h} = m^2 \int d\tau\, \mathcal{L}$$

$$m^2 = \frac{1}{2\pi\alpha'} .$$

Here $h = \det h_{ab}$ and $(h_{ab}) = (\partial_a x^\mu \partial_b x_\mu) = \begin{pmatrix} \dot{x}^2, & \dot{x}x' \\ x'\dot{x}, & x'^2 \end{pmatrix}$. τ and σ are parameters of the trajectory surface $x_\mu = x_\mu(\tau,\sigma)$ in Minkowski space.
σ is a "space-like" parameter:

$$x'^2 < 0 \qquad \left(x'_\mu \equiv \frac{\partial}{\partial\sigma} x_\mu(\tau,\sigma) \right).$$

τ is a time-like parameter:

$$\dot{x}^2 \geqslant 0 \qquad \left(\dot{x}_\mu \equiv \frac{\partial}{\partial \tau} x_\mu(\tau,\sigma) \right).$$

For every value of τ there exists a minimal period $\omega(\tau)$:

$$x_\mu(\tau, \sigma + \omega(\tau)) = x_\mu(\tau, \sigma).$$

This condition reflects the fact, that the string $\mathcal{C}(\cdot)$

$$0 \leq \sigma \leq \omega(\tau) \longrightarrow x_\mu = x_\mu(\tau, \sigma)$$

is closed.

The above action is invariant under arbitrary reparametrizations $\tau, \sigma \to \underset{\sim}{\tau}, \underset{\sim}{\sigma}$: $\tau = g(\underset{\sim}{\tau}, \underset{\sim}{\sigma})$, $\sigma = f(\underset{\sim}{\tau}, \underset{\sim}{\sigma})$. Moreover, it is invariant under the group of motions of the Minkowski space. The conserved quantities corresponding to this invariance are

$$P_\lambda, \ M_{\mu\nu}.$$

The description of the state of motion of the string by the data $x_\mu(\tau,\sigma)$, $\dot{x}_\mu(\tau,\sigma)$ τ fixed, $\sigma \in [0, \omega(\tau)]$ is highly redundant. This redundance is reflected in the canonical formalism in the non-invertibility of the defining equation for the momentum conjugate to x_μ:

$$p_\mu(\cdot,\sigma) \equiv \frac{\delta \mathcal{L}}{\delta \dot{x}^\mu(\cdot,\sigma)} = m^2 \frac{(\dot{x}x')x'_\mu - \dot{x}_\mu x'^2}{\sqrt{-h}}.$$

There exist two "first class" constraints in Dirac's classification

$$R(\cdot,\sigma) = x'(\cdot,\sigma)\cdot p(\cdot,\sigma) \approx 0,$$
$$L(\cdot,\sigma) = \tfrac{1}{2}\left[p^2(\cdot,\sigma) + m^4 x'^2(\cdot,\sigma)\right] \approx 0.$$

$p_\mu(\cdot,\sigma)$ has the interpretation of a momentum "density"

$$dP_\mu(\cdot,\sigma) = p_\mu(\cdot,\sigma)\, d\sigma \in \overline{V}^+$$

i.e. $dP(\cdot,\sigma)$ is interpreted as the momentum attached to the curve element $d\sigma$.

In the sequel, we shall set $m = 1$.

Under reparametrizations $\sigma \to \underset{\sim}{\sigma}$: $\sigma = f(\tau, \underset{\sim}{\sigma})$, x_μ and p_ν transform as follows

$$x_\mu(\cdot,\sigma) \longrightarrow \underset{\sim}{x}_\mu(\cdot, \underset{\sim}{\sigma}) = x_\mu(\cdot, f(\cdot, \underset{\sim}{\sigma}))$$
$$p_\nu(\cdot,\sigma) \longrightarrow \underset{\sim}{p}_\nu(\cdot, \underset{\sim}{\sigma}) = |f'(\cdot, \underset{\sim}{\sigma})|\, p_\nu(\cdot, f(\cdot, \underset{\sim}{\sigma})).$$

We impose the reparametrization invariant Poisson-brackets

$$\{x_\mu(\cdot,\sigma), p_\nu(\cdot,\sigma')\} = -g_{\mu\nu}\,\delta_{\omega(\cdot)}(\sigma-\sigma')\,, \quad \delta_{\omega(\cdot)} = \text{period. } \delta\text{-function}$$

$$\{x_\mu(\cdot,\sigma), x_\nu(\cdot,\sigma')\} = 0 = \{p_\mu(\cdot,\sigma), p_\nu(\cdot,\sigma')\}\,.$$

These can be reformulated with the help of the quantities

$$u^{\pm}_\mu(\cdot,\sigma) = x'_\mu(\cdot,\sigma) \pm p_\mu(\cdot,\sigma)$$

which transform covariantly under reparametrizations

$$u^{\pm}_\mu(\cdot,\sigma) \longrightarrow \underset{\sim}{u}^{\pm}_\mu(\cdot,\underset{\sim}{\sigma}) = |f'(\cdot,\underset{\sim}{\sigma})|\, u^{\pm}_\mu(\cdot, f(\cdot,\underset{\sim}{\sigma})):$$

$$\{u^{\pm}_\mu(\cdot,\sigma), u^{\pm}_\nu(\cdot,\sigma')\} = \begin{cases} -2g_{\mu\nu}\,\delta'_{\omega(\cdot)}(\sigma-\sigma') & \text{for } ++ \\ +2g_{\mu\nu}\,\delta'_{\omega(\cdot)}(\sigma-\sigma') & \text{for } -- \\ 0 & \text{otherwise.} \end{cases}$$

The "canonical" Hamiltonian $\mathcal{H} = p\dot{x} - \mathcal{L}$ vanishes identically. According to Dirac the evolution of the system in the parameter τ is given by the so-called total Hamiltonian formed of all first-class constraints.

$$\mathcal{H}_T = \int_0^{\omega(\cdot)} d\sigma \{\alpha(\cdot,\sigma) R(\cdot,\sigma) + \beta(\cdot,\sigma) L(\cdot,\sigma)\}$$

where α and β are quite arbitrary coefficient functions which can depend on anything like $\tau, \sigma, x_\mu(\cdot,\sigma), p_\nu(\cdot,\sigma')$ and even on more general (non-dynamical) variables like $\dot{x}_\lambda(\cdot,\sigma)$. It is however important that α and β are periodic functions of σ with period $\omega(\cdot)$.

The equations of motion are

$$\dot{x}_\mu = \alpha x'_\mu + \beta p_\mu \quad , \quad \dot{p}_\mu = (\alpha p_\mu + \beta x'_\mu)'\,.$$

We shall write them in the nearly equivalent form

$$\dot{u}^{\pm}_\mu = [(\alpha \pm \beta) u^{\pm}_\mu]'\,.$$

The solutions of these classical equations of motion describe minimal surfaces in Minkowski space. These surfaces have singular points in general.

$\mathcal{H}_T$ has no physical meaning. It should by no means be confused with the infinitesimal generator of translations in physical time: P_0, the energy.

1) The effect of the first term under the integral in the expression for $\mathcal{H}_T$:

$$\int_0^{\omega(\cdot)} d\sigma\, \alpha(\cdot,\sigma)\, R(\cdot,\sigma)$$

on a dynamical variable X parametrized with the help of some parameter σ just consists of a reparametrization of the quantity X. The condition $\{R(\cdot,\sigma), X\} = 0$ for the desired algebra guarantees that its elements are geometrical and observable objects.

2) Only the second part of $\mathcal{H}_T$

$$\int_0^{\omega(\cdot)} d\sigma\, \beta(\cdot,\sigma)\, L(\cdot,\sigma)$$

describes a change of the geometry as the parameter τ evolves. Vanishing commutators of our desired algebra $\mathcal{O}$ with this part of $\mathcal{H}_T$ mean that $\mathcal{O}$ consists only of conserved (observable) quantities.

The condition $L(\cdot,\sigma)|\Phi\rangle = 0$ can be viewed as a local version of the representation condition

$$[P^2 - m^2]|\Phi\rangle = 0 .$$

To see that, we pass from the above reparametrization-variant to the -invariant form

$$\left[\left(\frac{p(\cdot,\sigma)}{m\sqrt{-x'^2}}\right)^2 - m^2\right]|\Phi\rangle = 0$$

where for the time being we have reintroduced the mass m. Equivalently

$$\left[\frac{dP^\mu(\cdot,\sigma)}{ds}\,\frac{dP_\mu(\cdot,\sigma)}{ds} - m^2\right]|\Phi\rangle = 0$$

where $dP_\mu(\cdot,\sigma) = p_\mu(\cdot,\sigma)\, d\sigma$ = invariant quantity,

$$ds = m\sqrt{-x'^2}\, d\sigma = \quad " \quad " \quad .$$

Expanding $\frac{dP^\mu(\cdot,\sigma)}{ds}$ around its average value $\frac{P^\mu}{m\cdot L}$ gives the constant part ($L = L(\cdot)$ = Minkowski length of $\mathcal{C}(\cdot)$)

$$\langle \Phi | \left[P^2 - m^2 (m L)^2 \right] | \Phi \rangle \geq 0$$

and in addition a denumerable infinite collection of other representation conditions for the algebra of observable conserved quantities.

Certainly, among the elements of $\mathfrak{g}$ there are

$$P_\lambda = \oint d\sigma \, p_\lambda(\cdot, \sigma)$$

and

$$M_{\mu\nu} = \oint d\sigma \left\{ x_\mu(\cdot,\sigma) p_\nu(\cdot,\sigma) - x_\nu(\cdot,\sigma) p_\mu(\cdot,\sigma) \right\} .$$

The additional elements should rather refer to the "internal" state of the string, i.e. should be invariant under translations in space-time, i.e. should commute with P_λ , but not necessarily with $M_{\mu\nu}$.

Now we turn to the explicit construction of the desired algebra. The algebra is obtained with the help of a reparametrization invariant generating functional

$$U^{\pm}_{(A)} = U^{\pm}_{(A)[u^\pm]}(\cdot, \sigma)$$

where $U^{\pm}_{(A)}$ are two special pseudo-unitary $\infty \times \infty$ matrices which depend on τ, σ and on a set of real parameters $A^{\mu\alpha}$ $\mu = 1, \cdots, d, d+1$; $\alpha = 1, 2, 3, \cdots$: for a given μ: $A^{\mu\alpha} \neq 0$ only for finitely many values of α . Moreover - and that is absolutely essential - $U^{\pm}_{(A)}$ depend only on the dynamical variables $u^+_\mu(\cdot,\sigma)$ and $u^-_\nu(\cdot,\sigma')$ respectively: $u^\pm_{d+1} \equiv i u^\pm_0$.

The reparametrization invariant defining equations for $U^{\pm}_{(A)}$ are

$$U^{\pm}_{(A)}(\cdot,\sigma)' = u^\pm_\mu(\cdot,\sigma) A^{\mu\alpha} T_\alpha U^{\pm}_{(A)}(\cdot,\sigma) ,$$

where the T_α's are the infinitesimal generators of $su(\infty)$, e.g.

T_1, T_2, T_3 the infinitesimal generators of $su(2)$,

$T_1, \cdots, T_8$ the infinitesimal generators of $su(3)$,

$T_1, \cdots, T_{15}$ the infinitesimal generators of $su(4)$.

The equation for the τ-evolution of $U^{\pm}_{(A)}(\cdot,\sigma)$ is

$$U^{\pm}_{(A)}(\tau,\sigma)^{\cdot} = \left[\alpha(\tau,\sigma) \pm \beta(\tau,\sigma) \right] u^\pm_\mu(\tau,\sigma) A^{\mu\alpha} T_\alpha U^{\pm}_{(A)}(\tau,\sigma)$$

which is compatible with the previous defining equation due to the equations of motion.

Consider the monodromy matrix

$$V^{\pm}_{(A)} = V^{\pm}_{(A)}(\tau, \sigma_0(\tau)) = U^{\pm}_{(A)}(\tau, \sigma_0(\tau)+\omega(\tau)) \cdot U^{\pm}_{(A)}(\tau, \sigma_0(\tau))^{-1}$$

which does not depend on the parametrization of the curve $x_\mu = x_\mu(\tau, \sigma)$ $\sigma_0(\tau) \leq \sigma \leq \sigma_0(\tau) + \omega(\tau)$, τ fixed.

Its eigenvalues are independent of τ and $\sigma_0(\tau)$. Consequently all quantities

$$\mathrm{Tr}\left([V^{\pm}_{(A)} - \mathbb{1}]^k\right) \qquad k = 1, 2, \cdots$$

are reparametrization invariant conserved quantities. It is correct, that all the information is already contained in

$$\mathrm{Tr}\left(V^{\pm}_{(A)} - \mathbb{1}\right).$$

However, the higher powers $\mathrm{Tr}([V^{\pm}_{(A)} - \mathbb{1}]^k)$ - corresponding to the periods $k\,\omega(\tau)$, $k = 2, 3, \cdots$ - or rather the higher powers

$$\mathrm{Tr}\left(\ln^k V^{\pm}_{(A)}\right) \qquad k = 1, 2, \cdots$$

shall become important later.

The expansion of these traces in powers of the parameters $A^{\mu\alpha}$ yield "non-local" charges. (It is not clear whether in addition there is a "decent" set of infinitely many "local" observable conserved charges.)

With the help of the following matrices [10]

$$A^{\mu\alpha} T_\alpha = \mathrm{diag}(x_1^\mu, \cdots, x_n^\mu) \cdot T \quad , \quad T = \begin{pmatrix} 0 & 0 & \cdots & 0 & 1 \\ 1 & 0 & \cdots & 0 & 0 \\ \cdots & & & & \cdots \\ \cdots & & & & \cdots \\ 0 & 0 & \cdots & 1 & 0 \end{pmatrix}$$

it is possible to recover from the knowledge of

$$\mathrm{Tr}\left(V^{\pm}_{(A)} - \mathbb{1}\right)$$

as a function of the parameters $A^{\mu\alpha}$ all the cyclic sums $Z^{\pm}_{(\mu_1, \cdots, \mu_n)}$ of

$$R^{\pm}_{\mu_1,\cdots,\mu_n} \equiv \int_0^{\omega(\tau)} d\sigma_1 \int_0^{\sigma_1} d\sigma_2 \cdots \int_0^{\sigma_{n-1}} d\sigma_n \, u^{\pm}_{\mu_1}(\cdot,\sigma_1)\, u^{\pm}_{\mu_2}(\cdot,\sigma_2) \cdots u^{\pm}_{\mu_n}(\cdot,\sigma_n),$$

$$\mu_i \in [1,2,\cdots,d+1]$$

and vice versa: $\mathcal{Z}_{(\mu_1,\cdots,\mu_n)} = \mathrm{Z}\, R^{\pm}_{\mu_1,\cdots,\mu_n} \equiv R^{\pm}_{\mu_1,\cdots,\mu_n} + R^{\pm}_{\mu_2,\cdots,\mu_1} + \cdots$.

The $R^{\pm}_{\mu_1,\cdots,\mu_n}$ satisfy certain identities: $R^{\pm}_{\boxed{\mu_{i_1}\cdots\mu_{i_r}}\,\boxed{\mu_{j_1}\cdots\mu_{j_s}}} = R^{\pm}_{\mu_{i_1}\cdots\mu_{i_r}} \cdot R^{\pm}_{\mu_{j_1}\cdots\mu_{j_s}}$

Notation:
$$\sum_{\boxed{i_1\cdots i_r}\,\boxed{j_1\cdots j_s}} R^{\pm}_{\mu_1,\cdots,\mu_n} \equiv R^{\pm}_{\mu_1,\cdots,\mu_q \boxed{\mu_{i_1}\cdots\mu_{i_r}}\,\boxed{\mu_{j_1}\cdots\mu_{j_s}}\,\mu_{q+r+s+1},\cdots,\mu_n}$$

where the sum extends over all permutations of $\{q+1,\cdots,q+r+s\} = \{i_1,\cdots,i_r\} \cup \{j_1,\cdots,j_s\}$ which leave the order of the two disjoint sets $\{i_1,\cdots,i_r\}$; $\{j_1,\cdots,j_s\}$ separately invariant.

It is advantageous to pass to quantities with the same amount of information $R^{\pm t}_{\mu_1,\cdots,\mu_n}$ which differ from $R^{\pm}_{\mu_1,\cdots,\mu_n}$ by some products of R's with smaller rank which satisfy instead

$$R^{\pm t}_{\boxed{\mu_{i_1}\cdots\mu_{i_r}}\,\boxed{\mu_{j_1}\cdots\mu_{j_s}}} = 0 \qquad \text{for all partitions of } \{1,\cdots,n\}$$

These $R^{\pm t}$'s are just produced by $\mathrm{Tr}(\ln V^{\pm}_{(A)})$.

One consequence of these symmetries is

$$\mathrm{Z}\, R^{\pm t}_{\mu_1,\cdots,\mu_n} = 0 \quad , \text{ unless } n=1 \text{ for which case } P_{\mu_1} = \mathrm{Z}\, R^{t}_{\mu_1} = R_{\mu_1}.$$

Thus, the non-local charges we want to consider are produced by $\mathrm{Tr}(\ln^k V^{\pm}_{(A)})$ and are given by

$$\mathcal{Z}^{\pm n;k}_{(\mu_1,\cdots,\mu_n)} = \frac{1}{k!}\, \mathrm{Z} \sum_{0<i_1<\cdots<i_{k-1}<n} R^{\pm t}_{\mu_1,\cdots,\mu_{i_1}} R^{\pm t}_{\mu_{i_1+1},\cdots,\mu_{i_2}} \cdots R^{\pm t}_{\mu_{i_{k-1}},\cdots,\mu_n}$$

To give just two examples

$$\mathcal{Z}^{+3;2}_{(\mu_1,\cdots,\mu_3)} = P_{\mu_1} R^{+t}_{\mu_2,\mu_3} + \underset{123}{\mathrm{cycl}},$$

$$\mathcal{Z}^{+4;2}_{(\mu_1,\cdots,\mu_4)} = \frac{2}{3}\Big\{ P_{[\mu_1,} R^{+t}_{\mu_2][\mu_3,\mu_4]} + P_{[\mu_2,} R^{+t}_{\mu_3][\mu_4,\mu_1]} + P_{[\mu_3,} R^{+t}_{\mu_4][\mu_1,\mu_2]} + P_{[\mu_4,} R^{+t}_{\mu_1][\mu_2,\mu_3]} \Big\} + 2\Big\{ R^{+t}_{\mu_1,\mu_2} R^{+t}_{\mu_3,\mu_4} + R^{+t}_{\mu_2,\mu_3} R^{+t}_{\mu_4,\mu_1} \Big\}.$$

Alas, not all of those $\mathcal{Z}^{\pm n;k}$ are of this type, some containing products or even consisting of products of "smaller" $\mathcal{Z}$'s . The linear span of all those quantities $\mathcal{Z}^{\pm n;k}_{(\mu_1,\dots,\mu_n)}$ forms an algebra with an associative symmetric product, corresponding to the usual tensor product

$$\mathcal{Z}^{\pm r;k}_{(\mu_1,\dots,\mu_r)} \cdot \mathcal{Z}^{\pm s;l}_{(\nu_1,\dots,\nu_s)} = \sum \mathcal{Z}^{\pm r+s;k+l}_{(\mu_1,\dots,\mu_{i_1},\nu_{s-j},\mu_{i_1+1},\dots,\mu_{i_2},\nu_{s-j+1},\mu_{i_3},\dots)}$$

where the sum extends over all possible inequivalent (with respect to common cyclic permutations) distributions of the indices $\mu_1,\dots,\mu_r$; $\nu_1,\dots,\nu_s$ preserving the separate cyclic order of $\mu_1,\dots,\mu_r$ and $\nu_1,\dots,\nu_s$ respectively. Of course, the "tensor" product of a cyclic tensor $\mathcal{Z}_{(\mu_1,\dots,\mu_r)}$ and a second cyclic tensor $\mathcal{Z}^{\cdot\cdot}_{(\mu_{r+1},\dots,\mu_n)}$ in general does not lead to a cyclic tensor $\mathcal{Z}^{\cdot\cdot}_{(\mu_1,\dots,\mu_n)}$. For that it is necessary that the product is followed by cyclic symmetrization.

The linear span of the $\mathcal{Z}$'s closes also under the Poisson-bracket operation

$$\left\{\mathcal{Z}^{\pm}_{(\mu_1,\dots,\mu_r)}, \mathcal{Z}^{\pm}_{(\nu_1,\dots,\nu_s)}\right\} = \begin{cases} \pm\sum_{i=1}^{r}\sum_{j=1}^{s} 2g_{\mu_i\nu_j}\left(\mathcal{Z}^{\pm}_{\left(\mu_{i+1},\boxed{\begin{smallmatrix}\nu_{j+1}\cdots\nu_{j-1}\\ \mu_{i+2}\cdots\mu_{i-1}\end{smallmatrix}}\right)} - \mathcal{Z}^{\pm}_{\left(\boxed{\begin{smallmatrix}\nu_{j+1}\cdots\nu_{j-1}\\ \mu_{i+1}\cdots\mu_{i-2}\end{smallmatrix}},\mu_{i-1}\right)}\right) & \text{for } \pm\,\pm\,, \\ 0 & \text{for } \pm\,\mp\,. \end{cases}$$

This antisymmetric Lie product satisfies the Jacobi identity. It has all the properties of a derivation.
The Poisson-bracket for the quantities $\mathcal{Z}^{\pm n;k}_{(\mu_1,\dots,\mu_n)}$ can be given by a similar formula.

Now the idea is the following: do not attempt to represent in the quantum theory the original reparametrization variant Poisson-brackets by commutator brackets, but try less and represent only the Poisson-algebra of the above invariant quantities by a commutator algebra, the product being defined by the non-commutative, but associative operator version of the above "tensor-product".

Now in general this does not work unless the right hand side of the Poisson-brackets of some complete set of generating elements of the algebra does not give rise to ordering problems.

It is hardly surprising that indeed there will be ordering problems for the $\mathcal{Z}^{\pm}$'s as well as for the $\mathcal{Z}^{\pm n;k}$'s.

However, my collaborators and myself have checked for already fairly complicated situations that these are absent for certain filtered elements. For that one has to pass from the $\mathcal{Z}^{\pm n;k}_{(\mu_1,\dots,\mu_n)}$'s to truncated $\mathcal{Z}^{\pm n;k}_{(\mu_1,\dots,\mu_n)t}$ which differ from the $\mathcal{Z}^{\pm n;k}$'s only by linear combinations of products of lower rank $\mathcal{Z}^{\pm \cdot;\cdot}$'s and, moreover, have the property that all linear combinations which previously yielded products of $\mathcal{Z}^{\pm\cdot;\cdot}$'s vanish. Unfortunately, no simple generating functional (like the logarithm) for the $\mathcal{Z}^{\pm n;k}_{(\mu_1,\dots,\mu_n)t}$ has been found yet.

The $\mathcal{Z}^{\pm n;k}_{(\mu_1,\dots,\mu_n)t}$ define representation spaces for the Lorentz-group. The central projectors of the symmetric group commute, in particular, with the cyclic permutations. Hence the $\mathcal{Z}^{\pm n;k}_{(\mu_1,\dots,\mu_n)t}$ can be decomposed according to equivalent irred. representations of $SU(d,1)$ and finally of the Lorentz-group $SO(d,1)$.

The Poisson bracket operation obeys simple rules for the indices n, k and also for the symmetry content of $\mathcal{Z}^{\pm n;k}_{(\mu_1,\dots,\mu_n)t}$ i.e. there are selection rules. Under Poisson-bracket operation:

$$\{[n;k],[m;l]\} \longrightarrow [\leq n+m-2, k+l-1].$$

The possible representations of $SU(d,1)$ occuring on the r.h.s. follow from the Richardson-Littlewood rule. The inequality sign requires explanation: apart from the metric tensor $g_{\mu_i\nu_j}$ the rank of the r.h.s. of the Poisson-bracket is of course $n+m-2$. However, we allow that factors $P_{\mu_{i_1}}\cdots P_{\mu_{i_s}}$ be factored out of the elements on the r.h.s. , in other words, that the structure constants depend on the total momentum of the string. This is admissable since the P_μ's commute with all $\mathcal{Z}^{\pm n;k}_{(\;)t}$. As far as the <u>internal</u> algebra goes, the P_μ's can be regarded as numbers just as the Hamiltonian $\mathcal{H}$ enters the structure constants of the $O(4)$ group of the hydrogen atom. In fact, any such central element is acceptable in the structure constants on the r.h.s. of the Poisson-brackets.

Thus the total algebra $\mathfrak{g}$ is supposed to be generated from

$$so(d,1) \oplus (\{P_\mu \; \mu=0,1,\dots,d\} \oplus h_+ \oplus h_-)$$

where $h_\pm$ is the linear span of the elements $\mathcal{Z}^{+n;k}_{(\mu_1,\dots,\mu_n)t}$ and $\mathcal{Z}^{-n;k}_{(\mu_1,\dots,\mu_n)t}$ respectively.

The desired structure of the Poisson-brackets has been explicitely checked for all $\mathcal{Z}^{\pm n;2}_{(\mu_1,\cdots,\mu_n)t}$, $\mathcal{Z}^{\pm m;2}_{(\nu_1,\cdots,\nu_m)t}$ and moreover, for all $k,l,\ n+m \leq 10$. Heuristic arguments indicate that this structure holds true at least for an important infinite dimensional subalgebra of "dominant" charges.

Let me finish by expressing my conviction that here we have stumbled upon a remarkable structure of the classical string theory which should be taken into account when one passes to the quantum theory. The quantum theory should rather be defined as the implementation of a maximum of this structure in a commutator algebra, subject to the requirement that the energy be positive definite and the representation be "unitary".

References

[1] Y.Nambu: in "Symmetries and Quark Models" (R.Chand, Ed.), Gordon and Breach, New York, 1970
T.Goto: Progr.Theor.Phys. 46 (1971), 1560

[2] Y.Nambu: Phys.Lett. 80B (1979), 372

[3] M.Lüscher, K.Symanzik, P.Weisz: Nucl.Phys. B173 (1980), 365

[4] C.Rebbi: Phys.Rep. 12C (1974), 1
J.Scherk: Rev.Mod.Phys. 47 (1975), 123

[5] J.L.Gervais, B.Sakita: Phys.Rev. D4 (1971), 2291

[6] A.M.Polyakov: Phys.Lett. 103B (1981), 207

[7] B.Durhuus, P.Olesen, J.Petersen: Nucl.Phys. B198 (1982), 157; Nucl.Phys. B201 (1982), 176

[8] M.A.Virasoro: Phys.Rev. D1 (1970), 2933

[9] I.B.Frenkel, V.G.Kac: Inventiones Math. 62 (1980), 23

[10] K.Fredenhagen: private communication
The fact that these T^{μ}'s are not hermitean does not invalidate the argument.

YANG-BAXTER CHARGE ALGEBRAS IN INTEGRABLE CLASSICAL AND QUANTUM FIELD THEORIES.

H. Eichenherr
ETH Hönggerberg, Theoretical Physics
CH-8093 Zürich, Switzerland.

H.J. de Vega and J.M. Maillet
Laboratoire de Physique Théorique et Hautes Energies
(Laboratoire associé au CNRS n°LA280)
Université Pierre et Marie Curie
Tour 16 - ler étage, 4, place Jussieu
75230 Paris Cedex 05 (France).

Lecture given by one of us (H.E.) in the seminar series on "Non-linear Equations in Field Theory", Paris VI and Meudon (France), September 1983-May 1984.

Contents

I. Introduction

Integrable classical or quantum field theories possess by definition an infinite number of conserved charges commuting with each other. During the last decade, a considerable list of such models in two space-time dimensions has been established. The main tool for the construction and solution of them is the classical or quantum inverse scattering method, the algebraic backbone of which has turned out to be the Yang-Baxter equation. It was introduced to solve the one-dimensional gas of particles with Dirac delta interactions[1)] and the eight-vertex model[2)] in statistical mechanics, in particular, to prove the commutativity of transfer matrices $t(\lambda)$ at different values of the spectral parameter λ. The transfer matrix provides a link between vertex models and quantum mechanical spin chains: The XYZ Hamiltonian, for example, is a logarithmic derivative of the eight-vertex $t(\lambda)$ [2)]. It is a representative of a large class of theories to which the quantum inverse scattering method can be applied[3)]. One associates to such a model a local transition matrix $L_n(\lambda)$ and the monodromy operator $T(\lambda) = \prod_n L_n(\lambda)$, the trace of which is the transfer matrix $t(\lambda)$. Commutation relations for the various operator entries of $T(\lambda)$ are obtained from the Yang-Baxter algebra

$$R(\lambda,\mu)[T(\lambda)\otimes T(\mu)] = [T(\mu)\otimes T(\lambda)]R(\lambda,\mu) \qquad (1.1)$$

where the model dependent numerical matrix $R(\lambda,\mu)$ satisfies the consistency condition (Yang-Baxter equation)

$$R_{12}(\lambda,\mu)R_{13}(\lambda,\rho)R_{23}(\mu,\rho) = R_{23}(\mu,\rho)R_{13}(\lambda,\rho)R_{12}(\lambda,\mu) . \qquad (1.2)$$

In particular, $\log t(\lambda)$ is the generating functional of infinitely many commuting local conserved charges including the Hamiltonian. Moreover, the off-diagonal elements of $T(\lambda)$ provide creation and annihilation operators for the Bethe eigenstates of $t(\lambda)$.

These integrable quantum models have classical counterparts which are solvable by the classical inverse scattering method. They possess an associated system of linear differential equations (Lax pair) and its monodromy matrix $T(\lambda)$ which is the classical analogue of the quantum monodromy operator. Poisson brackets of the various matrix elements of $T(\lambda)$ on a finite spatial interval are obtained from the classical Yang-Baxter algebra[3)]

$$\{T(\lambda) \overset{\otimes}{,} T(\mu)\} = [r(\lambda,\mu), T(\lambda) \otimes T(\mu)] \tag{1.3}$$

where the model dependent numerical matrix $r(\lambda,\mu)$ satisfies the classical Yang-Baxter equation

$$[r_{12}(\lambda,\mu), r_{13}(\mu,\rho)] + [r_{13}(\mu,\rho), r_{23}(\rho,\lambda)] + [r_{23}(\rho,\lambda), r_{12}(\lambda,\mu)] = 0. \tag{1.4}$$

Usually, $R(\lambda,\mu)$ and $r(\lambda,\mu)$ are related through

$$R(\lambda,\mu) = P(1 - i\hbar\, r(\lambda,\mu) + O(\hbar^2)) \quad \text{as } \hbar \to 0, \tag{1.5}$$

P=permutation matrix (2.24). In the infinite volume limit (with suitable normalizations of $T(\lambda)$ and $r(\lambda,\mu)$) the trace of $T(\lambda)$ provides infinitely many commuting local conserved charges (action variables) including the Hamiltonian, whereas the off-diagonal elements furnish angle variables.

In contrast to this situation involving abelian charge algebras, these lecture notes deal with another class of models where the whole monodromy matrix (and not only its trace) is time independent and generates conserved charges satisfying non-abelian Poisson bracket and commutator algebras. This class contains two-dimensional field theories with non-abelian internal symmetry which classically are scale invariant. The currents associated to the internal symmetry group obey a flatness condition which allows the introduction of a Lax pair and the corresponding monodromy matrix. Examples are the non-linear σ models on symmetric spaces[4)] and fermionic theories like the chiral SU(N) and the O(2N) Gross-Neveu models[5)]. For these fermionic systems and their generalizations, the Poisson bracket algebra of the monodromy matrices has turned out to be a classical Yang-Baxter algebra[6)]. Due to conformal invariance, it is of "finite interval type", although we consider the system on the entire real axis. Thus the classical r matrix provides the structure constants of the canonical algebra of non-local charges. In particular, taking the trace of $T(\lambda)$, one obtains an infinite dimensional abelian subalgebra:

$$\{\mathrm{tr}\, T(\lambda), \mathrm{tr}\, T(\mu)\} = 0 \quad . \tag{1.6}$$

On the quantum level these theories exhibit dynamical generation of mass and asymptotic freedom. A sign of quantum integrability is the existence of a conserved quantum non-local charge $Q^{(1)}$. In certain cases, the renormalization of the classical $Q^{(1)}$ leads to anomalies. However, for the non-linear σ models on symmetric spaces G/H, the quantum $Q^{(1)}$ is time-independent provided H is simple[7]. In such a case, particle production is forbidden and the S matrix factorizes into two-body S matrices which (for consistency of the factorization) fulfill a Yang-Baxter equation (1.2) and can be calculated exactly[8]. Moreover, all higher non-local charges are expected to be conserved as well, so that there is a conserved quantum monodromy operator $T(\lambda)$ serving as their generating functional.

$T(\lambda)$ can be constructed from general properties, using $\mathcal{P}$, $\mathcal{T}$ and internal symmetries together with a factorization principle which relates the action of $T(\lambda)$ on k-particle states with its action on one-particle states[9][10]. As the result, the one-particle matrix element of $T_{ab}(\lambda)$ is expressed through the two-body physical S matrix as

$$\langle \theta\alpha | T_{ab}(\lambda) | \theta'\beta \rangle = \delta(\theta-\theta')\, S_{a\alpha,b\beta}(\theta+\gamma(\lambda)) \tag{1.7}$$

(a phase factor $\exp(i\phi(\lambda,\theta))$ cannot be ruled out completely). The quantum spectral parameter $\gamma(\lambda)$ is a real odd function of λ with the small λ behaviour

$$\gamma(\lambda) = c\lambda^{-1} + \mathcal{O}(\lambda)$$

where c is model dependent. From eq. (1.7) and the factorization principle one obtains closed expressions for all k-particle matrix elements of $T(\lambda)$ in terms of products of two-body S matrices. Their structure is identical to that of a k-sites monodromy operator in an inhomogeneous vertex model with statistical weights $S_{a\alpha,b\beta}(\theta_j+\gamma(\lambda))$, $1\leq j\leq k$. The Yang-Baxter equation (1.2) for $S_{a\alpha,b\beta}(\theta)$ finally implies that the monodromy operators $T(\lambda)$ satisfy the Yang-Baxter algebra (1.1) with the quantum R matrix

$$R(\lambda,\mu) = PS(\gamma(\lambda)-\gamma(\mu))$$

providing the structure constants. In particular, the trace of $T(\lambda)$ (i.e., the transfer matrix) furnishes an infinite dimensional abelian subalgebra:

$$[t(\lambda), t(\mu)] = 0 \quad , \quad t(\lambda) \equiv \sum_a T_{aa}(\lambda) .$$

In the limit of eq. (1.5) one recovers the classical Yang-Baxter charge algebras.

These lecture notes cover the content of refs. 6,9,10 in which the results on the non-abelian charge algebras have been explained. In section II, the construction of the classical monodromy matrices and non-local charges and the computation of their canonical algebras is described in general and illustrated by examples. Section II.4 contains some remarks on the rôle played in this context by Kac-Moody algebras. In section III, the general method for the derivation of Yang-Baxter quantum charge algebras for the class of models under consideration is explained and applied to the O(N)/O(N-1) non-linear σ model and to the SU(N) chiral and O(2N) Gross-Neveu models.

II. Classical Yang-Baxter charge algebras

II.1. Classical monodromy matrices and conserved charges

We consider 1+1 dimensional field theories with an internal non-abelian symmetry group G. The associated Noether current, $A_\mu(t,x)$ (taking values in the Lie algebra $\mathfrak{g}$ of G) is supposed to be conserved

$$\partial_\mu A^\mu = 0 \tag{2.1}$$

and flat:

$$F_{\mu\nu} \equiv \partial_\mu A_\nu - \partial_\nu A_\mu + [A_\mu, A_\nu] = 0 \ . \tag{2.2}$$

We assume the boundary conditions

$$A_\mu(t,x) \longrightarrow 0 \quad \text{as} \quad |x| \to \infty \ . \tag{2.3}$$

The main object of our interest is the monodromy matrix $T(X,Y|\lambda)$ which fulfills the equations[11) 4) 13)]

$$\frac{\partial}{\partial X^\mu} T(X,Y|\lambda) = -L_\mu(X,\lambda)\, T(X,Y|\lambda) \tag{2.4}$$

$$\frac{\partial}{\partial Y^\mu} T(X,Y|\lambda) = T(X,Y|\lambda)\, L_\mu(Y,\lambda) \tag{2.5}$$

$$T(X,X|\lambda) = \mathbb{1} \tag{2.6}$$

where

$$L_\mu(X,\lambda) = \frac{\lambda}{\lambda^2-1}\left(\lambda A_\mu(X) - \varepsilon_{\mu\nu} A^\nu(X)\right), \tag{2.7}$$

$$X = (t,x), \ Y = (t',y), \ g_{00} = -g_{11} = \varepsilon_{01} = 1 \ .$$

Eqs. (2.1), (2.2) are the compatibility conditions of eqs. (2.4), or equivalently, of eqs. (2.5). We also introduce

$$T^{+}(X,\lambda) = \lim_{y\to+\infty} T(X,Y|\lambda)$$
$$T^{-}(Y,\lambda) = \lim_{x\to-\infty} T(X,Y|\lambda) \qquad (2.8)$$
$$T(\lambda) = \lim_{x\to-\infty}\lim_{y\to+\infty} T(X,Y|\lambda) .$$

Since $L_\mu(X,\lambda)$ vanishes for $|x|\to\infty$, the monodromy matrix $T(X,Y|\lambda)$ becomes independent of t(t') as x(y) goes to infinity. In particular, $T(\lambda)$ is a time independent functional of the currents A_μ (thus, of the basic fields of the theory) and generates an infinite series of conserved non-local charges[12] $Q^{(n)}\in \mathcal{g}$:

$$T(\lambda) = \exp Q(\lambda)\ ,\quad Q(\lambda) = \sum_{1}^{\infty} \lambda^n Q^{(n)}\ ,\quad \frac{d}{dt}Q^{(n)} = 0$$
$$Q^{(0)} = -\int_{-\infty}^{+\infty} dx\, A_0(x) \qquad (2.9)$$
$$Q^{(1)} = -\frac{1}{4}\int_{-\infty}^{+\infty} dx \int_{-\infty}^{+\infty} dy\, \varepsilon(x-y)\,[A_0(x),A_0(y)] - \int_{-\infty}^{+\infty} dx\, A_1(x) \quad \text{etc.}$$

For later use, we note the transformation behaviour of currents and monodromy matrices under $\mathcal{P}$ (parity) and $\mathcal{T}$ (time reversal):

$$\mathcal{P}: \begin{cases} A_0(X) \to A_0(\tilde{X}) \\ A_1(X) \to -A_1(\tilde{X}) \\ L_0(X,\lambda) \to L_0(\tilde{X},-\lambda) \\ L_1(X,\lambda) \to -L_1(\tilde{X},-\lambda) \end{cases} \qquad \mathcal{P}\mathcal{T}: \begin{cases} A_0(X) \to -A_0(-X) \\ A_1(X) \to -A_1(-X) \\ L_0(X,\lambda) \to -L_0(-X,\lambda) \\ L_1(X,\lambda) \to -L_1(-X,\lambda) \end{cases}$$

with $\tilde{X} = (t,-x)$, and consequently from eqs. (2.4), (2.5), (2.6):

$$\mathcal{P}: T(X,Y|\lambda) \to T(\tilde{X},Y|-\lambda) \quad \text{and} \quad T(\lambda) \to T^{-1}(-\lambda) \qquad (2.10)$$
$$\mathcal{P}\mathcal{T}: T(X,Y|\lambda) \to T(-X,-Y|\lambda) \quad \text{and} \quad T(\lambda) \to T^{-1}(\lambda).$$

In several computations the following formula[14] will be used:

$$\frac{\delta T_{ab}(x,y|\lambda)}{\delta L_{1ij}(z,\lambda)} = T_{ai}(x,z|\lambda)\, T_{jb}(z,y|\lambda) \quad \text{for } x<z<y \qquad (2.11)$$

where the time arguments t=t' have been dropped. To prove it, we convert, say, eqs. (2.5), (2.6) into an integral equation

$$T(x,y|\lambda) = \mathbb{1} + \int_x^y dw\, T(x,w|\lambda)\, L_1(w,\lambda)$$

whence

$$\frac{\delta T_{ab}(x,y|\lambda)}{\delta L_{1ij}(z,\lambda)} = \int_z^y dw\, \frac{\delta T_{ac}(x,w|\lambda)}{\delta L_{1ij}(z,\lambda)}\, L_{1cb}(w,\lambda) + T_{ai}(x,z|\lambda)\,\delta_{jb}$$

which is solved by expression (2.11); analogously the x dependence is determined, starting from eq. (2.4).

II.2. Examples

II.2.1. The chiral fermionic models[6] are given through the Lagrangian

$$\mathcal{L} = i\bar{\psi}\partial\!\!\!/\psi + g(\bar{\psi}\gamma_\mu I_\alpha \psi)(\bar{\psi}\gamma^\mu I^\alpha \psi) \tag{2.12}$$

Here the γ^u are the usual Dirac matrices, the $I_\alpha = -I_\alpha^+$ ($\alpha = 1\ldots\dim G$) provide a unitary representation of the (compact) internal symmetry group G,

$$[I_\alpha, I_\beta] = f^\gamma_{\alpha\beta} I_\gamma \;, \qquad I^\alpha = k^{\alpha\beta} I_\beta$$

where $K_{\alpha\beta} = \mathrm{tr}(I_\alpha I_\beta) = (K^{-1})^{\alpha\beta}$ is the Killing form of G. $\psi_{a\mu}$ ($a=1\ldots N$; $\mu=1,2$) is a fermion field in two-dimensional space-time. We shall deal with both commuting and anticommuting spinors. If we set G = SU(N) in the fundamental representation, we recover the chiral Gross-Neveu model. The Hamiltonian of the theory reads

$$H = \int_{-\infty}^{+\infty} dx \left[-i\psi^+\gamma^5\partial_1\psi - g(\bar{\psi}\gamma_\mu I_\alpha\psi)(\bar{\psi}\gamma^\mu I^\alpha\psi)\right] \tag{2.13}$$

where $\gamma^5 = \gamma^o \gamma^1$, and for the case of commuting spinors, the Poisson brackets are the usual ones:

$$\{\psi_{a\mu}(t,x), \psi_{b\nu}(t,y)\} = i\delta_{ab}\delta_{\mu\nu}\delta(x-y) \tag{2.14}$$

If $\psi_{a\mu}(t,x)$ is a Grassmann variable, however, we use

$$\{A,B\} = i\int_{-\infty}^{+\infty} dx \sum_{a,\mu} A\left(\frac{\overleftarrow{\delta}}{\delta\psi^+_{a\mu}(x)}\frac{\overrightarrow{\delta}}{\delta\psi_{a\mu}(x)} + \frac{\overleftarrow{\delta}}{\delta\psi_{a\mu}(x)}\frac{\overrightarrow{\delta}}{\delta\psi^+_{a\mu}(x)}\right)B \tag{2.15}$$

where $\frac{\overleftarrow{\delta}}{\delta\psi_{a\mu}(x)}$ and $\frac{\overleftarrow{\delta}}{\delta\psi_{a\mu}(x)}$ denote the right and left derivative[15] in the Grassmann algebra, respectively. For even functions A, B this Poisson bracket is antisymmetric and fulfills the Jacobi identity.

It is easy to show from the equations of motion that the current

$$A_\mu = \sum_{\alpha=1}^{\dim G} A^\alpha_\mu I_\alpha \,,$$
$$A^\alpha_\mu = -4ig(\bar{\psi}\gamma_\mu I^\alpha\psi) \quad, \quad \mu = 0,1 \tag{2.16}$$

verifies eqs. (2.1) and (2.2). Moreover, its classical current algebra is given by

$$\{A_\mu(t,x), A_\nu(t,y)\} = 4g\,\delta(x-y)\left[1\otimes A_{|\mu-\nu|}(t,x), \Pi\right] \tag{2.17}$$

where

$$\Pi = \sum_{\alpha=1}^{\dim G} I_\alpha \otimes I^\alpha \tag{2.18}$$

and we have used the tensor product notation

$$(A\otimes B)_{ac,bd} = A_{ab}B_{cd}, \quad \{A \overset{\otimes}{,} B\}_{ac,bd} = \{A_{ab}, B_{cd}\}.$$

Both assertions hold for commuting as well as anticommuting spinors.

To control the conservation of $T(\lambda)$, we compute from eqs.(2.14) or (2.15)

$$\{H \overset{\otimes}{,} L_1(x,\lambda)\} = \partial_1 L_0(x,\lambda) - [L_0(x,\lambda), L_1(x,\lambda)]$$

(cf. eq. (2.2)), and using eqs. (2.11) and (2.4), (2.5):

$$\begin{aligned}\{H \overset{\otimes}{,} T(\lambda)\} &= \int_{-\infty}^{+\infty} dx \sum_{ij} \{H, L_{1ij}(x,\lambda)\} \frac{\delta T(\lambda)}{\delta L_{1ij}(x,\lambda)} \\ &= \int_{-\infty}^{+\infty} dx\, T^-(x,\lambda)\left(\partial_1 L_0(x,\lambda) - [L_0(x,\lambda), L_1(x,\lambda)]\right) T^+(x,\lambda)\end{aligned}$$

= 0 upon partial integration.

II.2.2. <u>The Gross-Neveu model</u> with the Lagrangian

$$\mathcal{L} = \sum_{a=1}^{N} i\bar{\phi}_a \partial\!\!\!/ \phi_a + g\Big(\sum_{a=1}^{N} \bar{\phi}_a \phi_a\Big)^2 \tag{2.19}$$

provides another example. The symmetry group of the model depends on whether $\phi_{a\mu}$ $(a = 1,\ldots N;\ \mu = 1,2)$ is a commuting or anticommuting spinor.

a) <u>Commuting spinors</u>

We shall employ the notation[16)]

$$\begin{aligned}(u^a) &= \sqrt{2}\,(\mathrm{Re}\,\phi_{11}\ldots\mathrm{Re}\,\phi_{N1}, \mathrm{Im}\,\phi_{11}\ldots\mathrm{Im}\,\phi_{N1})^T, \qquad a = 1\ldots 2N \\ (v_a) &= \sqrt{2}\,(\mathrm{Re}\,\phi_{12}\ldots\mathrm{Re}\,\phi_{N2}, \mathrm{Im}\,\phi_{12}\ldots\mathrm{Im}\,\phi_{N2})^T, \\ u_a &= \varepsilon_{ab}u^b, \quad v^a = \varepsilon^{ab}v_b, \\ (\varepsilon_{ab}) &= \begin{pmatrix} 0 & 1_N \\ -1_N & 0\end{pmatrix} = -(\varepsilon^{ab}).\end{aligned}$$

The canonical formalism is given by the Hamiltonian

$$H = -\int_{-\infty}^{+\infty} dx \left[\frac{1}{2} \sum_{a=1}^{2N} (u^a \partial_1 u_a - v^a \partial_1 v_a) + g \left(\sum_{a=1}^{2N} u^a v_a \right)^2 \right] \tag{2.20}$$

and the Poisson brackets

$$\begin{aligned} \{u^a(t,x), u^b(t,y)\} &= \{v^a(t,x), v^a(t,y)\} = \varepsilon^{ab} \delta(x-y) \\ \{u^a(t,x), v^b(t,y)\} &= 0 \quad . \end{aligned} \tag{2.21}$$

Eqs. (2.20) and (2.21) exhibit the symplectic group Sp(N) as the symmetry group of the commuting Gross-Neveu model.

The conserved and curvatureless current reads

$$\begin{aligned} (A_0)^a{}_b &= -2g(u^a u_b + v^a v_b) \\ (A_1)^a{}_b &= -2g(u^a u_b - v^a v_b) \end{aligned} \tag{2.22}$$

and the current algebra is found to be

$$\{A_\mu(t,x) \overset{\otimes}{,} A_\nu(t,y)\} = -2g\, \delta(x-y) \left[A_{|\mu-\nu|}(t,x) \otimes 1, P + \Sigma \right] \tag{2.23}$$

where

$$P^{ac}{}_{bd} = \delta^a{}_d \delta^c{}_b \quad , \quad \Sigma^{ac}{}_{bd} = \varepsilon^{ac} \varepsilon_{bd} \tag{2.24}$$

(P is a permutation operator: $P(A \otimes B)P = B \otimes A$, it coincides with Π (eq. (2.18)) if G is U(N) in the fundamental representation.)

b) <u>Anticommuting spinors</u>

Going over from the complex spinors $\phi_{a\mu}$ (a = 1...N) to Majorana spinors $\psi_{a\mu}$ (a = 1,...,2N) we have the Hamiltonian

$$H = \int_{-\infty}^{+\infty} dx \left[-i \sum_{a=1}^{2N} \psi_a^+ \gamma^5 \partial_1 \psi_a - g \left(\sum_{a=1}^{2N} \bar\psi_a \psi_a \right)^2 \right] \tag{2.25}$$

and the Poisson brackets

$$\{A, B\} = i \int_{-\infty}^{+\infty} dx \sum_{a,\alpha} A \left(\frac{\overleftarrow{\delta}}{\delta \psi_{a\alpha}(x)} \frac{\overrightarrow{\delta}}{\delta \psi_{a\alpha}(x)} \right) B \quad , \tag{2.26}$$

showing the O(2N) symmetry of the anticommuting Gross-Neveu model.

The current

$$(A_\mu)_{ab} = -8ig \left(\bar\psi_a \gamma_\mu \psi_b \right) \tag{2.27}$$

is conserved and flat; the current algebra is

$$\{A_\mu(t,x) \overset{\otimes}{,} A_\nu(t,y)\} = 8g\,\delta(x-y)\left[A_{|\mu-\nu|}\otimes 1,\ P-\Delta\right] \tag{2.28}$$

with

$$\Delta_{ac,bd} = \delta_{ac}\delta_{bd} \quad . \tag{2.29}$$

Notice that P-Δ coincides with Π in eq. (2.18) for the fundamental representation of O(2N): Choosing

$$I_\alpha \equiv I_{k\ell} \quad , \quad (I_{k\ell})_{ab} = \delta_{ka}\delta_{\ell b} - \delta_{kb}\delta_{\ell a} \quad , \quad 1 \le k < \ell \le 2N ,$$

$$I^{k\ell} = -\tfrac{1}{2} I_{k\ell} \quad ,$$

we have

$$\Pi = -\tfrac{1}{2} \sum_{1\le k<\ell\le 2N} I_{k\ell}\otimes I_{k\ell} = P-\Delta \quad .$$

Thus, the O(2N) Gross-Neveu model as well as the chiral fermionic models in II.2.1. provide explicit realizations of the Lie Poisson bracket construction in ref. 3 (Faddeev). The above classical current algebra holds also for the O(2N+1) case (whereas the properties of the quantum models might be different, due to different features of the representations of O(N) for odd and even N).

The current algebras (2.17) and (2.23) are called <u>ultralocal</u> because they contain the δ function but not its derivatives.

II.2.3. <u>The chiral σ model</u> involves a field g(t,x) taking values in a Lie group G. The Lagrangian is

$$\mathcal{L} = -\tfrac{1}{2}\,\mathrm{tr}\, A_\mu A^\mu \tag{2.30}$$

where

$$A_\mu = g^{-1}\partial_\mu g \tag{2.31}$$

is the conserved and flat current. The symmetry group is actually GxG. G can be replaced by a Riemannian symmetric space G/H, leading to the non-linear σ model on G/H [4]. For (2.30), the Hamiltonian [17]

$$H = \tfrac{1}{2}\int_{-\infty}^{+\infty} dx\, \mathrm{tr}\left(-\pi^T g\, \pi^T g + \partial_x g\, \partial_x (g^{-1})\right)$$

and the Poisson brackets for g and $\pi = \partial_0 (g^{-1})^T$

$$\{q(t,x) \overset{\otimes}{,} \pi^T(t,y)\} = \delta(x-y)\, P$$
$$\{q(t,x) \overset{\otimes}{,} q(t,y)\} = \{\pi(t,x) \overset{\otimes}{,} \pi(t,y)\} = 0$$

lead to the non-ultralocal current algebra [9]

$$\begin{aligned} \{A_0(t,x) \overset{\otimes}{,} A_0(t,y)\} &= \delta(x-y)\,[\mathbb{1} \otimes A_0(t,x), P] \\ \{A_1(t,x) \overset{\otimes}{,} A_0(t,y)\} &= \delta(x-y)(\mathbb{1} \otimes A_1(t,x))P - \delta'(x-y)(g^{-1}(x)g(y) \otimes \mathbb{1})P \\ \{A_1(t,x) \overset{\otimes}{,} A_1(t,y)\} &= 0\,. \end{aligned} \tag{2.32}$$

The δ' function in (2.32) (caused by the spatial derivative in $A_1(t,x)$) is a source of trouble in the determination of the charge algebra for the chiral models.

II.3. Computation of the charge algebras

The Poisson bracket of any two charges $Q^{(n)}$ can be obtained from the Poisson bracket of the monodromy matrices. To determine it, we start from the chain rule:

$$\{T(\lambda) \overset{\otimes}{,} T(\mu)\} = \int_{-\infty}^{+\infty} dx \int_{-\infty}^{+\infty} dy \left(\frac{\delta T(\lambda)}{\delta L_{1ij}(x,\lambda)} \otimes \frac{\delta T(\mu)}{\delta L_{1kl}(y,\mu)} \right) \{L_{1ij}(x,\lambda), L_{1kl}(y,\mu)\}\,. \tag{2.33}$$

For the fermionic models (sects. II.2.1. and II.2.2.) we have

$$\{L_1(x,\lambda) \overset{\otimes}{,} L_1(y,\mu)\} = \delta(x-y)\,[r(\lambda,\mu), L_1(x,\lambda) \otimes \mathbb{1} + \mathbb{1} \otimes L_1(y,\mu)] \tag{2.34}$$

where the classical r matrix is given by

$$r(\lambda,\mu) = \begin{cases} \dfrac{4g}{\lambda^{-1}-\mu^{-1}}\,\Pi & \text{for models (2.12)} \quad (2.35) \\[2ex] \dfrac{2g}{\lambda^{-1}-\mu^{-1}}(P+\Sigma) & \text{for model (2.19), commuting} \quad (2.36) \\[2ex] -\dfrac{8g}{\lambda^{-1}-\mu^{-1}}(P-\Delta) & \text{for model (2.19), anticommuting.} \quad (2.37) \end{cases}$$

To derive eq. (2.34) from the current algebras (2.17), (2.23), (2.28), respectively, one uses the formulas

$$[X \otimes \mathbb{1} + \mathbb{1} \otimes X, r(\lambda,\mu)] = 0 \qquad X \in \mathfrak{g}\,,$$
$$A_1(x) = \frac{1}{\lambda-\mu}\,(\alpha(\lambda)L_1(x,\lambda) - \alpha(\mu)L_1(x,\mu))$$

$$A_0(x) = \frac{1}{\lambda-\mu}\left(\mu\alpha(\lambda)L_1(x,\lambda)-\lambda\alpha(\mu)L_1(x,\mu)\right) \quad , \quad \alpha(\lambda)=\frac{\lambda^2-1}{\lambda} .$$

Inserting eqs.(2.11), (2.34) into (2.33) and using eqs. (2.4), (2.5), we find

$$\begin{aligned} \{T(\lambda) \overset{\otimes}{,} T(\mu)\} &= -\int_{-\infty}^{+\infty} dx\, \partial_x \left[\left(T^-(x,\lambda) \otimes T^-(x,\mu)\right) r(\lambda,\mu) \left(T^+(x,\lambda) \otimes T^+(x,\mu)\right)\right] \\ &= \left[r(\lambda,\mu) , T(\lambda) \otimes T(\mu)\right] . \end{aligned} \tag{2.38}$$

The result (2.38) shows that the canonical algebra of the monodromy matrices is a quadratic algebra: the Poisson bracket of two T's is not a linear, but a quadratic expression in the T's.

The algebra (2.38) is called the classical Yang-Baxter algebra. Its structure constants (as well as those of the current algebra) are provided by the classical r matrix. The antisymmetry of the Poisson bracket (2.38)

$$P\{T(\lambda) \overset{\otimes}{,} T(\mu)\}P = -\{T(\mu) \overset{\otimes}{,} T(\lambda)\}$$

is assured because of

$$[r(\lambda,\mu) , P] = 0 .$$

An important property of the r matrices (2.35) - (2.37) is that they satisfy the classical Yang-Baxter equation

$$[r_{12}(\lambda,\mu), r_{13}(\mu,\rho)] + [r_{13}(\mu,\rho), r_{23}(\rho,\lambda)] + [r_{23}(\rho,\lambda), r_{12}(\lambda,\mu)] = 0 \tag{2.39}$$

In eq. (2.39), the $r_{ij}(\lambda,\mu)$ act in a threefold tensor product space rather than in a twofold one, the indices i,j indicating those spaces on which the action is nontrivial:

$$r_{12}(\lambda,\mu) \equiv r(\lambda,\mu) \otimes 1_3$$

for example. The classical Yang-Baxter equation is a sufficient condition for the algebra (2.38) to satisfy the Jacobi identity.

Inserting the expansion

$$T(\lambda) = \sum_0^{\infty} \lambda^n T^{(n)} \tag{2.40}$$

into eq. (2.38), we obtain

$$\{T^{(n)} , T^{(m)}\} = \sum_{\ell=0}^{m-1} [\Gamma , T^{(\ell+n)} \otimes T^{(m-\ell-1)}] \tag{2.41}$$

where $\Gamma = \begin{cases} -4g\pi & \text{for models (2.12)} \\ -2g(P+\Sigma) & \text{for model (2.19), commuting,} \\ 8g(P-\Delta) & \text{for model (2.19), anticommuting.} \end{cases}$

Eq. (2.41) gives, in particular,

$$\begin{aligned} \{T^{(n)} \overset{\otimes}{,} T^{(1)}\} &= [\Gamma, T^{(n)} \otimes 1] \\ \{T^{(2)} \overset{\otimes}{,} T^{(2)}\} &= [\Gamma, T^{(3)} \otimes 1 + T^{(2)} \otimes T^{(1)}] \\ \{T^{(3)} \overset{\otimes}{,} T^{(2)}\} &= [\Gamma, T^{(4)} \otimes 1 + T^{(3)} \otimes T^{(1)}] \end{aligned} \tag{2.42}$$

From eq. (2.41) one can obtain the algebra of the charges $Q^{(n)}$ (2.9) upon expanding the logarithm. For the SU(N) chiral Gross-Neveu model one gets, for example,

$$\{Q^{(1)} \overset{\otimes}{,} Q^{(1)}\} = [P, (Q^{(2)} - \tfrac{1}{12}(Q^{(0)})^3) \otimes 1 + \tfrac{1}{4}(Q^{(0)})^2 \otimes Q^{(0)}] .$$

Writing

$$T^{(1)} = Q^{(0)} = \sum_{\alpha=1}^{\dim G} Q_\alpha^{(0)} I^\alpha ,$$

the first of eqs. (2.42) yields

$$\{Q_\alpha^{(0)}, Q_\beta^{(0)}\} = k f_{\alpha\beta}^{\gamma} Q_\gamma^{(0)} ,$$

$k = \begin{cases} -4g & \text{for models (2.2)} \\ -4g & \text{for model (2.19), commuting} \\ 8g & \text{for model (2.19), anticommuting,} \end{cases}$

so the $Q_\alpha^{(o)}$ are the generators of the internal symmetry group G.

All the considerations in sect. II.3 rely on the current algebra being ultralocal. For the non-ultralocal case of the chiral σ models, the δ' function in the current algebra produces singularities in the Poisson brackets of non-local objects: for example, the two integrations in the r.h.s. of eq. (2.33) do not commute. As a consequence, one does not obtain a classical r matrix solving the classical Yang-Baxter equation, hence, the Jacobi identity for the charge algebra fails. A detailed analysis of this problem can be found in ref. 9.

II.4. Kac-Moody current algebras

The formulas (2.17), (2.28) for the (equal time) current algebra

$$\{A_\mu(x) \overset{\otimes}{,} A_\nu(y)\} = c\,\delta(x-y)\left[1 \otimes A_{|\mu-\nu|}(x), \Pi\right] \qquad (2.43)$$

(c = fct. (coupling constant g)) can be rewritten in terms of

$$A_\pm(x) = A_0(x) \pm A_1(x)$$

as follows:

$$\{A_\pm(x) \overset{\otimes}{,} A_\pm(y)\} = 2c\,\delta(x-y)\left[1 \otimes A_\pm(x), \Pi\right]$$

$$\{A_+(x) \overset{\otimes}{,} A_-(y)\} = 0$$

For the components of $A_\pm(x)$ with respect to a basis of generators I_α,

$$A_\pm(x) = A_\pm^\alpha(x)\, I_\alpha$$

we find

$$\begin{aligned} \{A_\pm^\alpha(x), A_\pm^\beta(y)\} &= 2c\, f^{\alpha\beta}_{\ \ \gamma}\, \delta(x-y)\, A_\pm^\gamma(x) \\ \{A_+^\alpha(x), A_-^\beta(y)\} &= 0\,. \end{aligned} \qquad (2.44)$$

Taking x-space to be a circle, $x \in [0,2\pi]$, and inserting the Fourier decomposition

$$A_\pm^\alpha(x) = \frac{1}{2\pi} \sum_{n=-\infty}^{+\infty} A_\pm^\alpha(n)\, e^{inx}$$

we have

$$\{A_\pm^\alpha(n), A_\pm^\beta(m)\} = 2c\, f^{\alpha\beta}_{\ \ \gamma}\, A_\pm^\gamma(n+m)\,. \qquad (2.45)$$

Eqs. (2.44) or (2.45) show that the current algebra (2.43) is built up of two mutually commuting Kac-Moody algebras [19] with vanishing central charge. The result of sect. II.3 tells us that, going over to the monodromy matrix $T(\lambda)$ as functional of the A_μ's, we obtain a representation of the Yang-Baxter algebra induced by the representations (2.44) of the Kac-Moody algebra.

It is well known that a quantization of the current algebra in general leads to extra terms in the commutation relations (2.44), the so called Schwinger terms, which can be computed perturbatively. In two dimensions they turn out to be finite. Taking the currents to be made up of free fermion fields which is equivalent to the large N limit one gets [18] [6]

$$[A_\mu(x) \overset{\otimes}{,} A_\nu(y)] = -ic\,\delta(x-y)\left[1 \otimes A_{|\mu-\nu|}(x), \Pi\right] + \frac{ic^2}{2\pi}\, \delta_{|\mu-\nu|,1}\, \delta'(x-y)\, \Pi \qquad (2.46)$$

or, for the redefined current

$$B_\pm^\alpha(x) = -\frac{1}{c} A_\pm^\alpha(x) \quad :$$

$$[B^{\alpha}_{\pm}(x), B^{\beta}_{\pm}(y)] = 2i f^{\alpha\beta}_{\gamma} \delta(x-y) B^{\gamma}_{\pm}(x) \pm \frac{i}{\pi} \delta'(x-y) K^{\alpha\beta} \tag{2.47}$$

and for the Fourier components of $B^{\alpha}_{\pm}(x)$:

$$[B^{\alpha}_{\pm}(n), B^{\beta}_{\pm}(m)] = 2i f^{\alpha\beta}_{\gamma} B^{\gamma}_{\pm}(n+m) \mp 2n \delta_{n,-m} K^{\alpha\beta} \tag{2.48}$$

Formulas (2.47) or (2.48) show that, in the approximation indicated above, the quantum current algebra consists of two mutually commuting Kac-Moody algebras with central charge $\pm$ 1. At present, it is not known whether one can construct, in analogy to the classical case, a functional of the quantum currents providing a representation of the quantum Yang-Baxter algebra. The method for the computation of the quantum charge algebra explained in the next chapter avoids a direct use of currents.

There is another representation of the Kac-Moody algebra with zero central charge going along with the classical models considered in section II. Here the generators are realized as infinitesimal symmetry transformations of the field equations[20]. This representation, which sometimes was wrongly apostrophized as algebra of non-local charges, is in fact an infinitesimal version of the Riemann-Hilbert transformation method to compute solutions of the field equations. Its drawback is that the transformations are not canonical (not extendable to phase space)[17] and cannot be taken over to the quantum models.

Our conclusion is that the Kac-Moody algebra has its place at the level of current algebras, whereas the non-local charges generate a Yang-Baxter algebra, in the classical as well as in the quantum case to be discussed in the next section.

III. Quantum Yang-Baxter charge algebras

III.1 General method

The existence of a conserved monodromy operator $T(\lambda)$ in the quantum field theory involves a non-trivial renormalization problem since $T(\lambda)$ is built of products of arbitrary numbers of fields at the same point.More precisely, the n-th charge $Q^{(n)}$ contains up to n+1 fields at the same point. It is possible to construct a renormalized and conserved $Q^{(1)}$ for the O(N)/O(N-1) non-linear σ model [21] and for the Gross-Neveu models [10] through short distance operator product expansion. The conservation of $Q^{(1)}$ implies the absence of particle production and ensures the factorizability of the S matrix [21]. In general the renormalization procedure can lead to anomalies [7].

In principle, it is possible to extend this procedure to the higher charges $Q^{(n)}$ ($n \geq 2$). Practically, this seems impossible, so we shall instead follow a simpler and more direct way. Assuming the existence of $T(\lambda)$ as a time independent operator in the quantum theory, we develop a method that provides an explicit construction for it, giving closed form expressions for all matrix elements of $T(\lambda)$ in asymptotic states.

The quantum $T(\lambda)$ will be constructed from the following general properties [9] [10]:

a) $T(\lambda)$ exists as a quantum operator and it commutes with energy-momentum:

$$[T_{ab}(\lambda), P^{\mu}] = 0 \tag{3.1}$$

b) $T(\lambda)$ fulfills a quantum factorization principle (eqs.(3.4),(3.5).

c) $\mathcal{P}$, $\mathcal{T}$ and internal symmetries hold in the quantum theory.

To understand the meaning of (b), let us consider a classical field configuration formed by two separated lumps:

$$A_{\mu}(X) = \begin{cases} A_{\mu}^{(1)}(X) , & x \leq a \\ 0 , & a \leq x \leq b \\ A_{\mu}^{(2)}(X) , & b \leq x \end{cases} . \tag{3.2}$$

From eqs. (2.4),(2.5) follows the factorization formula

$$T_{ab}(\lambda, A_{\mu}) = \sum_{c} T_{ac}(\lambda, A_{\mu}^{(1)}) \, T_{cb}(\lambda, A_{\mu}^{(2)}) \tag{3.3}$$

for the classical monodromy matrix. Eq. (3.3) admits a very useful quantum generalization [22]: Let us denote by $|\theta,\alpha\rangle$ the one-particle asymptotic states of the theory, where θ stands for the rapidity and α for the set of internal quantum numbers. We assume here all particles to be massive. The quantum version of eq. (3.3) relates the action of $T_{ab}(\lambda)$ on k-particle asymptotic states to a product of $T_{ab}(\lambda)$ applied to one-particle states:

$$T_{ab}(\lambda)|\theta_1\alpha_1\ldots\theta_k\alpha_k\rangle_{out} = \sum_{a_1\ldots a_{k-1}} T_{aa_1}(\lambda)|\theta_1\alpha_1\rangle T_{a_1a_2}(\lambda)|\theta_2\alpha_2\rangle\ldots T_{a_{k-1}b}(\lambda)|\theta_k\alpha_k\rangle \tag{3.4}$$

and

$$T_{ab}(\lambda)|\theta_1\alpha_1\ldots\theta_k\alpha_k\rangle_{in} = \sum_{a_1\ldots a_{k-1}} T_{a_1b}(\lambda)|\theta_1\alpha_1\rangle T_{a_2a_1}(\lambda)|\theta_2\alpha_2\rangle\ldots T_{aa_{k-1}}(\lambda)|\theta_k\alpha_k\rangle \tag{3.5}$$

where $\theta_i > \theta_j$ for $i > j$. Eqs. (3.1),(3.4) or (3.5) and the orthogonality of k-particle states imply that matrix elements of $T_{ab}(\lambda)$ between states of different particle numbers vanish:

$$\langle\theta_1\alpha_1\ldots\theta_k\alpha_k|[T_{ab}(\lambda),P^\mu]|\theta'\alpha'\rangle = 0 \Rightarrow \begin{cases}\sum_1^k \cosh\theta_j = \cosh\theta' \\ \sum_1^k \sinh\theta_j = \sinh\theta'\end{cases}$$

Taking the difference of the squares, we have

$$k^2 \leq k + \sum_{j\neq\ell}\cosh(\theta_j-\theta_\ell) = 1 ,$$

thus either k = 1, or the matrix element vanishes. The one-particle matrix element reads

$$\langle\theta\alpha|T_{ab}(\lambda)|\theta'\beta\rangle = \delta(\theta-\theta')\, T_{a\alpha,b\beta}(\lambda,\theta) . \tag{3.6}$$

The asymptotic states of the theory being connected as usual by the S matrix through

$$|in\rangle = S|out\rangle$$

we have the identity

$$\langle in|S\,T_{ab}(\lambda)|in\rangle = \langle out|T_{ab}(\lambda)\,S|out\rangle . \tag{3.7}$$

With the help of eqs. (3.4),(3.5) this gives for two-particle states

$$\begin{aligned}&\sum_{c,\delta_1,\delta_2} S_{\gamma_1\gamma_2,\delta_1\delta_2}(\theta_2-\theta_1)\,T_{a\delta_2,c\gamma_2'}(\lambda,\theta_2)\,T_{c\delta_1,b\gamma_1'}(\lambda,\theta_1) = \\ &= \sum_{c,\delta_1,\delta_2} T_{a\gamma_1,c\delta_1}(\lambda,\theta_1)\,T_{c\gamma_2,b\delta_2}(\lambda,\theta_2)\,S_{\delta_1\delta_2,\gamma_1'\gamma_2'}(\theta_2-\theta_1)\end{aligned} \tag{3.8}$$

where $S_{\alpha_1\alpha_2,\alpha'_1\alpha'_2}(\theta_2-\theta_1)$ stands for the two-particle S matrix

$$_{out}\langle\theta_1\alpha_1,\theta_2\alpha_2|\theta_1'\alpha_1',\theta_2'\alpha_2'\rangle_{in} = \delta(\theta_1-\theta_1')\delta(\theta_2-\theta_2')\,S_{\alpha_1\alpha_2,\alpha_1'\alpha_2'}(\theta_2-\theta_1) \pm \begin{pmatrix}\theta_1'\leftrightarrow\theta_2' \\ \alpha_1'\leftrightarrow\alpha_2'\end{pmatrix}$$

(here + holds for bosons and - for fermions). Eq. (3.8) can be recasted in the form of a matrix product on one-particle Fock indices:

$$R(\theta_2-\theta_1) \sum_c T_{ac}(\lambda,\theta_2) \bullet T_{cb}(\lambda,\theta_1) = \sum_c T_{ac}(\lambda,\theta_1) \bullet T_{cb}(\lambda,\theta_2) R(\theta_2-\theta_1) \tag{3.9}$$

where

$$R_{\alpha\beta,\gamma\delta}(\theta) = (S(\theta)P)_{\alpha\beta,\gamma\delta} = S_{\alpha\beta,\delta\gamma}(\theta) \tag{3.10}$$

with P the permutation operator (eq.(2.24)), and

$$(T_{ac}(\lambda,\theta))_{\gamma\delta} = T_{a\gamma,b\delta}(\lambda,\theta) \quad .$$

Eq. (3.9) shows that $T_{a\alpha,b\beta}(\lambda,\theta)$ is a representation of the (quantum) Yang-Baxter algebra [3] associated to $R(\theta)$, acting on an N-dimensional space (a,b = 1...N) with θ as spectral parameter.

Following (c), $T_{ab}(\lambda)$ is further restricted by the invariance of the quantum theory under $\mathcal{P}$ and $\mathcal{T}$. The quantum analogue of eqs. (2.10) is the existence of a unitary operator $\mathcal{P}$ and of an antiunitary operator $\tau = \mathcal{P}\mathcal{T}$ such that

$$\tau\, T(\lambda)\, \tau^{-1} = T(\lambda)^{-1}$$

or

$$T_{ac}(\lambda)\, \tau\, T_{cb}(\lambda)\, \tau^{-1} = \delta_{ab} \mathbb{1} \tag{3.11}$$

and

$$\mathcal{P}\, T(\lambda)\, \mathcal{P}^{-1} = T(-\lambda)^{-1}$$

or

$$T_{ac}(-\lambda)\, \mathcal{P}\, T_{cb}(\lambda)\, \mathcal{P}^{-1} = \delta_{ab} \mathbb{1} \quad . \tag{3.12}$$

In eqs. (3.11),(3.12) T^{-1} stands for the inverse operator in Fock space and for the inverse matrix in the N-dimensional auxiliary space. This gives on one-particle states

$$T_{a\alpha,c\gamma}(\lambda,\theta)\, T^{*}_{c\gamma,b\beta}(\lambda,\theta) = \delta_{ab}\delta_{\alpha\beta} \tag{3.13}$$

and

$$T_{a\alpha,c\gamma}(-\lambda,-\theta)\, T_{c\gamma,b\beta}(\lambda,\theta) = \delta_{ab}\delta_{\alpha\beta} \quad . \tag{3.14}$$

Since $|\theta,\alpha\rangle$ (α = 1...N) are possible particle states in the theory (they usually correspond to the fundamental fields) we can consider the S matrix $S_{a\alpha,b\beta}(\theta)$ which satisfies the factorization equations [8]

$$S_{i_1 i_2, k_1 k_2}(\theta)\, S_{i_3 k_1, k_3 j_1}(\theta+\theta')\, S_{k_2 k_3, j_2 j_3}(\theta') = \tag{3.15}$$
$$= S_{i_2 i_3, k_2 k_3}(\theta')\, S_{i_1 k_3, k_1 j_3}(\theta+\theta')\, S_{k_1 k_2, j_1 j_2}(\theta) ,$$

the unitarity relation

$$S_{a\alpha, c\gamma}(\theta)\, S_{c\gamma, b\beta}(\theta^*) = \delta_{ab}\, \delta_{\alpha\beta} \tag{3.16}$$

and the real analyticity condition

$$S^*_{a\alpha, b\beta}(\theta^*) = S_{a\alpha, b\beta}(-\theta) . \tag{3.17}$$

This provides an explicit solution of eqs. (3.8),(3.13),(3.14):

$$(T_{ab}(\lambda,\theta))_{\alpha\beta} = S_{a\alpha, b\beta}(\theta+\gamma(\lambda))\, e^{i\phi(\lambda,\theta)} \tag{3.18}$$

where the quantum spectral parameter γ is a function of λ only, and $\phi(\lambda,\theta)$ is a real phase. Eqs. (3.13),(3.14) and (3.16),(3.17) imply

$$\gamma(\lambda)^* = \gamma(\lambda)$$
$$\gamma(-\lambda) = -\gamma(\lambda) \tag{3.19}$$

and

$$\phi(-\lambda,-\theta) = -\phi(\lambda,\theta) .$$

When a continuous symmetry is present like in the non-linear σ model, SU(N) chiral Gross-Neveu and O(2N) Gross-Neveu models, an expansion of $T(\lambda)$ and $S(\theta)$ into irreducible channels (see refs. 9,10) shows that (3.18),(3.19) provide the general solution of the problem.

All matrix elements of $T_{ab}(\lambda)$ can now be computed using the one-particle matrix elements (3.18) and the factorization principle (3.4), (3.5):

$$_{out}\langle\theta_1'\alpha_1' \ldots \theta_k'\alpha_k' | T_{ab}(\lambda) | \theta_1\alpha_1 \ldots \theta_\ell\alpha_\ell\rangle_{out} =$$
$$= \delta_{k\ell} \prod_{i=1}^{k} \delta(\theta_i-\theta_i') \sum_{a_1 \ldots a_{k-1}} S_{a\alpha_1', a_1\alpha_1}(\theta_1+\gamma(\lambda)) \ldots S_{a_{k-1}\alpha_k', b\alpha_k}(\theta_k+\gamma(\lambda))\, e^{i\{\sum_{j=1}^{k}\phi(\lambda,\theta_j)\}} \tag{3.20}$$
$$\equiv T_{a\{\alpha'\}, b\{\alpha\}}(\lambda,\{\theta\})\, \delta_{k\ell} \prod_{i=1}^{k} \delta(\theta_i-\theta_i')$$

where $\theta_i > \theta_j$ for $i > j$. This formula displays the formal coincidence of $T_{a\{\alpha'\},b\{\alpha\}}(\lambda,\{\theta\})$ with the monodromy matrix in the statistical mechanics of an inhomogeneous vertex model on a line of k sites, the statistical weights being $S_{a_{j-i}\alpha'_j, a_j\alpha_j}(\theta_j+\gamma(\lambda))$:

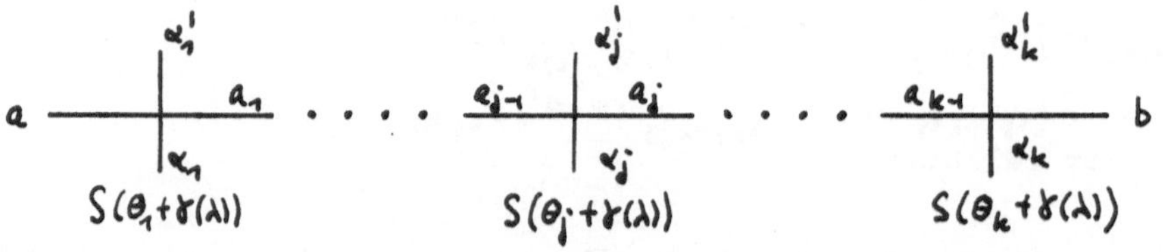

The algebra of quantum monodromy operators follows from (3.20) and the Yang-Baxter equation (3.15):

$$S_{ab,cd}(\gamma(\lambda)-\gamma(\mu))\left[T(\lambda)\otimes T(\mu)\right]_{cd,ef} = \left[T(\mu)\otimes T(\lambda)\right]_{ba,dc} S_{cd,ef}(\gamma(\lambda)-\gamma(\mu)) \qquad (3.21)$$

Here, the product over Fock (greek) indices is understood, and the tensor product is over matrix (latin) indices. Since eq. (3.21) takes the same form in all k-particle subspaces, it holds as an operator equation in the whole Fock space. It can be written in the more compact form

$$R(\lambda,\mu)\left[T(\lambda)\otimes T(\mu)\right] = \left[T(\mu)\otimes T(\lambda)\right]R(\lambda,\mu) \qquad (3.22)$$

where

$$R(\lambda,\mu) = P\, S(\gamma(\lambda)-\gamma(\mu)) \ . \qquad (3.23)$$

So we find that the quantum R matrix is (up to a permutation of indices) equal to the two-body S matrix for states $|a\theta,b\theta'\rangle$ at $\theta = \gamma(\lambda), \theta' = \gamma(\mu)$; the phase $\phi(\lambda,\theta)$ drops out of the algebra.

Quadratic algebras (3.22) of Yang-Baxter type are characteristic of integrable field theories and of integrable statistical models [3]. However, all the $T_{ab}(\lambda)$ are conserved here and not just the trace

$$t(\lambda) = \sum T_{aa}(\lambda) \quad ,$$

the latter generating an abelian subalgebra of the charge algebra:

$$[t(\lambda), t(\mu)] = 0 \ .$$

The classical monodromy matrix is invariant under Lorentz transformations

$$(x \pm t) \rightarrow (x' \pm t') = e^{\pm\varepsilon}(x \pm t) \ .$$

However, since the rapidity transforms as

$$\theta \rightarrow \theta' = \theta + \varepsilon \ ,$$

the quantum spectral parameter $\gamma(\lambda)$ carries a representation of the Lorentz group

$$\lambda \rightarrow \lambda' = \lambda(\varepsilon) \ , \quad \gamma(\lambda) \rightarrow \gamma' = \gamma(\lambda) + \varepsilon$$

leading to a nontrivial transformation behaviour of the charges.

III.2. Examples

III.2.1. The O(N)/O(N-1) non-linear σ model

The spectrum of this model contains an N-plet of massive particles transforming under the fundamental representation of O(N). Their scattering is governed by a factorized S matrix; the two-body S matrix reads [8)]

$$S_{c_1c_2,c_1'c_2'}(\theta) = \delta_{c_1c_2}\delta_{c_1'c_2'}\sigma_1(\theta) + \delta_{c_1c_1'}\delta_{c_2c_2'}\sigma_2(\theta) + \delta_{c_1c_2'}\delta_{c_2c_1'}\sigma_3(\theta) \tag{3.24}$$

where

$$\sigma_1(\theta) = -\frac{2\pi i\Delta}{i\pi-\theta}\sigma_2(\theta) \quad , \quad \sigma_3(\theta) = -\frac{2\pi i\Delta}{\theta}\sigma_2(\theta)$$

$$\sigma_2(\theta) = \frac{\Gamma(\Delta+\frac{\theta}{2\pi i})\Gamma(\frac{1}{2}+\frac{\theta}{2\pi i})\Gamma(\Delta+\frac{1}{2}-\frac{\theta}{2\pi i})\Gamma(1-\frac{\theta}{2\pi i})}{\Gamma(\frac{\theta}{2\pi i})\Gamma(\Delta+\frac{1}{2}+\frac{\theta}{2\pi i})\Gamma(\frac{1}{2}-\frac{\theta}{2\pi i})\Gamma(\Delta+1-\frac{\theta}{2\pi i})} \quad , \Delta = (N-2)^{-1}. \tag{3.25}$$

The first two renormalized charges read explicitly [21)]

$$Q^{(0)}_{ab}|\theta_1c_1\ldots\theta_kc_k\rangle_{\substack{in\\out}} = \sum_{\{d\}}|\theta_1d_1\ldots\theta_kd_k\rangle_{\substack{in\\out}}\sum_{j=1}^{k} 2i\,(I^{(j)}_{ab})_{\{d\}\{c\}} \; ,$$

$$Q^{(1)}_{ab}|\theta_1c_1\ldots\theta_kc_k\rangle_{\substack{in\\out}} = \sum_{\{d\}}|\theta_1d_1\ldots\theta_kd_k\rangle_{\substack{in\\out}} \cdot$$

$$\cdot 2\left\{\mp\sum_{i<j}(I^{(i)}_{ac}I^{(j)}_{bc} - I^{(i)}_{bc}I^{(j)}_{ac}) + \frac{N-2}{i\pi}\sum_{j=1}^{N}\theta_j I^{(j)}_{ab}\right\}_{\{d\}\{c\}} \tag{3.26}$$

with

$$(I^{(j)}_{ab})_{\{d\}\{c\}} = (\delta_{ad_j}\delta_{bc_j} - \delta_{bd_j}\delta_{ac_j})\prod_{i\neq j}\delta_{d_ic_i} \tag{3.27}$$

From these formulas one can check that our basic assumptions (a),(b) and (c) are fulfilled to orders λ^1 and λ^2. Applying the general formalism in section III.1, we identify [9)]

$$T_{ad,bc}(\lambda,\theta) = S_{ad,bc}(\theta+\gamma(\lambda))\,e^{i\phi(\lambda,\theta)} \quad . \tag{3.28}$$

To get information on the functions $\gamma(\lambda)$ and $\phi(\lambda,\theta)$, we expand both sides of eq. (3.28): On the l.h.s. we get from eqs. (3.26),(3.27):

$$T_{ad,bc}(\lambda,\theta) = (1-2\lambda^2)\delta_{ab}\delta_{cd} - 2i\lambda(1-\frac{\lambda}{\pi}(N-2)\theta)\delta_{ac}\delta_{bd}$$
$$+2i\lambda(1+\frac{\lambda}{\pi}(N-2)(i\pi-\theta))\delta_{ad}\delta_{bc} + \mathcal{O}(\lambda^3) \; . \tag{3.29}$$

Expanding the r.h.s. of eq. (3.28) around $\gamma^{-1} = 0$, Stirling's formula yields

$$S_{ad,bc}(\theta+\gamma) = \left(1-2\left(\frac{\pi}{N-2}\right)^2\gamma^{-2}\right)\delta_{ab}\delta_{cd} + \frac{2\pi i}{N-2}\left(-\gamma^{-1}+\theta\gamma^{-2}\right)\delta_{ac}\delta_{bd} + \frac{2\pi i}{N-2}\left(\gamma^{-1}+(i\pi-\theta)\gamma^{-2}\right)\delta_{ad}\delta_{bc} + \mathcal{O}(\gamma^{-3}) \; . \tag{3.30}$$

Both expressions match if

$$\gamma(\lambda) = \frac{\pi}{N-2}\lambda^{-1} + \mathcal{O}(\lambda) \quad , \quad \phi(\lambda,\theta) = \mathcal{O}(\lambda^3) \tag{3.31}$$

as $\lambda \to 0$. Inserting eqs. (3.24),(3.25) into the Yang-Baxter algebra (3.21), we finally get the commutator algebra for the monodromy operators $T_{ab}(\lambda)$:

$$[T_{ac}(\lambda), T_{bd}(\mu)] = \frac{2\pi i}{N-2}\cdot\frac{1}{\gamma(\lambda)-\gamma(\mu)}\left(T_{bc}(\lambda)T_{ad}(\mu) - T_{bc}(\mu)T_{ad}(\lambda)\right) + \frac{2\pi i}{N-2}\frac{1}{\gamma(\lambda)-\gamma(\mu)-i\pi}\left(\delta_{cd}T_{be}(\mu)T_{ae}(\lambda) - \delta_{ab}T_{ec}(\lambda)T_{ed}(\mu)\right) . \tag{3.32}$$

The Yang-Baxter equation (3.15) guarantees that the algebra (3.32) satisfies the Jacobi identity, unlike the classical case.

III.2.2 The O(2N) Gross-Neveu model

This model (2.25) has a rich spectrum containing kinks, elementary fermions and bound states. The physical S matrix of the elementary fermions (taking generalized statistics into account) coincides with the O(2N)σ model S matrix up to a CDD pole [5]:

$$S^{GN}_{ab,cd}(\theta) = \frac{\sinh\theta + i\sin\frac{\pi}{N-1}}{\sinh\theta - i\sin\frac{\pi}{N-1}}\, S^{NL\sigma}_{ab,cd}(\theta) \quad .$$

The charges $Q^{(0)}$ and $Q^{(1)}$ of both models are equal up to a (g-dependent) normalization factor, hence, our assumptions (a) - (c) are verified up to $\mathcal{O}(\lambda^3)$. We get for the matrix elements of $T_{ab}(\lambda)$ between states of elementary fermions

$$\langle\theta d|T_{ab}(\lambda)|\theta' c\rangle = \delta(\theta'-\theta)\, S^{GN}_{ad,bc}(\theta+\gamma(\lambda))\, e^{i\phi(\lambda,\theta)} \tag{3.33}$$

with

$$\gamma(\lambda) = \frac{\pi}{8g(N-1)}\lambda^{-1} + \mathcal{O}(\lambda^3) \quad , \quad \phi(\lambda,\theta) = \mathcal{O}(\lambda^3) \; . \tag{3.34}$$

The commutator algebra looks as in eq. (3.32), the factors $\frac{2\pi i}{N-2}$ being replaced by $\frac{i\pi}{N-1}$. Comparing with the classical charge algebra (2.38), (2.37)

$$\{T_{ac}(\lambda), T_{bd}(\mu)\} = \frac{8g}{\lambda^{-1}-\mu^{-1}} \big(T_{bc}(\lambda) T_{ad}(\mu) - T_{bc}(\mu) T_{ad}(\lambda) + \delta_{cd} T_{be}(\mu) T_{ae}(\lambda) - \delta_{ab} T_{ec}(\lambda) T_{ed}(\mu) \big) \tag{3.35}$$

we see that both algebras cannot be isomorphic, due to the $i\pi$ in the second term of (3.32). However, in the classical limit ($g \to 0$) we recover the classical result (3.35) through the correspondance $\{\ ,\ \} \longleftrightarrow -i[\ ,\ ]$.

III.2.3. The SU(N) chiral Gross-Neveu model

This model is given by the Hamiltonian (2.13) with G = SU(N) in the fundamental representation. Its spectrum contains an N-plet of massive particles transforming under the fundamental representation $\{N\}$ and an N-plet of their antiparticles transforming under $\{\bar{N}\}$. Multiparticle states are denoted as

$$|\theta_1 c_1^{\alpha_1} \dots \theta_k c_k^{\alpha_k}\rangle \tag{3.36}$$

with $\alpha_i = +(-)$ standing for a particle (antiparticle) SU(N) label. The S matrix factorizes into the two-body S matrices [23)]

$$_{out}\langle \theta_1 c_1^+, \theta_2 c_2^+ | \theta_1' d_1^+, \theta_2' d_2^+\rangle_{in} = \delta(\theta_1-\theta_1')\delta(\theta_2-\theta_2') S_{c_1c_2,d_1d_2}(\theta_2-\theta_1) - \begin{pmatrix}\theta_1 \leftrightarrow \theta_2 \\ c_1 \leftrightarrow c_2\end{pmatrix}$$

$$_{out}\langle \theta_1 c_1^+, \theta_2 c_2^- | \theta_1' d_1^+, \theta_2' d_2^-\rangle_{in} = \delta(\theta_1-\theta_1')\delta(\theta_2-\theta_2') F_{c_1c_2,d_1d_2}(\theta_2-\theta_1)$$

where

$$\begin{aligned} S_{c_1c_2,d_1d_2}(\theta) &= u_1(\theta)\,\delta_{c_1d_1}\delta_{c_2d_2} + u_2(\theta)\,\delta_{c_1d_2}\delta_{c_2d_1} \\ F_{c_1c_2,d_1d_2}(\theta) &= t_1(\theta)\,\delta_{c_1d_1}\delta_{c_2d_2} + t_2(\theta)\,\delta_{c_1c_2}\delta_{d_1d_2} \\ u_1(\theta) &= \frac{\Gamma(1-\frac{\theta}{2\pi i})\Gamma(\frac{\theta}{2\pi i}-\frac{1}{N})}{\Gamma(1-\frac{\theta}{2\pi i}-\frac{1}{N})\Gamma(\frac{\theta}{2\pi i})} = t_1(i\pi-\theta), \quad u_2(\theta) = -\frac{2\pi i}{N\theta}u_1(\theta) \\ & \qquad\qquad t_2(\theta) = -\frac{2\pi i}{N(i\pi-\theta)}t_1(\theta). \end{aligned} \tag{3.37}$$

The action of $Q^{(0)}$ and $Q^{(1)}$ on the states (3.36) is [10)]

$$Q_{ab}^{(0)}|\theta_1 c_1^{\alpha_1}\dots\theta_k c_k^{\alpha_k}\rangle_{\substack{in\\out}} = \sum_{\{d\}} |\theta_1 d_1^{\alpha_1}\dots\theta_k d_k^{\alpha_k}\rangle_{\substack{in\\out}}\, 4g \sum_{i=1}^{k} (J_{ab}^{(i)})_{\{d^\alpha\}\{c^\alpha\}},$$

$$\begin{aligned} Q_{ab}^{(1)}|\theta_1 c_1^{\alpha_1}\dots\theta_k c_k^{\alpha_k}\rangle_{\substack{in\\out}} &= \sum_{\{d\}} |\theta_1 d_1^{\alpha_1}\dots\theta_k d_k^{\alpha_k}\rangle_{\substack{in\\out}} \cdot \\ &\cdot 8g^2\Big\{\mp\sum_{i>j}\big(J_{ae}^{(i)}J_{eb}^{(j)} - J_{ae}^{(j)}J_{eb}^{(i)}\big) + \frac{N}{\pi}\sum_{i=1}^{k}\theta_i J_{ab}^{(i)}\Big\}_{\{d^\alpha\}\{c^\alpha\}}. \end{aligned} \tag{3.38}$$

where

$$\left(J^{(i)}_{ab}\right)_{\{d^\alpha\}\{c^\alpha\}} = \begin{cases} (M_{ab})_{d_i c_i} \prod_{\ell \neq i} \delta_{d_\ell c_\ell} & , \quad \alpha_i = + \\ (M^+_{ab})_{d_i c_i} \prod_{\ell \neq i} \delta_{d_\ell c_\ell} & , \quad \alpha_i = - \end{cases} \tag{3.39}$$

and

$$(M_{ab})_{cd} = i\left(\delta_{ac}\delta_{bd} - \frac{1}{N}\delta_{ab}\delta_{cd}\right) .$$

Again, to orders λ^1 and λ^2 our assumptions are fulfilled. The method of sect. III.1 gives, taking the SU(N) transformation behaviour of the various labels into account [10],

$$\begin{aligned} \langle \theta d^+ | T_{ab}(\lambda) | \theta' c^+ \rangle &= \delta(\theta-\theta')\, F_{ad,bc}(\theta+\gamma(\lambda))\, e^{i\phi(\lambda,\theta)} \\ \langle \theta d^- | T_{ab}(\lambda) | \theta' c^- \rangle &= \delta(\theta-\theta')\, S_{ad,bc}(\theta+\gamma(\lambda))\, e^{i\phi(\lambda,\theta)} \end{aligned} \tag{3.40}$$

From eqs. (3.38),(3.39) and (3.37), we find the expansions

$$\begin{aligned} \langle \theta d^+ | T_{ab}(\lambda) | \theta' c^+ \rangle = \delta(\theta-\theta') \Big\{ &\Big(1 - \frac{4ig\lambda}{N} - 8(g\lambda)^2\Big(\frac{1}{N^2} - \frac{i\theta}{\pi}\Big)\Big)\delta_{ab}\delta_{cd} \\ &+ \Big(4ig\lambda + 8(g\lambda)^2\Big(\frac{2}{N} - \frac{iN\theta}{\pi} - N\Big)\Big)\delta_{ad}\delta_{bc} + O(\lambda^3)\Big\} \end{aligned} \tag{3.41}$$

and

$$\begin{aligned} F_{ad,bc}(\theta+\gamma) = e^{-\frac{i\pi}{N}\varepsilon(\theta+\gamma)} \Big\{ &\Big(1 - \frac{2\pi i}{N^2\gamma} + \frac{2\pi}{N^2\gamma^2}\Big(i\theta - \frac{\pi}{N^2}\Big)\Big)\delta_{ab}\delta_{cd} \\ &+ \Big(\frac{1}{\gamma} + \frac{2\pi i}{N\gamma^2}\Big(i\pi - \theta - \frac{2\pi i}{N^2}\Big)\Big)\delta_{ad}\delta_{bc} + O(\gamma^{-3})\Big\} \end{aligned} \tag{3.42}$$

Eqs. (3.41) and (3.42) coincide provided

$$\gamma(\lambda) = \frac{\pi}{2gN\lambda} + O(\lambda) \tag{3.43}$$

$$\phi(\lambda,\theta) = \frac{\pi}{N}\varepsilon(\lambda) + O(\lambda^3) . \tag{3.44}$$

It has been argued in ref. 5 that the physical particles in this model are more naturally described by field operators obeying generalized statistics. The corresponding physical states $|\theta \hat{c}^\alpha\rangle$ are connected by physical scattering amplitudes

$$\begin{aligned} \hat{S}_{c_1c_2,d_1d_2}(\theta) &= e^{\frac{i\pi}{N}\varepsilon(\theta)}\, S_{c_1c_2,d_1d_2}(\theta) \\ \hat{F}_{c_1c_2,d_1d_2}(\theta) &= e^{\frac{i\pi}{N}\varepsilon(\theta)}\, F_{c_1c_2,d_1d_2}(\theta) \end{aligned} \tag{3.45}$$

This shows that the one-particle matrix elements of $T_{ab}(\lambda)$ are given by the physical two-body S matrix without any phase (up to $O(\lambda^3)$):

$$\begin{aligned} \langle \theta \hat{d}^+ | T_{ab}(\lambda) | \theta' \hat{c}^+ \rangle &= \hat{F}_{ad,bc}(\theta+\gamma(\lambda))\, \delta(\theta-\theta') \\ \langle \theta \hat{d}^- | T_{ab}(\lambda) | \theta' \hat{c}^- \rangle &= \hat{S}_{ad,bc}(\theta+\gamma(\lambda))\, \delta(\theta-\theta') \end{aligned} \tag{3.46}$$

The commutator algebra of the quantum monodromy operators follows from eqs. (3.21),(3.37):

$$[T_{ac}(\lambda), T_{bd}(\mu)] = \frac{2\pi i}{N} \frac{1}{\gamma(\lambda)-\gamma(\mu)} (T_{bc}(\lambda) T_{ad}(\mu) - T_{bc}(\mu) T_{ad}(\lambda)) .$$

Comparing with the classical algebra (2.38),(2.35)

$$\{T_{ac}(\lambda), T_{bd}(\mu)\} = \frac{4g}{\lambda^{-1}-\mu^{-1}} (T_{bc}(\lambda) T_{ad}(\mu) - T_{bc}(\mu) T_{ad}(\lambda)) ,$$

we see that if the expression (3.43) for $\gamma(\lambda)$ is exact, then the classical and quantum charge algebras are isomorphic with respect to $\{\ ,\ \} \longleftrightarrow -i[\ ,\ ]$.

References

1. C.N. Yang, Phys. Rev. 19 (1967) 1312.
2. R.J. Baxter, Exactly solved models in statistical mechanics, Academic Press, London (1982).
3. L.D. Faddeev, Les Houches Lectures (1982), Saclay preprint T-82-76. P.P. Kulish and E.K. Sklyanin, in Tvärminne Lectures, Eds. J. Hietarinta and C. Montonen, Springer Lecture Notes in Physics vol. 151 (1982).
4. H. Eichenherr and M. Forger, Nucl. Phys. B 155 (1979) 381, A.V. Mikhailov and V.E. Zakharov, Sov. Phys. JETP 47 (1978) 1017.
5. D.J. Gross and A. Neveu, Phys. Rev. D10 (1974) 3235
   R. Köberle, V. Kurak, J.A. Swieca, Phys. Rev. D20, (1979) 897
   M. Karowski and H.J. Thun, Nucl. Phys. B190(FS3) (1981) 61.
6. H.J. de Vega, H. Eichenherr and J.M. Maillet, Phys.Lett. 132B (1983) 337.
7. E. Abdalla, M. Forger and M. Gomes, Nucl. Phys. B210(FS6)(1982)181.
8. A.B. Zamolodchikov and Al.B.Zamolodchikov,Ann.Phys.NY120 (1979)253.
9. H.J. de Vega, H. Eichenherr and J.M. Maillet, Commun. Math. Phys. 92 (1984) 507.
10. H.J. de Vega, H. Eichenherr and J.M. Maillet, preprint PAR-LPTHE 84/05, to appear in Nucl. Phys. B.
11. K. Pohlmeyer, Commun. Math. Phys. 46 (1976) 207.
12. M. Lüscher und K. Pohlmeyer, Nucl. Phys. B137 (1978) 46.
13. H.J. de Vega, Phys. Lett. 87B (1979) 233.
14. A.G. Jzergin and V.E. Korepin, Commun. Math. Phys. 79 (1981) 303.
15. F.A. Berezin, The Method of Second Quantization, Academic Press, New York 1966.
16. A. Neveu and N.Papanicolaou, Commun.Math.Phys 58 (1978) 31.
17. M.C. Davies, P.J. Houston, J.M. Leinaas, A.J. MacFarlane, Phys. Lett. 119B (1982) 187.
18. H.J. de Vega and H.O. Girotti, Nucl. Phys. B79 (1974) 77.
19. V.G. Kac, Math. USSR Izv. 2 (1968) 1271, R. Moody, J. Algebra 10 (1968) 211.
20. L. Dolan, Phys. Rev. Lett. 47 (1981) 1371, K. Ueno, Kyoto University preprint RIMS-374 (1981).
21. M. Lüscher, Nucl. Phys. B135 (1978) 1 .
22. Al.B. Zamolodchikov, Dubna preprint E2-11485 (1978).
23. B. Berg, M. Karowski, P. Weisz and V. Kurak, Nucl. Phys. B134 (1978) 125.

THE QUANTUM TODA CHAIN

E.K. SKLYANIN*
Laboratoire de Physique Théorique
et Hautes Eneriges
Université Pierre et Marie Curie
Tour 16 - 1er étage
4, place Jussieu 75230 PARIS CEDEX 05/FRANCE
(Laboratoire associé au CNRS LA 280)

ABSTRACT :

The exact equations for the spectrum of the finite and infinite quantum Toda chains are derived. The method used does not need the existence of the pseudovacuum (reference) state and provides thus an alternative to the Bethe ansatz method. The results obtained are in good agreement with those of B. Sutherland and M. Gutzwiller.

0. INTRODUCTION

The Toda chain that is the one-dimensional system of equal particles with the exponential interaction of nearest neighbors was introduced by M. Toda in 1967 [1] and since that time is one of the most popular models in the theory of the completely integrable systems.

Among the papers concerning the Toda chain in classical mechanics one must mention [2] where the complete integrability was proven and an exhaustive investigation was carried out of the classical infinite Toda chain and also the papers [3] where the action-angle variables for the finite periodic chain were constructed.

The investigation of the quantum Toda chain in contrast with the classical case is far yet from being completed. The complete integrability of the finite quantum Toda chains with free or periodic boundary conditions was probably first proven in [4] . The first important

---

(*) *On leave from Steklov Mathematical Institute, Leningrad, USSR.*

step in investigating the quantum infinite Toda chain was made by S. Sutherland [5] who has noted that the Toda chain represents a degenerate case of another well known completely integrable model, namely the one-dimensional system of equal particles interacting in pairs via the potential $1/\mathrm{sh}\,(x_i - x_j)$ [6]. This fruitful idea was not unfortunately developed by him in its full extent.

A completely different approach was proposed by M. Gutzwiller [7] who had succeeded in transferring the ideas of the above mentioned papers [3] to the quantum case. Having rewritten the Hamiltonian of the periodic Toda chain in the base of the eigenfunctions of the open chain, Gutzwiller has shown that the problem of finding the spectrum of the Hamiltonian and higher integrals of motion for the N-particle periodic quantum chain reduces to solving a system of N-1 (the total momentum being fixed) identical spectral problems for functions of one variable. Such a reduction of the N-dimensional problem to the one dimensional one can be considered as a version of separation of variables. Lacking efficient formulas for the open chain eigenfunctions for arbitrary N, Gutzwiller could investigate only the cases N = 3 and N = 4 though the formulation of his results can easily be generalized to arbitrary N.

After creation of the quantum spectral transform method (QSTM) [8] which seems at present to be the most universal and powerful method of treating the quantum completely integrable systems several attempts were made to apply it to the quantum Toda chain [9]. However only a little progress has been achieved up to now because of the absence in case of the Toda chain of the so-called pseudovacuum (see below § 3) which makes the direct application of the Bethe-ansatz impossible.

The present article originates as a result of combining the ideas of the paper [7] and QSTM [8,9] which has brought double profit. First, for the periodic quantum chain we have succeeded to separate the variables in a form close to that of Gutzwiller but for arbitrary N. The application of the so-called R-matrix formalism was extremely useful here resulting in a drastic reduction of the calculations. Second, and maybe the more important result is that the arsenal of QSTM is enriched now with a new powerful tool which hopefully will permit to increase the number of solved models.

The paper is organized as follows. The first paragraph contains the description of the model and fixes the notation. The paragraph 2 deals with the classical Toda chain. Though it reproduces already known results, the way of presentation based on the R-matrix formalism is new and clarifies the analogy between the classical and quantum cases. The third paragraph is crucial for the whole article. It contains the derivation of the equations determining the spectrum. The paragraph 4 is devoted to the infinite quantum chain. It is shown there that the determination of the spectrum reduces in this case to solving integral Fredholm equations of a form which is very close to the usual Bethe-ansatz integral equations. It is interesting that the equations obtained are almost the same that those derived previously by B. Sutherland on the basis of a totally different approach.

I would like to express here my gratitude to I.V. Komarov who has influenced my understanding of the papers 3,7 and has given me an impetus to write the present article. I would like also to thank L.D. Faddeev, A.G. Reyman, Yu. N. Reshetikhin, M.A. Semenov-Tian-Shansky L.A. Takhtajan, O. Babelon and H. de Vega for their interest in the work and useful discussions. I am grateful to LPTHE for hospitality.

1. *DESCRIPTION OF THE MODEL*

Consider the chain (finite or infinite) of one-dimensional non-relativistic particles of equal masses m which are described by the quantum mechanical coordinate q_n and momentum p_n operators satisfying the canonical commutation relation $[p_m, q_n] = -i\hbar\delta_{mn}$. The energy of the system is the sum of the kinetic energy for each particle $p_n^2/2m$ and of the potential interaction of the nearest neighbours $V\exp(q_{n+1}-q_n)/d$.

Since the system is characterized by the only dimensionless parameter $\eta = \hbar m^{-1/2} V^{-1/2} d^{-1}$, we shall choose, following [7], the quantities m, V and d to be the units. Note that then $\eta = \hbar$. It will be also convenient for us to use the quantities p_n and $e_n^{\pm} = \exp(\pm q_n)$ satisfying the commutation relation

$$[p_m, e_n^{\pm}] = \mp i\eta e_n^{\pm} \delta_{mn} \tag{1.1}$$

and the constraint

$$e_n^+ e_n^- = 1 \tag{1.2}$$

as the principal dynamical variables instead of p_n and q_n.

In the classical limit $\eta \to 0$ the commutation relation (1.1) is replaced by the Poisson brackets relation

$$\{ p_m , e_n^\pm \} = \pm e_n^\pm \delta_{mn} \tag{1.3}$$

Our final goal is the investigation of the infinite Toda chain whose energy, taking in consideration the above conventions, is

$$E_\infty = \sum_{n=-\infty}^{\infty} \left(\frac{1}{2} p_n^2 + e_{n+1}^+ e_n^- - \mathcal{E} \right) \tag{1.4}$$

where the energy density

$$\mathcal{E} = \left\langle \frac{1}{2} p_n^2 + e_{n+1}^+ e_n^- \right\rangle \tag{1.5}$$

is subtracted in order to cancel the volume divergence. Here and below $\langle X \rangle \equiv \langle 0 | X | 0 \rangle$ stands for the expectation value of the observable X on the vacuum $|0\rangle$.

Concerning the vacuum state $|0\rangle$ we assume the following hypotheses.

a) Normalization. $\langle 0|0 \rangle = 1$.

b) Translational invariance.

For some positive Ξ there exists the unitary translation operator U defined by

$$U p_n U^{-1} = p_{n+1} \tag{1.6a}$$

$$U e_n^\pm U^{-1} = \Xi^{\mp 1} e_{n+1}^\pm \tag{1.6b}$$

The vacuum $|0\rangle$ is required to be invariant under U :

$$U|0\rangle = |0\rangle \qquad (1.7)$$

Comment 1. In particular, it follows from (1.6) and (1.7) that

$$\langle e_n^{\pm}\rangle = \Xi^{\pm n}\langle e_0^{\pm}\rangle$$

Since we always have the freedom of unitary transformation of the form $\rho_n \longmapsto \rho_n$, $e_n^{\pm} \longmapsto x^{\pm 1} e_n^{\pm}$, $x > 0$ under which the quantity $\nu = \langle e_0^+\rangle\langle e_0^-\rangle$ is invariant, it is reasonable to fix x in some way, for example, requiring that $\langle e_0^+\rangle = \langle e_0^-\rangle = \nu^{1/2}$

Comment 2. By virtue of the translational invariance, the vacuum energy density $\mathcal{E}$ (1.5) does not depend on n.

Comment 3. The definition of the translation operator U including a constant Ξ which must not necessarily be 1 may seem not quite usual but it gives the opportunity to describe the effects of expansion and contraction of the crystal lattice due to changes of pressure or temperature. The quantity $(-\ln \Xi)$ plays here the role of the volume per particle.

Comment 4. In the variables $f_n^{\pm} = \Xi^{\mp n} e_n^{\pm}$ the operator U acts in the usual way $U f_n^{\pm} = f_{n+1}^{\pm}$. The term $f_{n+1}^{+} f_n^{-}$ in the energy aquires however the coefficient Ξ .

c) Superselection rule.

We shall assume that the Hilbert space $\mathcal{H}$ of the states of the system is expanded into the direct integral of spaces $\mathcal{H}_{\xi}$, $\xi \in (0,\infty)$ which are invariant for all the observables and in each of which the representation of the algebra (1.1) of the observables $\rho_n, e_n^{\pm}$ is scalar. It is assumed also that the vacuum vector $|0\rangle$ belongs to the subspace $\mathcal{H}_1$ corresponding to $\xi = 1$. As we shall see later, the quantity $(-\ln \xi)$ plays the role of the total volume of the system.

d) Minimal enthalpy principle.

From the physical point of view it is most natural to consider the infinite Toda chain under a constant pressure $\mathcal{P}$ which is defined as the vacuum expectation.

$$\mathcal{P} = \langle e_{n+1}^{+} e_{n}^{-} \rangle \quad > 0$$

It is obvious that due to the translational invariance $\mathcal{P}$ as ε (1.5) does not depend on n. It is supposed that the vacuum $|0\rangle$ corresponds to the state with the minimal enthalpy density

$$h = \epsilon - \mathcal{P} \ln \Xi \tag{1.8}$$

in the class of the translationally invariant vacua corresponding to various values of $\Xi > 0$. Moreover, it must correspond to the minimal enthalpy

$$H = E - \mathcal{P} \ln \xi \tag{1.9}$$

among all the states corresponding to given Ξ . One sees from (1.8) and (1.9) that the quantities $(-\ln \Xi)$ and $(-\ln \xi)$, as mentioned above, play the roles of the volume per particle and the total volume respectively.

The validity of the assumptions made is quite obvious in the classical case. In fact, the translationally invariant vacuum $|0\rangle$ corresponds in classical mechanics to the state

$$p_n = 0 \quad , \quad e_n^{\pm} = \Xi^{\pm n} \tag{1.10}$$

(the normalization $e_0^{\pm} = 1$ is taken). The energy density is $\varepsilon = \Xi$. The enthalpy density $h = \Xi - \mathcal{P} \ln \Xi$ reaches its minimum when $\Xi = \mathcal{P}$.

The quantitiy ξ is defined in the classical case as the limit

$$\xi = \lim_{N_{\pm} \to \pm\infty} e_{N_+}^{+} e_{N_-}^{-} \Xi^{-(N_+ - N_-)} = \left[\lim_{N_{\pm} \to \pm\infty} e_{N_+}^{-} e_{N_-}^{+} \Xi^{(N_+ - N_-)} \right]^{-1} \tag{1.11}$$

The physical sense of the parameter ξ is simple. It determines the asymptotic shift

$$\ln \xi = \lim_{N_\pm \to \pm\infty} (q_{N_+} - q_{N_-}) \tag{1.12}$$

The phase space of the system is thus foliated into the submanifolds ξ = const on which the Poisson bracket is nondegenerate (that is there is no functional other than a constant commuting with p_n and $e_n^\pm$ $\forall n$). One can show that the enthalpy H(1.9) constructed with ξ defined by (1.11) or (1.12) is in fact positive for all excited states. The necessity of considering the states with non zero shift ($\xi \neq 1$) follows, for instance, from the fact that the one-soliton states [1,2] are of this type.

As for a rigorous proof of the hypotheses a)-d) in the quantum case we hope that it will be available soon due to the recent progress in calculating the Green's functions in the quantum completely integrable models [10].

Besides the infinite Toda lattice we shall also deal with the variants of the finite Toda chain which are described below.

Consider the system of N one-dimensional particles which are labelled by the number $n = N_-, N_-+1, \ldots, N_+$, supposing that

$$N_+ - N_- + 1 = N \tag{1.13}$$

As previously, we describe the particles by the dynamical variables p_n, $e_n^\pm$ (1.1-1.2).

The Hamiltonian of the <u>quasiperiodic</u> Toda chain is given by

$$E_N(x) = \sum_{n=N_-}^{N_+} \frac{1}{2} p_n^2 + \sum_{n=N_-}^{N_+-1} e_{n+1}^+ e_n^- + x e_{N_-}^+ e_{N_+}^- \tag{1.14}$$

where $x > 0$ is a constant. For $x = 1$ the Hamiltonian (1.14) describes the <u>periodic</u> and for $x = 0$ the <u>open</u> Toda chain.

Let us define the unitary translation operator U by

$$U\, p_n\, U^{-1} = p_{n+1} \quad , \quad U e_n^{\pm}\, U^{-1} = x^{\mp 1/N} e_{n+1}^{\pm} \tag{1.15a}$$

for $n = N_-, \ldots, N_+ - 1$ and

$$U\, p_{N_+}\, U^{-1} = p_{N_-} \quad , \quad U e_{N_+}^{\pm}\, U^{-1} = x^{\pm (N-1)/N} e_{N_-}^{\pm} \tag{1.15b}$$

One verifies easily that $U^N = 1$ and $U E_N(x)\, U^{-1} = E_N(x)$

As for the infinite chain, the quasiperiodic Toda chain rewritten in the variables $f_n^{\pm} = x^{\mp n/N} (n = N_-, \ldots N_+)$ reduces to the periodic one but with a different strength $V = x^{1/N}$.

To terminate the paragraph let us discuss the connection between the quasiperiodic chain (1.14) and the infinite one (1.4). Consider first the classical case.

Let us substitute in (1.14)

$$x = \xi\, \Xi^N \tag{1.16}$$

and take the limit $N_\pm \to \pm\infty$. We shall show now that the limit being taken, the quasiperiodic Toda chain becomes the infinite chain with the density and the asymptotic shift determined by the parameters Ξ and ξ respectively.

To begin with, note that the equilibrium state (vacuum) for the quasiperiodic chain (1.14) is

$$p_n = 0 \quad , \quad e_n^{\pm} = x^{\pm n/N} \quad , \quad n = N_-, \ldots N_+ \tag{1.17}$$

(the normalization $e_0^{\pm} = 1$ is assumed) and upon substituting (1.16) for $\xi = 1$ it becomes in the limit $N_\pm \to \pm\infty$ the vacuum (1.10) for the infinite chain. It seems reasonable to suppose that as $N_\pm \longrightarrow \pm\infty$ the configurations corresponding to the excited states would be spatially localized and the variables p_n, $e_n^{\pm}$ would thus go to their asymptotic values

$$p_\nu = 0 \quad , \quad e_\nu^{\pm} = x^{\pm \nu / N} = \xi^{\pm \nu / N} \Xi^{\pm \nu} \quad , \quad \nu = \pm N_{\pm} \tag{1.18}$$

as $n \longrightarrow \pm \infty$.

Upon substituting (1.18) into (1.11) and taking in consideration (1.13) we obtain an identity. The definition of Ξ and ξ by (1.16) is therefore equivalent to the previous definitions (1.6b) and (1.11).

One needs also to check up the correspondance between the energies of the quasiperiodic chain (1.14) and of the infinite one (1.4). Using (1.18) and the hypothesis of the locality of the excitations one obtains that the contribution $x^{1/N} = \xi^{1/N} \Xi$ of the term $x\, e_{N-}^{+} e_{N+}^{-}$ to the energy $E_N(x)$ of the quasiperiodic chain cancels the contribution Ξ of the same term to the vacuum energy $E_{N,vac.}(x)$ in the limit $N \to \infty$. The following relations are thus valid

$$\varepsilon = \lim_{N \to \infty} \frac{1}{N} E_{N, vac.} (\Xi^N) \quad , \tag{1.19}$$

$$E_\infty = \lim_{N \to \infty} \left[E_N (\xi \Xi^N) - E_{N, vac.} (\Xi^N) \right] \tag{1.20}$$

The question of the rigorous justification of the similar limit transitions in the quantum case goes beyond the scope of the present paper. We shall limit ourselves here by simply postulating the relations (1.19-1.20) and taking (1.16) as definition of Ξ and ξ .

2. THE CLASSICAL TODA CHAIN

In this paragraph mainly well known results [1,2,3] concerning the complete integrability of the classical Toda chain are reproduced in the framework of the r-matrix formalism.

Let us consider the classical quasiperiodic Toda chain, and following [9,11] put into correspondance to the n-th particle (n = N-,..., N+) the 2x2 matrix $L_n(u)$ depending on a spectral parameter u.

$$L_n(u) = \begin{pmatrix} u - p_n & -e_n^+ \\ e_n^- & 0 \end{pmatrix} \quad (2.1)$$

Let us introduce also the monodromy matrix

$$T_N(u) = L_{N_+}(u) \ldots L_{N_-}(u) = \begin{pmatrix} A_N(u) & B_N(u) \\ C_N(u) & D_N(u) \end{pmatrix}, \quad (2.2)$$

and its weighted trace

$$t_N(u,x) = x^{-1/2} A_N(u) + x^{1/2} D_N(u) = t_r \, X \, T_N(u) \quad (2.3)$$

where $X = \begin{pmatrix} x^{-1/2} & 0 \\ 0 & x^{1/2} \end{pmatrix}$,

the constant $x > 0$ being the same as in the Hamiltonian (1.14) of the quasiperiodic chain.

Note that by virtue of (1.2) the matrix $L_n(u)$ and consequently $T_N(u)$ are unimodular

$$\det L_n(u) = 1,$$

$$\det T_N(u) = A_N(u) D_N(u) - B_N(u) C_N(u) = 1 \quad (2.4)$$

The r-matrix method is based on the following remarkable fact. Let $(L_n(u))^a_b$ be a matrix element of $L_n(u)$. Then the Poisson bracket between $(L_m(u))^{a_1}_{b_1}$ and $(L_n(v))^{a_2}_{b_2}$ can be represented as the commutator

$$\{(L_m(u))^{a_1}_{b_1}, (L_n(v))^{a_2}_{b_2}\} = [r(u-v), L_m(u) \otimes L_n(v)]^{a_1 a_2}_{b_1 b_2} \delta_{mn} \quad (2.5)$$

of the quantities

$$r(u-v)^{a_1 a_2}_{b_1 b_2} = -(u-v)^{-1} \delta^{a_1}_{b_2} \delta^{a_2}_{b_1} \tag{2.6}$$

and $$\left(L_m(u) \otimes L_n(v) \right)^{a_1 a_2}_{b_1 b_2} = \left(L_m(u) \right)^{a_1}_{b_1} \left(L_n(v) \right)^{a_2}_{b_2}$$

considered as 4x4 matrices [8,9].

It is easy to see that the identity (2.5) for the L-operator (2.1) entrains the analogous identity for the monodromy matrix $T_N(u)$ (2.2) which can be written in matrix notation [8,9] as

$$\{ T_N(u) \overset{\otimes}{,} T_N(v) \} = [r(u-v), T_N(u) \otimes T_N(v)] \tag{2.7}$$

The complete list of the relations contained in (2.7) is given in the Appendix A.

Noting that the identity holds

$$[r(u-v), X \otimes X] = 0 \tag{2.8}$$

for the matrix X (2.3) and the r-matrix (2.6) and using the equality (2.7) one obtains that the quantities $t_N(u,x)$ are in involution

$$\{ t_N(u,x), t_N(v,x) \} = 0 \tag{2.9}$$

From the definitions (2.1), (2.2) and (2.3) it follows immediately that $t_N(u,x)$ is a polynomial in u of degree N with the coefficient $x^{-1/2}$ at the highest power of u

$$t_N(u,x) = x^{-1/2}(u^N + t_1 u^{N-1} + t_2 u^{N-2} + \dots t_N) \tag{2.10}$$

In particular,

$$t_1 = - P_N , \tag{2.11a}$$

$$t_2 = \frac{1}{2} P_N^{\,2} - E_N(x) \,, \tag{2.11b}$$

where

$$P_N = \sum_{n=N_-}^{N_+} p_n \tag{2.12}$$

is the total momentum and $E_N(x)$ is the energy (1.14) of the quasi-periodic Toda chain.

By virtue of (2.9) the coefficients t_j; ($j = 1,\ldots, N$) of the polynomial $t_N(u,x)$ are in involution. Moreover, one can show [3] that they are functionally independent. Since the quasiperiodic Toda chain has exactly N degrees of freedom and its Hamiltonian $E_N(x) = \frac{1}{2}t_1^2 - t_2$ can be expressed in terms of t_j, the complete integrability of the system in the Liouville's sense is thus established.

It is more convenient to use the quantity

$$\tau_N(u,x) = \ln t_N(u,x) \tag{2.13}$$

which due to (2.10) expands as $u \to \infty$ like

$$\tau_N(u,x) = N \ln u - \frac{1}{2} \ln x + \tau_1 u^{-1} + \ldots + \tau_j u^{-j} + \ldots \tag{2.14a}$$

where

$$\tau_1 = -P_N \,, \quad \tau_2 = -E_N(x) \tag{2.14b}$$

Using the methods of [11] one can show that the coefficients τ_j being like t_j integrals of motion are local that is representable as sum of the terms (densities) containing p_n, $e_n^{\pm}$ for no more than j neighbor particles.

For those accustomed to the traditional approach [2,3] let us notice that it is equivalent to ours. In fact, the first-order difference linear problem [9,11]

$$\Phi_{n+1} = L_n(u)\Phi_n \quad , \quad \Phi_n = \begin{pmatrix} \varphi_n \\ \psi_n \end{pmatrix} \tag{2.15}$$

lying in the base of our approach is easily transformed by using the explicit expression (2.1) for the L-operator into the second-order difference linear problem

$$u\varphi_n = \varphi_{n+1} + p_n\varphi_n + e_n^+ e_{n-1}^- \varphi_{n-1} \tag{2.16}$$

which coincides up to a non essential change of notation with that used in [2,3]. Moreover, one can verify that the trace t(u) (2.3) of the monodromy matrix (2.2) for the linear problem (2.15) coincides with the determinant of the jacobian matrix corresponding to (2.16).

Let us describe now a construction of the action-angle variables for the quasiperiodic Toda chain. The construction given below represents essentially an adaptation of the approach of the papers [3] to the linear problem (2.15). The utilization of the r-matrix formalism adopted here has the following advantages. First, less calculation is needed in treating the Poisson brackets. Second, the whole construction can immediately be trasferred to the quantum case. And third, the construction can be generalized to other models permitting the application of the r-matrix method.

Let us go now to the construction itself. Following [12] we define the variables u_j as the roots of the polynomial C(u) of the (N-1)-th power

$$C(u_j) = 0 \quad , \quad j = 1, \dots, N-1 \tag{2.17}$$

One can show [3] that all the u_j are real and, consequently, one can order them, for example like $u_1 < u_2 < \dots < u_{N-1}$.

We define also the variables $\lambda_j^{\pm}$ by

$$\lambda_j^- = A_N(u_j) \quad , \quad \lambda_j^+ = D_N(u_j) \quad , \quad j = 1, \dots, N-1 \tag{2.18}$$

Note that, by virtue of (2.4)

$$\lambda_j^- \lambda_j^+ = A(u_j)\, D(u_j) = \det h(u_j) = 1 \qquad (2.19)$$

The central point of the whole construction is the calculation of the following Poisson brackets

$$\begin{aligned} \{u_j, u_k\} &= \{\lambda_j^{\pm}, \lambda_k^{\pm}\} = 0 \ , \\ \{u_j, \lambda_k^{\pm}\} &= \pm \lambda_j^{\pm} \delta_{jk} \end{aligned} \qquad (2.20)$$

Let us show, for example, how to calculate the Poisson bracket $\{u_j, \lambda_k^+\}$. Using (A11, A12) one obtains

$$\{u_j, \lambda_k^+\} = \{u_j, D(u_k)\} = \{u_j, u_k\} D'(u_k) + \{u_j, D(v)\}\Big|_{v=u_k}$$

$$= -\frac{1}{C'(u_j)} \{C(u), D(v)\}\Big|_{\substack{u=u_j \\ v=u_k}} = \frac{1}{C'(u_j)} \frac{C(u)D(v) - D(u)C(v)}{u-v}\Big|_{\substack{u=u_j \\ v=u_k}}$$

$$= \lambda_j^+ \delta_{jk}$$

The rest of equalities (2.20) are verified in the same manner.

The equalities (2.20) give the opportunity to represent $\lambda_j^{\pm}$ as $\lambda_j^{\pm} = \exp(\mp v_j)$ where v_j are the momenta canonically conjugated to u_j that is $\{v_j, u_k\} = \delta_{jk}$ It follows from the reality of u_j and the obvious relation $\overline{T(u)} = T(\bar{u})$ that v_j are also real.

Adding two variables $P \equiv P_N$ and $Q = -\ln e_{N+}^-$ to the set u_j, v_j we obtain a complete set of canonical variables on the phase space of the quasiperiodic Toda chain (one can easily verify using (2.7) and the asymptotics (2.21) that P and Q commute with all the u_j, v_j). The completeness means that any observable can be expressed in terms of P, Q, u_j, v_j. The general proof can be found in [3] . Here we shall illustrate this statement on the example of the elements $A_N(u)$, $B_N(u)$, $C_N(u)$, $D_N(u)$ of the monodromy matrix (2.2). In fact, the polynomials $C_N(u)$, $A_N(u)$, $D_N(u)$ are unambiguously determined from their values at $u = u_j$ and their asymptotics at $u \to \infty$

$$C_N(u) = e_{N+}^{-} u^{N-1} + O(u^{N-2}) \tag{2.21a}$$

$$A_N(u) = u^N - P u^{N-1} + O(u^{N-2}) \tag{2.21b}$$

$$D_N(u) = O(u^{N-2}) \tag{2.21c}$$

Using the Lagrange's interpolation formula one obtains

$$C_N(u) = e_{N+}^{-} \prod_{j=1}^{N-1} (u - u_j) , \tag{2.22a}$$

$$A_N(u) = (u - P + u_1 + \dots + u_{N-1}) \prod_{j=1}^{N-1} (u - u_j) + \sum_{j=1}^{N-1} \left(\prod_{\substack{k=1 \\ k \neq j}}^{N-1} \frac{u - u_k}{u_j - u_k} \lambda_j^{-} \right) , \tag{2.22b}$$

$$D_N(u) = \sum_{j=1}^{N-1} \left(\prod_{\substack{k=1 \\ k \neq j}}^{N-1} \frac{u - u_k}{u_j - u_k} \lambda_j^{+} \right) \tag{2.22c}$$

The expression for $B_N(u)$ can then be obtained from (2.4).

The variables P, Q, u_j, v_j as mentioned above form a complete set of canonical variables. Though they are not exactly the action-angle variables the latter are obtained from them by a standard procedure which is described in detail in [3]. So, we shall not more deal with them and proceed now to the quantum case.

3. THE QUANTUM QUASIPERIODIC CHAIN

Let us consider the quantum quasiperiodic Toda chain defined by the Hamiltonian (1.14) written in terms of the dynamical variables p_n, $e_n^{\pm}$ satisfying the commutation relations (1.1) and the constraint (1.2). In order to preserve parallelism in treating the quantum and classical cases and to avoid rewriting many classical formulas which survive also in the quantum case we do not change the notation and denote the majority of the quantum objects by the same letters that the corresponding classical ones.

To begin with, let us introduce the quantum L-operator, the quantum monodromy matrix and its weighted trace by the same formulas (2.1), (2.2) and (2.3) respectively as in the classical case paying attention that p_n, $e_n^{\pm}$ are now the quantum operators. The relation (2.5) is replaced in the quantum case by [8]

$$R(u-v)\,\tilde{L}_n(u)\,\tilde{\tilde{L}}_n(v) = \tilde{\tilde{L}}_n(v)\,\tilde{L}_n(u)\,R(u-v) \tag{3.1}$$

where $\tilde{L}_n(u) = L_n(u) \otimes I$, $\tilde{\tilde{L}}_n(u) = I \otimes L_n(u)$ and

$$\left[R(u-v)\right]^{a_1 a_2}_{b_1 b_2} = u\,\delta^{a_1}_{b_1}\,\delta^{a_2}_{b_2} - i\eta\,\delta^{a_1}_{b_2}\,\delta^{a_2}_{b_1} \,. \tag{3.2}$$

As in the classical case the analogous identity holds for the monodromy matrix

$$R(u-v)\,\tilde{T}(u)\,\tilde{\tilde{T}}(v) = \tilde{\tilde{T}}(v)\,\tilde{T}(u)\,R(u-v) \tag{3.3}$$

The complete list of relations contained in (3.3) is given in Appendix B.

The identity (3.3) plays in the quantum case the same role as (2.7) in the classical case providing the commutativity of the operator-valued functions $t_N(u,x)$

$$\left[\,t_N(u,x)\,,\ t_N(v,x)\,\right] = 0 \tag{3.4}$$

As in the classical case the functions $t_N(u,x)$ and $\tau_N(u,x) = \ln t_N(u,x)$ have the expansions in powers of u (2.10-211) and (2.14) respectively. The function $\tau_N(u,x)$ can thus be considered as a generating function of the local integrals of motion.

The principal problem in studying the quantum quasiperiodic Toda chain is to determine the spectrum of its Hamiltonian (1.14) and higher integrals of motion or, which is the same, to determine the spectrum of $t_N(u,x)$. It is worth mentioning that from the mathematical point of view the problem is mostly an algebraic one. The matrix elements A, B, C D of $T_N(u)$ for different N can be considered as representations of the same infinite-dimensional associative algebra with the quadratic constraints given by (3.3), the R-matrix (3.2) being the set of the structure constants of the algebra [13]. There is strong resemblance between the commuting family $t_N(u,x)$ and the Cartan subalgebra in the theory of the Lie algebras. The quantum determinant (B.17 - B.18) can also be considered as the set of the Casimir operators of the algebra. For more information on the algebras with quadratic constraints and their applications to the quantum completely integrable models see [14].

The usual way to find the spectrum of $t_N(u,x)$ known as the algebraic Bethe ansatz [8] is as follows. One needs first to find a vector Ω (analogue of the lowest weight in the Lie groups representation theory) such that $C(u)\,\Omega = 0 \ \forall u$. The eigenvectors f of t(u) are then written in the form $f = B(u_1)\ldots B(u\ )$. A simple algebraic manipulation using (3.3) provides then a system of equations for determining values of u_i (i = 1,...,) as well as the corresponding eigenvalue of $t_N(u)$.

Unfortunately, in the case of the Toda chain the method does not work. The reason is the absence of the vector Ω with the needed properties. It follows for instance from the absence of zero eigenvalue for the operator e_{N+}^{-} which determines the asymptotics (2.21a) of C(u) as $u \to \infty$.

The main idea of our approach is borrowed from the paper [7] of M. Gutzwiller who realized that the proper quantum analogue of the Flaschka-Mc-Laughlin-Kac-van Moerbeke construction [3] consists in expanding the operator $t_N(u,x)$ in the eigenfunctions of the (N-1)- particle open chain (the eigenvalue of the total momentum P_N being fixed). In our language it means expanding in eigenfunctions of $C_N(u)$. In fact, it follows from (B.11) that $[C_N(u), C_N(v)] = 0$ and from (2.2) that $C_N(u) = e_{N+}^{-} A_{N-1}(u)$ where $A_{N-1}(u)$ due to (2.3) for x = 0 contains integrals of motion for the open (N-1) particle chain (this observation belongs to I.V. Komarov [12]).

Now, let us suppose by analogy with the classical case that the formula (2.22a) is valid also in the quantum case, $\{u_j\}_{j=1}^{N-1}$ being now a commutative family of self-adjoint operators. Let us introduce the operators $\lambda_j^{\pm}$ by the formulas

$$\lambda_j^- = A_N(u \rightleftarrows u_j) \quad , \quad \lambda_j^+ = D_N(u \rightleftarrows u_j) \quad , \quad j = 1, \dots, N-1 \qquad (3.5)$$

where the sign $\rightleftarrows$ means that the operators u_j are substituted into the operator-valued polynomials $A_N(u)$ and $D_N(u)$ to the left, that is, for example a polynomial $X(u) = \sum_k x_k u^k$ is replaced by the operator $X(u \rightleftarrows u_j) = \sum_k u_j^k x_k$. This convention is essential since the operators u_j and $A(u)$, $D(u)$ generally speaking do not commute.

As in the classical case we supplement the set of variables u_j, $\lambda_j^{\pm}$ with the variables $P = P_N$ and e_{N+}^{-}.

Using the formulas (B1-16) and the definitions (2.22a), (3.5) one can derive the commutation relations between P, e_{N+}^{-}, u_j, $\lambda_j^{\pm}$:

$$[P, u_j] = 0 = [P, \lambda_j^{\pm}] = [e_{N+}^{-}, u_j] = [e_{N+}^{-}, \lambda_j^{\pm}] \qquad (3.6a)$$

$$[u_j, u_k] = [\lambda_j^{\pm}, \lambda_k^{\pm}] = 0 \, , \qquad (3.6b)$$

$$\lambda_j^{\pm} u_k = (u_k \pm i\eta\,\delta_{kj})\,\lambda_j^{\pm} \qquad (3.6c)$$

$$e_{N+}^{-} P = (P - i\eta)\, e_{N+}^{-} \qquad (3.6d)$$

In addition, it follows from (B19) and (B28) that

$$\lambda_j^- \lambda_j^+ = \lambda_j^+ \lambda_j^- = 1 \qquad (3.7)$$

Let us show, for example, the derivation of the commutation relation (3.6c) between λ_j^- and u_k. One rewrites first the equality (B3) as

$$(u-v)A(u)\,C(v) = (u-v-i\eta)\,C(v)A(u) + \\ + i\eta\,C(u)\,A(v). \tag{3.8}$$

and substitues $u \rightleftarrows u_j$. Using the definition (3.5) and the fact that by virtue of (B11) $[\,C(v), u_j\,] = 0$ and $C(u \rightleftarrows u_j) = 0$ one obtains

$$(u_j - v)\,\lambda_j^-\,C(v) = (u_j - v - i\eta)\,C(v)\,\lambda_j^- \tag{3.9}$$

Substituting the expression (2.22a) for C(v) into (3.9) and cancelling from the left the factor $e_{N\mp}^-\,(v-u_j)$ which can always be done for v lying out of the spectrum of u_j one arrives at

$$\lambda_j^-(v-u_1)\,\ldots\,(v-u_j)\,\ldots\,(v-u_{N-1}) = \\ = (v-u_1)\,\ldots\,(v-u_j+i\eta)\ldots\,(v-u_{N-1})\,\lambda_j^- \tag{3.10}$$

From (3.10) it follows immediately the equality

$$\lambda_j^-\,\mathcal{L}(u_1,\ldots,u_j,\ldots,u_{N-1}) = \\ = \mathcal{L}(u_1,\ldots,u_j-i\eta,\ldots,u_{N-1})\,\lambda_j^- \tag{3.11}$$

which holds for any symmetric function $\mathcal{L}(u_1,\ldots,u_{N-1})$ since any symmetric function of (N-1) variables can be expressed in terms of the elementary symmetric polynomials which are nothing but the coefficients at the powers of v in the polynomial $(v-u_1)\ldots(v-u_{N-1})$. We shall see later that it will be enough to understand the commutation relation (3.6c) in the weak sense of (3.11).

The rest of the relations (3.6) are proved in quite analogous way. As for the relations (3.7), their proof contains more tedious calculations (see analogous proof in [15]).

The expressions (2.22) for the operators $C_N(u)$, $A_N(u)$, $D_N(u)$ in terms of P, $e_{N\mp}^-$, u_j, $\lambda_j^\pm$ are valid also in the quantum case with the only difference that attention must be paid to the operator ordering. The operator $B_N(u)$ can be found from the identities (B19, B22, B23, B26).

Looking at the commutation relations (3.6) and the constraint (3.7) it is natural to represent the operators P, u_j as the multiplication operators

$$(Pf)(k;v_1,\dots,v_{N-1}) = k\, f(k;v_1,\dots v_{N-1})\,, \tag{3.12a}$$

$$(u_j f)(k;v_1,\dots,v_{N-1}) = v_j\, f(k;v_1,\dots,v_{N-1})\,, \tag{3.12b}$$

and e_{N+}^{-}, $\lambda_j^{\pm}$ as the shift operators

$$(e_{N+}^{-} f)(k;v_1,\dots,v_{N-1}) = f(k-i\eta;v_1,\dots,v_{N-1}) \tag{3.13a}$$

$$(\lambda_j^{\pm} f)(k;v_1,\dots,v_j,\dots,v_{N-1}) = i^{\pm N} f(k;v_1,\dots v_j \pm i\eta,\dots v_{N-1}) \tag{3.13b}$$

acting on functions of N scalar arguments $k;v_1,\dots,v_{N-1}$

The coefficients $i^{\pm N}$ are introduced in (3.13b) in order to simplify some subsequent calculations. The representation (3.12-3.13) is in fact the only (up to equivalent transformations), irreducible representation of the Heisenberg algebra (3.6).

Since the expressions (2.22) for the operators $C_N(u)$, $A_N(u)$, $D_N(u)$ are invariant under permutations of u_j and $\lambda_j^{\pm}$, one can restrict the representation to the functions $f(k\,;\,v_1,\dots,v_{N-1})$ which are symmetric in $v_1,\dots v_{N-1}$. The restriction corresponds precisely to the fact mentioned above that the commutation relations (3.6) are valid only for the symmetric functions of u_j. One must mention also that it is only for the symmetric functions $f(k,v_1,\dots,v_{N-1})$ the Lagrange's interpretation polynomials in (2.22) are cancelled.

Strictly speaking, the operators u_j (3.12b) and $\lambda_j^{\pm}$ (3.13b) are not defined on the symmetrical functions since they break the symmetry. One can, nevertheless, use them in formal calculations provided the final result is symmetric.

Up to the moment we did not study the conjugation properties of the operators u_j and $\lambda_j^{\pm}$. They are extracted easily from the obvious relation

$$T_N^*(u) = T_N(\bar{u}) \tag{3.14}$$

which follows immediately from the definitions (2.1) and (2.2) and the selfadjointness of p_n, e_n^+. The conjugation $*$ acts here on the quantum operators only without transposition of the 2x2 matrix T. Using (3.14) and (2.22) one arrives after same manipulations at the relations

$$u_j^* = u_j \tag{3.15a}$$

$$(\lambda_j^\pm)^* = \prod_{\substack{k \neq j \\ k=1}}^{N-1} \frac{(u_j - u_k \pm i\eta)}{u_j - u_k} \lambda_j^\pm \tag{3.15b}$$

Let us look for the scalar product for functions $f(k\,;\,v_1,\ldots,v_{N-1})$ in the form

$$\langle f,g\rangle = \int_{-\infty}^{\infty} dk\, dv^{N-1} \mu(v_1,\ldots,v_{N-1})\, \bar{f}(k;v_1,\ldots,v_{N-1})\, g(k;v_1,\ldots,v_{N-1}) \tag{3.16}$$

where the variables k, v_j are assumed to be real and μ is some positive function. Assuming that the functions f, g and μ have an analytical continuation into $v_j < \eta + \epsilon$ and using (3.12b) and (3.13b) one obtains for the measure μ the recurrence relation

$$\mu(v_1,\ldots,v_j + i\eta,\ldots,v_{N-1}) = \prod_{\substack{k \neq j \\ k=1}}^{N-1} \frac{(v_k - v_j - i\eta)}{v_j - v_k} \mu(v_1,\ldots,v_j,\ldots,v_{N-1})$$

which has the solution

$$\mu(v_1,\ldots,v_{N-1}) = \prod_{j<k=1}^{N-1} \frac{1}{\left|\Gamma\left(\frac{u_j - u_k}{i\eta}\right)\right|^2} \tag{3.17}$$

Let us collect now the facts obtained and formulate the following

Theorem. The two representations of the algebra defined by the generators A(u), B(u), C(u), D(u) and the quadratic relations (3.3) which are given, first, by the formula (2.1) and the definition (2.2) of the monodromy matrix as the product of L-operators, and, second, by the formulas (2.22),(3.12-3.13) in the space of the functions $f(k\ ;\ v_1,\ldots,v_{N-1})$ symmetric in v_j with the scalar product (3.16), (3.17), are unitary equivalent.

The previous reasoning is, of course, a heuristical one and cannot be considered as a rigorous proof. Nevertheless, we have presented it to show how one can get a guess which can be proven afterwards by different tools. As for the complete and rigorous proof of the Theorem, a version based on the induction in N, will be published in a separate paper.

Let us attack now the problem of finding the spectrum of the operators $t_N(u,x)$. By virtue of the Theorem one can replace the operators $A_N(u)$ and $D_N(u)$ in the definition (2.3) of $t_N(u,x)$ by their expressions (2.22bc) in terms of u_j and $\lambda_j^{\pm}$ given by (3.12b-3.13b). Since the total momentum P commute with $t_N(u,x)$ their common eigenfunctions $f(k\ ;\ v_1,\ldots,v_{N-1})$ have the trivial dependence on k that is $f(k\ ;\ v_1,\ldots,v_{N-1}) = \delta(k-p)\ \Phi(v_1,\ldots,V_{N-1})$ where p is a specified eigenvalue of P. Denoting the eigenvalue of $t_N(u,x)$ by the same letter one obtains the following eigenvalue problem

$$t_N(u,x)\Phi(v_1,\ldots,v_{N-1}) = (u-p+v_1+\ldots+v_{N-1})\prod_{j=1}^{N-1}(u-v_j)\,\cdot$$
$$\cdot\,\Phi(v_1,\ldots,v_{N-1}) + \sum_{j=1}^{N-1}\prod_{\substack{k\neq j\\ k=1}}^{N-1}\Big[x^{-1/2}i^{N}\Phi(v_1,\ldots v_j+i\eta,\ldots v_{N-1}) + \tag{3.18}$$
$$+\,x^{1/2}i^{-N}\Phi(v_1,\ldots,v_j-i\eta,\ldots v_{N-1})\Big]$$

To reduce the equation (3.18) to the system of one-dimensional equations we shall use a trick due to M. Gutzwiller [7]. Note that $t_N(u,x)$ is a polynomial of N-th degree in u and thus can be determined from its values in (N-1) points, say u = v , and its asymptotics (2.10). Substituting $u = v_j$ into (3.18) we arrive at (N-1) identical equations.

$$t_N(v_j,x)\Phi(v_1,\ldots,v_{N-1}) = x^{-1/2}i^{N}\Phi(v_1,\ldots,v_j+i\eta,\ldots,v_{N-1}) + \tag{3.19}$$
$$+\,x^{1/2}i^{-N}\Phi(v_1,\ldots,v_j-i\eta,\ldots,v_{N-1})$$

Looking for the solution $\Phi(v_1, \ldots, v_{N-1})$ of (3.19) in the form

$$\Phi(v_1, \ldots, v_{N-1}) = \prod_{j=1}^{N-1} \varphi(v_j) \tag{3.20}$$

and recollecting the asymptotical condition (2.10) for $t_N(u,x)$ we obtain, finally, the linear problem for $\varphi(v_j)$

$$t_N(u,x)\ \varphi(u) = x^{-1/2} i^N \varphi(u+i\eta) + x^{1/2} i^{-N} \varphi(u-i\eta) \tag{3.21a}$$

$$t_N(u,x) = x^{-1/2} u^N + O(u^{N-1}),\quad u \to \infty \tag{3.21b}$$

which is equivalent to the multidimensional problem (3.18).

One needs to do now several comments.

Comment 1. To obtain (3.19) it is not necessary to use the explicit expressions (2.22). One can also apply directly the definitions (3.5) of $\lambda_j^{\pm}$.

Comment 2. Generally speaking there might be no one-to-one correspondance between the solutions of (3.21) and the eigenfunctions of $t_N(u,x)$. If the eigenvalue $t_N(u,x)$ of the problem (3.21) is M-times degenerate, that is there exist M linearly independent functions $\varphi_k(u)$ satisfying (3.21) with the same $t_N(u,x)$, then the corresponding eigenvalue $t_N(u,x)$ of the problem (3.18) has the degeneracy $(N+M-2)!/(N-1)!\,(M-1)!$ that is the number of the symmetrical polynomials in φ_k of order (N-1). The cases of degeneracy seem to be however extremely rare it they exist at all.

Comment 3. The equation (3.21) for the function $\varphi(u)$ has the same form as the equation for the eigenvalues of the Baxter's [16] operator Q(u). No connection between the wave function $\varphi(u)$ and the operator Q(u) is known yet.

To sum up, we have shown that the problem of finding the spectrum of the N-particle quasiperiodic quantum Toda chain is reduced for all N to a one-dimensional problem of almost the same form as that obtained in [7] for N = 3 and 4. Such a reduction can be treated as a kind of seperation of variables.

M. Gutzwiller has given in [7] a solution of the problem (3.21) in terms of the Hill's determinant. In the next paragraph we propose another approach which seems to be more suitable for the subsequent transition to the infinite chain.

4. THE INFINITE QUANTUM CHAIN

To attack the problem of finding solutions to (3.21) let us make once more the analogy with the Baxter's Q-operator [16]. Consider, for example, Baxter's equation for the XXX-magnetic chain of spin ℓ (cf. also[17])

$$\begin{cases} t_N(u)Q(u) = (u+i\ell\eta)^N Q(u-i\eta)+(u-i\ell\eta)^N Q(u+i\eta) \\ t_N(u) \simeq 2u^N + O(u^{N-1}), \quad u\to\infty \end{cases} \tag{4.1}$$

The usual way of solving (4.1) is to substitute for Q(u) the polynomial $Q(u)=\prod_{k=1}^{M}(u-u_k)$ and to put then $u=u_k$. The left-hand-side of (4.1) being zero, one obtains for u_k the set of equations

$$\left(\frac{u_k-i\ell\eta}{u_k+i\ell\eta}\right)^N = \prod_{j=1}^{M}\frac{u_k-u_j-i\eta}{u_k-u_j+i\eta} \tag{4.2}$$

In case of the Toda chain the polynomial ansatz for φ(u), however, does not work in (3.21) since it contradicts the asymptotic condition for $t_N(u,x)$. This fact seems to be related to the absence of the reference (pseudovacuum) state Ω for the Toda chain discussed in the paragraph 1.

We are led thus to look for the solutions of (3.21) among functions more general than polynomials, for example, among holomorphic functions having infinitely many zeroes. To get an idea of what kind of functions it may be, let us derive the asymptotic conditions for φ(u). Substituting for φ(u) the ansatz φ(u) = and taking the leading orders in u at $u\to\infty$, one obtains

$$\varphi(u) \simeq \exp\left(-\frac{\pi}{2}\frac{N}{\eta}u\right)\exp\left(-i\frac{\ln x}{2\eta}u\right)\exp\left[\pm i\frac{u}{\eta}\left(N\ln\frac{u}{e}-\frac{1}{2}\ln x\right)\right]$$

Taking a combination of two solutions we are led to the asymptotics

$$\varphi(u) \sim \exp\left(-\frac{\pi}{2}\frac{N}{\eta}u\right)\exp\left(-i\frac{\ln x}{2\eta}u\right)\sin\left[\frac{u}{\eta}\left(N\ln\frac{u}{e}-\frac{1}{2}\ln x\right)\right]$$

which implies that $\varphi(u)$ has infinitely many zeroes going to the infinity according to the asymptotics

$$u_n\left(N\ln\frac{u_n}{e}-\frac{1}{2}\ln x\right) \sim \pi\eta n. \qquad (4.3)$$

One can look for $\varphi(u)$ roughly speaking as the infinite (divergent) product

$$\varphi(u) = \prod_n (u-u_n) \qquad (4.4)$$

where the zeroes u_n have the asymptotics (4.3). Leaving the correct definition of the product (4.4) for the further investigation, let us now substitute quite formally (4.4) into (3.21a) and put $u = u_n$. The result is quite similar to the equation (4.2) for the ferromagnet

$$(-1)^N x = \prod_m \frac{u_n - u_m + i\eta}{u_n - u_m - i\eta} \qquad (4.5)$$

with the difference that the left-hand-side does not depend on u and the right-hand-side is now an infinite product.

The equations for the infinite chain ($N \to \infty$) are obtained now from (4.5) by a standard procedure [18]. Consider first the vacuum state. We assume that the zeroes of $\varphi(u)$ condensate as $N \to \infty$ and fill continuously the real axis except the interval $[-\mu, \mu]$ where $\mu > 0$ is a parameter to be determined subsequently. The basis for such an assumption is given by the classical limit (see Appendix C). Introducing the density of zeroes [18]

$$\rho_+(u) = \pi\eta \lim_{N\to\infty}\frac{1}{N}\left\{\frac{\text{number of } u_n \text{ in } du}{du}\right\} \qquad (4.6)$$

and using the equation (4.5) and the asymptotic conditions (1.16) and

(4.3) one obtains in the usual way [18] the integral equation and the asymptotic condition for $\rho_+(u)$

$$\begin{cases} \rho_+(u) - \dfrac{1}{\pi}\displaystyle\int_{|v|>\mu} \frac{\eta\,\rho_+(v)\,dv}{(u-v)^2+\eta^2} = 0 \,, \quad |u|>\mu & (4.7a) \\[2ex] \rho_+(u) = \ln|u| - \dfrac{1}{2}\ln \Xi + O(u^{-1}) \,, \quad u\to\infty & (4.7b) \end{cases}$$

Since the integral equation (4.7a) is homogenious, $\rho_+(u)$ is the eigenfunction (belonging to the continuous spectrum) of the integral operator determined by the asymptotic condition. To avoid difficulties connected with non compact integral operators it is convenient to reduce the problem to a Fredholm type integral equation [19].

To this end, let us define $\rho_+(u) = 0$ for $|u|<\mu$ and introduce the function $\rho_-(u)$ by the formulas

$$\rho_-(u) = \begin{cases} 0 \,, & |u| \geq \mu \\ -\dfrac{1}{\pi\eta}\left[\rho_+(u) - \dfrac{1}{\pi}\displaystyle\int_{|v|>\mu} \frac{\eta\,\rho_+(v)\,dv}{(u-v)^2+\eta^2}\right] , & |u|<\mu \end{cases} \tag{4.8}$$

The equation (4.7a) and the formula (4.8) can be rewritten then as the integral equation on the whole real axis

$$\pi\eta\,\rho_- + (1 - K_+)\,\rho_+ = 0 \tag{4.9}$$

where K_+ is the integral operator

$$(K_+\rho)(u) = \int_{-\infty}^{\infty} K_+(u-v)\,\rho(v)\,dv \tag{4.10}$$

with the kernel

$$K_+(u) = \frac{1}{\pi}\,\frac{\eta}{u^2+\eta^2} \tag{4.11}$$

Through the constant function is the eigenfunction of the integral operator K_+ with the eigenvalue 1 the inverse operator

$$1 - K_- = (1 - K_+)^{-1}$$

can be defined on the functions vanishing at infinity ($\rho_-(u)$ is by definition such a function). Due to the degeneracy of the operator $1-K_+$, the kernel $K_-(u)$ is defined up to an additive constant. Its explicit form can be easily found by means of the Fourier transform. It reads

$$K_-(u) = \frac{1}{\pi\eta}\left[\Psi\left(1+i\frac{u}{\eta}\right) + \Psi\left(1-i\frac{u}{\eta}\right)\right] + \Delta \qquad (4.12)$$

where $\Psi(u) = \Gamma'(u)/\Gamma(u)$ is the digamma function [20] . In this paragraph we take the constant Δ in (4.12) to be zero. The asymptotics of the kernel $K_-(u)$ then will be the following

$$K_-(u) = \frac{1}{\pi\eta}\ln\frac{|u|}{\eta} + O(u^{-1}) \quad , \quad u \to \infty \qquad (4.13)$$

We can transform now the equation (4.9) into the equivalent one

$$\pi\eta\,(1-K_-)\,\rho_- + \rho_+ = c \qquad (4.14)$$

where the constant c which appears in the right-hand-side of (4.14) due to the degeneracy of $(1-K_+)$ should be determined from the asymptotic condition (4.7b).

Using the asymptotics (4.7b) and (4.13) one obtains then from (4.14) that $c = \ln\eta - \frac{1}{2}\ln\mathcal{E}$ and $\int \rho_-(u)\,du = 1$.

Recalling then that $\rho_+(u) = 0$ for $|u| < \mu$ one obtains finally the system of equations for $\rho_-(u)$

$$\begin{cases} \pi\eta\left[\rho_-(u) - \int_{-\mu}^{\mu} K_-(u-v)\,\rho(v)\,dv\right] = \ln\eta - \frac{1}{2}\ln\mathcal{E} & (4.15a) \\ \int_{-\mu}^{\mu} \rho_-(v)\,dv = 1 & (4.15b) \end{cases}$$

which is equivalent to the system (4.7) for $\rho_+(u)$.

It is clear from the system (4.15) how to fix the parameter μ . The equation (4.15a) determines $\rho_-(u;\mathcal{E},\mu)$ and the normalization condition (4.15b) determines μ as a function of $\mathcal{E}$.

To finish the study of the vacuum state one needs to calculate the integrals of motion and, in particular, the energy density of the ground state.

To this end let us consider once more, equation (3.21a) and using (4.4) rewrite it as

$$t_N(u,x) = \Lambda_N\left(u+i\frac{\eta}{2}\right) + \Lambda_N^{-1}\left(u-i\frac{\eta}{2}\right) \qquad (4.16)$$

where

$$\Lambda_N(u) = x^{-1/2}\, i^N \prod_n \frac{u-u_n+i\eta/2}{u-u_n-i\eta/2} \qquad (4.17)$$

One can show that in the limit $N \to \infty$ the first term in the expression (4.16) dominates for $\mathrm{Im}\,u > 0$ and the second term for $\mathrm{Im}\,u < 0$.

Let us define the generating function for the vacuum densities of the integrals of motion as (cf. (1.19))

$$\tau(u) = \lim_{N\to\infty} \frac{1}{N} \ln t_N(u, \Xi^N) = \qquad (4.18)$$

$$= \ln u - P u^{-1} - \epsilon u^{-2} + O(u^{-3})$$

Using (4.18), (4.16) and (4.6) one obtains for the derivative of $\tau(u)$ the expression

$$\tau'(u) = \begin{cases} \xi\left(u+i\frac{\eta}{2}\right), & \mathrm{Im}\,u > 0 \\ \xi\left(u-i\frac{\eta}{2}\right), & \mathrm{Im}\,u < 0 \end{cases} \qquad (4.19)$$

where

$$\xi(u) = \frac{1}{i}\int_{|v|>u} dv \frac{\eta\,\rho_+(v)}{(u-v)^2+\eta^2/4} \qquad (4.20)$$

With the help of (4.8) one can now express $\tau'(u)$ in terms of $\rho_-(u)$:

$$\tau'(u) = \int_{-\mu}^{\mu} \frac{\rho_-(u)\,dv}{u-v} \qquad (4.21)$$

Expanding (4.21) in powers of u and using the expansion

$$\tau'(u) = u^{-1} + P u^{-2} + 2\epsilon u^{-3} + \dots \qquad (4.22)$$

which follows from (2.14) one obtains for the density of the momentum

$$\mathcal{P} = \int_{-\mu}^{\mu} v \rho_-(v)\, dv = 0$$

(due to the symmetry of the ground state) and

$$\epsilon = \int_{-\mu}^{\mu} \frac{v^2}{2} \rho_-(v)\, dv \tag{4.23}$$

Note that the first term u^{-1} in the expansion (4.22) conforms with the condition (4.15b).

The pressure $\mathcal{P}$ being given, the parameters $\bar{\Sigma}$ and $\mu(\bar{\Sigma})$ are defined now from the minimum enthalpy principle (see paragraph 1).

Let us discuss briefly the excitations. As usually [18] to obtain the excited states one needs to add a finite number of zeroes of the function φ(u) into the lacune $[-\mu, \mu]$ and to remove a finite number of zeroes from the vacuum distribution outside of $[-\mu, \mu]$. Let the additional zeroes be in the points $\{u_p\}$ (p for particle) and the missing ones in $\{u\}$ (h for holes). The comparison with the classical limit shows that the first type excitations (particles) correspond to the phonons and the second ones (holes) to solitons of the Toda chain.

Omitting the standard calculations let us present the final result. For the relative density $\sigma_+(u)$ defined by

$$\sigma_+(u) = \begin{cases} \pi\eta \sum_{\{u_p\}} \delta(u - u_p), & |u| < \mu, \ |u_p| < \mu \\ \pi\eta \lim_{N\to\infty} \left[\frac{1}{u_{n+1} - u_n} - \frac{1}{u_{n+1}^{vac} - u_n^{vac}}\right], & |u| > \mu \end{cases} \tag{4.24}$$

and its dual density defined by

$$\sigma_-(u) = \begin{cases} -\frac{1}{\pi\eta}\left[\sigma_+(u) - \frac{1}{\pi}\int_{|v|>\mu} \frac{\eta\, \sigma_+(v)\, dv}{(u-v)^2 + \eta^2}\right], & |u| < \mu \\ \sum_{\{u_n\}} \delta(u - u_n), & |u| > \mu, \ |u_n| > \mu \end{cases} \tag{4.25}$$

one obtains the integral equation with the boundary condition determined by the asymptotic shift Σ (see paragraph 1)

$$\begin{cases} \pi\eta\sigma_- + (1-K_+)\sigma_+ = 0 & (4.26a) \\ \sigma_+(u) = -\frac{1}{2}\ln\xi + O(u^{-1}),\ u \to \infty & (4.26b) \end{cases}$$

which, being restricted on $|u| > \mu$ gives a consistent equation for the smooth part of $\sigma_+(u)$.

The dual equation is

$$\begin{cases} \pi\eta(1-K_-)\sigma_- + \sigma_+ = -\frac{1}{2}\ln\xi & (4.27a) \\ \int_{-\infty}^{\infty}\sigma_-(v)\,dv = 0 & (4.27b) \end{cases}$$

Being restricted on $|u|<\mu$ it provides a Fredholm type equation for the smooth part of $\sigma_-(u)$. The normalization condition (4.27b) defines ξ as an implicit function of $\{u_p, u_n\}$.

One can show that the generating fucntion of the integrals of motion

$$\begin{aligned} \tau(u) &= \lim_{N\to\infty}[\ln t_N(u,\xi\Xi^N) - \ln t_N^{vac}(u,\xi\Xi^N)] \\ &= \frac{1}{2}\ln\xi - P u^{-1} - E u^{-2} + \dots \end{aligned} \tag{4.28}$$

is expressed in terms of $\sigma_-(u)$ like

$$\tau(u) = \frac{1}{2}\ln\xi + \int_{-\infty}^{\infty}\ln(u-v)\,\sigma_-(v)\,dv \tag{4.29}$$

and, therefore,

$$P = \int_{-\infty}^{\infty} v\,\sigma_-(v)\,dv, \tag{4.30a}$$

$$E = \int_{-\infty}^{\infty}\frac{v^2}{2}\sigma_-(v)\,dv \tag{4.30b}$$

Let us make now a comparison of our results and the Sutherland's ones [5]. The equations (4.15), (4.27) can be considered as the usual Bethe-ansatz equations for the one-dimensional Bose-gas corresponding to the two-particle S-matrix

$$S(u) = \frac{\Gamma(1-iu/\eta)}{\Gamma(1+iu/\eta)} \tag{4.31}$$

which corresponds to the exponential potential between the particles. The logarithmic derivative of the S-matrix (4.31) is the kernel K_- (4.12) and it is this kernel which has been proposed by Sutherland in [5]. A slight difference in the normalization with [5] seems to be due to the fact that we consider here the Toda chain under a constant pressure, so its volume determined by ξ may vary. Such a good correspondance with Sutherland's results is rather surprising since his approach is valid striclty speaking only for low densities [21] and, in fact, the exact equations (3.21) for the finite Toda chain can by no means be represented as Bethe-ansatz-like equations with the S-matrix (4.31).

5. CONCLUSION

To conclude, I would like to stress that the success in finding the spectrum of the Toda chain is due to the application of a new method consisting in constructing some special representations of the algebra (3.3) determined by the R-matrix. The method itself seems to be quite general and to be applicable to many other models. It has been applied already to the Goryachev-Chaplygin top [15] and the author hopes to apply it to the sine-Gordon model.

Concerning the Toda chain, there are also many problems to be solved such as a rigorous proof of all the assumptions made, a proper investigation of the integral equations, the construction of the thermodynamics, Green's functions, etc. The author hopes that progress in these areas will be made in the next few years.

APPENDIX A

Below the complete list of the matrix elements of the relation (2.7) is given. Using (2.6) this can be rewritten as

where according to (2.2)

$$T^1_1(u) = A_N(u) \; , \quad T^1_2(u) = B_N(u) \; , \quad T^2_1(u) = C_N(u) \; , \quad T^2_2(u) = D_N(u.$$

In the relations given below the values of the indices $(a_1 a_2 b_1 b_2)$ are given on the left margin. For the sake of brevity the index N is omitted.

(1111) : $$\{A(u), A(v)\} = 0 \qquad \text{(A1)}$$

(1112) : $$\{A(u), B(v)\} = -(u-v)^{-1}\left[A(u)B(v) - B(u)A(v)\right] \qquad \text{(A2)}$$

(1211) : $$\{A(u), C(v)\} = -(u-v)^{-1}\left[-A(u)C(v) + C(u)A(v)\right] \qquad \text{(A3)}$$

(1212) : $$\{A(u), D(v)\} = -(u-v)^{-1}\left[\quad (u)B(v) - B(u)C(v)\right] \qquad \text{(A4)}$$

(1121) : $$\{B(u), A(v)\} = -(u-v)^{-1}\left[B(u)A(v) - A(u)B(v)\right] \qquad \text{(A5)}$$

(1122) : $$\{B(u), B(v)\} = 0 \qquad \text{(A6)}$$

(1221) : $$\{B(u), C(v)\} = -(u-v)^{-1}\left[D(u)A(v) - A(u)D(v)\right] \qquad \text{(A7)}$$

(1222) : $$\{B(u), D(v)\} = -(u-v)^{-1}\left[-B(u)D(v) + D(u)B(v)\right] \qquad \text{(A8)}$$

(2111) : $$\{C(u), A(v)\} = -(u-v)^{-1}\left[-C(u)A(v) + A(u)C(v)\right] \qquad \text{(A9)}$$

(2112) : $$\{C(u), B(v)\} = -(u-v)^{-1}\left[A(u)D(v) - D(u)A(v)\right] \qquad \text{(A10)}$$

(2211) : $$\{C(u), C(v)\} = 0 \qquad \text{(A11)}$$

(2212) : $$\{C(u), D(v)\} = -(u-v)^{-1}\left[C(u)D(v) - D(u)C(v)\right] \qquad \text{(A12)}$$

(2121) : $$\{D(u), A(v)\} = -(u-v)^{-1}\left[B(u)C(v) - C(u)B(v)\right] \qquad \text{(A13)}$$

(2122) : $\{D(u), B(v)\} = -(u-v)^{-1}[-D(u)B(v) + B(u)C(v)]$ (A14)

(2221) : $\{D(u), C(v)\} = -(u-v)^{-1}[D(u)C(v) - C(u)D(v)]$ (A15)

(2222) : $\{D(u), D(v)\} = 0$ (A16)

APPENDIX B

Using (3.2) one can rewrite the relation (3.3) as

$$(u-v)\, T^{a_1}_{b_1}(u)\, T^{a_2}_{b_2}(v) - i\eta\, T^{a_2}_{b_1}(u)\, T^{a_1}_{b_2}(v) =$$

$$=(u-v)\, T^{a_2}_{b_2}(v)\, T^{a_1}_{b_1}(u) - i\eta\, T^{a_2}_{b_1}(v)\, T^{a_1}_{b_2}(u)$$

The complete list of the matrix elements of (3.3) is given below. As in the Appendix A the values of the indices $(a_1a_2b_1b_2)$ are given on the left margin and the index N is omitted.

(1111) : $(u-v-i\eta)A(u)A(v) = (u-v-i\eta)A(v)A(u)$ (B1)

(1112) : $(u-v-i\eta)A(u)B(v) = (u-v)B(v)A(u) - i\eta A(v)B(u)$ (B2)

(1211) : $(u-v)A(u)C(v) - i\eta C(u)A(v) = (u-v-i\eta)C(v)A(u)$ (B3)

(1212) : $(u-v)A(u)D(v) - i\eta C(u)B(v) = (u-v)D(v)A(u) - i\eta C(v)B(u)$ (B4)

(1121) : $(u-v-i\eta)B(u)A(v) = (u-v)A(v)B(u) - i\eta B(v)A(u)$ (B5)

(1122) : $B(u)B(v) = B(v)B(u)$ (B6)

(1221) : $(u-v)B(u)C(v) - i\eta D(u)A(v) = (u-v)C(v)B(u) - i\eta D(v)A(u)$ (B7)

(1222) : $(u-v)B(u)D(v) - i\eta D(u)B(v) = (u-v-i\eta)D(v)B(u)$ (B8)

(2111) : $(u-v)C(u)A(v) - i\eta A(u)C(v) = (u-v-i\eta)A(v)C(u)$ (B9)

(2112) : $(u-v)C(u)B(v) - i\eta A(u)D(v) = (u-v)B(v)C(u) - i\eta A(v)D(u)$ (B10)

(2211) : $C(u)C(v) = C(v)C(u)$ (B11)

$$(2212) : (u-v-i\eta)\, C(u)\, D(v) = (u-v)\, D(v)\, C(u) - i\eta\, C(v)\, D(u) \tag{B12}$$

$$(2121) : (u-v)\, D(u)\, A(v) - i\eta\, B(u)\, C(v) = (u-v)\, A(v)\, D(u) - i\eta\, B(v)\, C(u) \tag{B13}$$

$$(2122) : (u-v)\, D(u)\, B(v) - i\eta\, B(u)\, D(v) = (u-v-i\eta)\, B(v)\, D(u) \tag{B14}$$

$$(2221) : (u-v-i\eta)\, D(u)\, C(v) = (u-v)\, C(v)\, D(u) - i\eta\, D(v)\, C(u) \tag{B15}$$

$$(2222) : (u-v-i\eta)\, D(u)\, D(v) = (u-v-i\eta)\, D(v)\, D(u) \tag{B16}$$

In the main text of the paper some formulas are used connected with notion of the quantum determinant [8]. Using the explicit expression (2.1) for the L-operator one can verify that the following relation holds

$$\sigma_2\, L(u)\, \sigma_2\, L(u-i\eta) = 1 \tag{B17}$$

as well as the relation

$$L(u)\, \sigma_2\, L(u+i\eta)\, \sigma_2 = 1 \tag{B18}$$

where $\sigma_2 = \begin{pmatrix} 0 & -i \\ i & 0 \end{pmatrix}$

Below the matrix elements of (B17) are written down

$$D(u)\, A(u-i\eta) - B(u)\, C(u-i\eta) = 1 \tag{B19}$$

$$D(u)\, B(i-i\eta) - B(u)\, D(u-i\eta) = 0 \tag{B20}$$

$$A(u)\, C(u-i\eta) - C(u)\, A(u-i\eta) = 0 \tag{B21}$$

$$A(u)\, D(u-i\eta) - C(u)\, B(u-i\eta = 1 \tag{B22}$$

and of (B18)

$$A(u)\, D(u+i\eta) - B(u)\, C(u+i\eta) = 1 \tag{B23}$$

$$B(u)\, A(u+i\eta) - A(u)\, B(u+i\eta) = 0 \tag{B24}$$

$$C(u)\, D(u+i\eta) - D(u)\, C(u+i\eta) = 0 \tag{B25}$$

$$D(u)\,A(u+i\eta) - C(u)\,B(u+i\eta) = 1 \qquad \text{(B26)}$$

APPENDIX C

Here we consider the classical limits of the equations obtained in the paragraph 4.

Let us consider the vacuum equations. The kernel $K_+(u)$ has the following expansion as $\eta \to 0$

$$K_+(u) = \delta_+(u) + \frac{\eta}{\pi}\,\frac{1}{u^2} + O(\eta^2) \qquad \text{(C1)}$$

The right-hand-side of (C1) should be understood in the sense of distributions, $\frac{1}{u^2}$ being the standard renormalization of u^{-2} [22]

$$\int_{-\infty}^{\infty} \frac{f(u)}{u^2}\,du = \int_0^{\infty} u^{-2}\,[f(u) + f(-u) - 2f(0)]\,du$$

Due to (C1) the vacuum equation (4.7) becomes in the classical limit (cf. the analogous treatment of the non linear Schrödinger equation in 23)

$$\begin{cases} -\dfrac{1}{\pi^2} \displaystyle\int_{|v|>\mu} \dfrac{\rho_+(v)\,dv}{(u-v)^2} = 0\,, \quad |u| > \mu & \text{(C2a)} \\[2ex] \rho_+(u) = \ln|u| - \dfrac{1}{2}\ln \Xi + O(u^{-1}),\ u \to \infty & \text{(C2b)} \end{cases}$$

The constant μ being fixed it has the unique solution

$$\Xi = \frac{\mu^2}{4} \qquad \text{(C3a)}$$

$$\rho_+(u) = \ln \frac{u + \sqrt{u^2 - \mu^2}}{\mu} \qquad \text{(C3b)}$$

In order to perform the limit $\eta \to 0$ in the equation (4.15) it is convenient to use another normalization of the kernel K_- . Let us choose the constant Δ in (4.12) to be $\ln\eta/\pi\eta$. Using the new kernel $\tilde{K}_-$ the equation (4.15) can be rewritten as

$$\begin{cases} \pi\eta\left[\rho_-(u) - \int_{-\mu}^{\mu} \tilde{K}(u-v)\rho_-(v)\,dv\right] = -\frac{1}{2}\ln\Xi & \text{(C4a)} \\ \int_{-\mu}^{\mu} \rho_-(v)\,dv = 1 & \text{(C4b)} \end{cases}$$

Using the asymptotics of the kernel $\tilde{K}_-$

$$\tilde{K}_-(u) = \frac{1}{\pi\eta}\ln|u| + O(\eta) \quad , \quad \eta \to 0$$

one obtains in the classical limit

$$\begin{cases} -\int_{-\mu}^{\mu} \ln|u-v|\,\rho_-(v)\,dv = -\frac{1}{2}\ln\Xi \quad , \quad |u| < \mu & \text{(C5a)} \\ \int_{-\mu}^{\mu} \rho_-(v)\,dv = 1 & \text{(C5b)} \end{cases}$$

The unique solution to the equations (C5) is

$$\Xi = \frac{\mu^2}{4} \tag{C6a}$$

$$\rho_-(u) = \frac{1}{\pi}(\mu^2 - u^2)^{-1/2} \tag{C6b}$$

The expressions (4.21) for $\tau'(u)$ and (4.23) for the energy do not change in the classical limit. One obtains

$$\tau(u) = \ln\frac{u + \sqrt{u^2 - \mu^2}}{\mu}$$

$$\varepsilon = \frac{\mu^2}{4} = \Xi$$

which coincides with the results of the paragraph 1.

The classical limit in the equations (4.26-27) for the excited states can be performed in the same manner and leads to the well known classical results (cf.[5]).

REFERENCES

1 M. Toda - J. Phys. Soc. Ipn. 23, 501 (1967).
M. Toda : Theory of Nonlinear Lattices, Springer (1981).

2 S.V. Manakov, Zh. Eksp. Teor. Fiz. 67, 543 (1974) Sov. Phys_ JETP, 67, 269 (1974)
H. Flaschka, Phys. Rev. B9, 1924 (1974).
H. Flaschka, Progr. Theor. Phys. 51, 703 (1974).

3 H. Flaschka, D.W. McLaughlin, Progr. Theor. Phys. 55, 438 (1976).
M. Kac, P. van Moerbeke, Proc. Nat. Acad. Sci. USA, 72, 1627 (1975) ; ibid. 72, 2879 (1975).

4 M.A. Olshanetsky, A.M. Perelomov, Lett. Math. Phys. 2, 7 (1977).

5 B. Sutherland, Rocky Mountain J. Math., 8, 413 (1978).

6 M.A. Olshanetsky, A.M. Perelomov, Phys. Rep. 71, 313 (1981) ; ibid. 94, 313 (1983).

7 M.C. Gutzwiller, Ann. Phys., 133, 304 (1981).

8 L.D. Faddeev, Les Houches Lectures 1982, Elsevier Sci. Publ. (1984).
P.P. Kulish, E.K. Sklyanin, in : Lecture Notes in Physics, 151, 61, Springer (1982).

9 L.D. Faddeev, Soviet Sci. Rev., Sect. C : Math. Phys. Rev. 1, 107, Harwood Academic, Chur (1980).
V.E. Korepin, Zap. Nauch. Semin. LOMI, 101, 90 (1980) J. Sov. Math. 23, 2429 (1983)
M. Gaudin : La fonction de l'onde de Bethe pour les modèles exacts de la mécanique statistique, ch. 14, Sur la chaîne de Toda, Coll. du Comm. à l'en. at Série sci. (1983).

10 A.G. Izergin, V.E. Korepin, Comm. Math. Phys., 94, 67 (1984).
V.E. Korepin, Comm. Math. Phys., 94, 93 (1984).
P.P. Kulish, F.A. Smirnov, Phys. Lett., 90A, 74 (1982).

11 L.D. Faddeev, L.A. Takhtajan, The Hamiltonian Approach to the Soliton Theory, Springer, to be published.

12 I.V. Komarov, private communication (1983).

13 A.G. Izergin, V.E. Korepin, Lett. Math. Phys. 8, 259 (1984).

14 E.K. Sklyanin, Funct. Anal. Appl. 16, 263 (1982) ; ibid. 17, 273 (1983).

15 E.K. Sklyanin, Zapiski nauch semin. LOMI, 133, 236 (1984), to be translated in J. Sov. Math. .

16 R.J. Baxter, Ann. Phys. 70, 193 (1972).

17 N.Yu. Reshetikhin, Lett. Math. Phys. 7, 205 (1983).

18 C.N. Yang, C.P. Yang, J. Math. Phys. 10, 1115 (1969).

19 P.M. Morse, H. Feshbach, Methods of Theoretical Physics, v.1, ch. 8, McGraw-Hill (1953).

20 A. Erdélyi, Higer Transcendental Functions, v.1, McGraw-Hill (1953).

21 N. Theodorakopoulos, Phys. Rev. Lett. 53, 871 (1984).

22 I.M. Gelfand, G.E. Shilov : Generalized functions, vol.1, Properties and operations, Academic Press, NY (1964).

23 N.Yu. Reshetikhin, F.A. Smirnov, Zap. Nauch. Semin. LOMI, 131, 128 (1983) (to be translated in J. Sov. Math.).

ON SOLUBLE CASES OF STAGGERED ICE-RULE ON A SQUARE LATTICE

T.T. Truong
Institut für Theoretische Physik
Freie Universität Berlin
Arnimallee 14, D-1000 Berlin 33
W. Germany

Abstract. We discuss a solubility condition for staggered vertex systems with ice-rule and its impacts on the Ashkin-Teller, Potts and layered Potts models.

§1. Introduction

Contemporary equilibrium statistical mechanics deals with a wide variety of problems pertaining to critical behavior of physical systems and their phase structure and numerous are the methods available to achieve these goals. However, simple models made to describe real systems appear always attractive for they provide, if successful, a clear cut understanding of the phenomenon. The Ising model /1/ and its solution is certainly the most outstanding example of effort in this direction. Generalisations of the Ising model have been subsequently proposed by Ashkin-Teller /2/ and R.B. Potts /3/ with models carrying their respective names, but these models have failed to be soluble up to now. But in the meantime progress in modeling ferroelectric systems in two dimensions had been successful and their complete solution was obtained by E.H. Lieb /4/ and later generalised by R.J. Baxter /5/. It is therefore not surprising to witness efforts by many workers to convert the Ashkin-Teller and Potts models to vertex models.But one obtains their staggered versions /19/ whereby two systems of vertices coexist on interpenetrating sublattices of the main lattice. Unfortunately, the problem of staggered vertex models still represents a considerable challenge to the theoretician for it has eluded complete resolution up to now.

In what follows we propose to reach a partial solution to the problem . The earliest class of solutions for staggered ice-rule is of Pfaffian type and was discovered by F.Y. Wu and collaborators /6/ who also extended their findings to other lattices /7/ and considered the staggering of several lattices /8/. Recently reconsidering the most general ferroelectric model on a square lattice he solved in 1971 /9/ Baxter discovered that a special staggered system is hidden in the class of soluble inhomogeneous systems, which he identified to the critical antiferromagnetic q-state Potts model /10/. This discovery is an important landmark for it sweeps away the belief that staggered vertex models do not seem to admit other solution than Pfafians and consequently raises the question when does a staggered vertex system is integrable by a Bethe-ansatz type of solution.

To this end we shall elaborate on some ideas of R.Z. Bariev regarding the treatment of vertex systems /11/. More precisely, restricting ourselves first to the simple homogeneous zero-field six-vertex model we shall show that Bariev's diagonal-to-diagonal transfer matrix admits eigenfunctions of the Bethe-ansatz type and the partition function has the usual product form. Now staggered vertex systems appear in this "diagonal representation" of the lattice as described by the product of two-diagonal-to-diagonal transfer matrices. Hence if these matrices, under certain restrictions, admit the same set of Bethe-ansatz wave functions as eigenstates we would have obtained thus soluble cases of staggered vertex systems. These solubility conditions yield in turn solubility lines for the Ashkin-Teller and Potts models. Conversely these spin models do have themselves special soluble cases, which will reappear on the staggered vertex systems as new sets of soluble points. Some of the solubility lines but not all are in fact critical lines. Finally we sketch some gneralisations of this construction for other systems.

§2. Methodology for the ice-rule on a square lattice

2.1 Conventions. We consider a square lattice of sites labelled by (I,J) with periodic boundary conditions, i.e. site (I+N,J+M) is identified to site (I,J) where M and N are the numbers of sites on the horizontal and vertical directions. Pairs of sites are connected by vertical and inclined edges as shown in fig. 1. The four edges round a site form a vertex, each edge carries an "arrow" variable. Let us agree that a "down-arrow" by the value -1 of σ (resp. τ). The so-called "ice-rule" which originates from the relative position of a hydrogen atom on an edge in a planar ice molecule model is then representated by the constraint: $\sigma + \tau = \sigma' + \tau'$. See figure 2. Let C be a configuration of edge variables (σ,τ) obeying the ice-rule on the

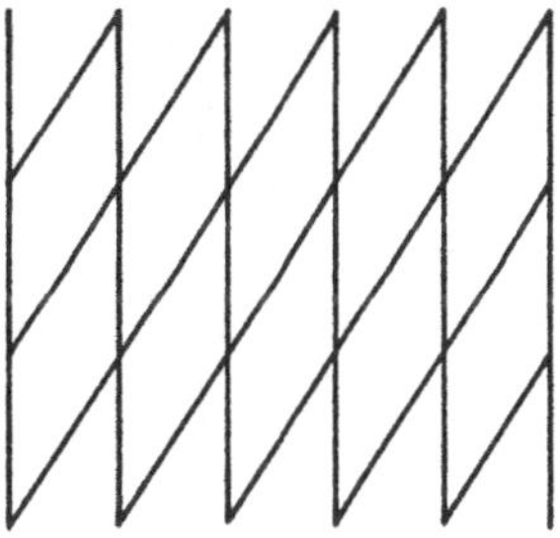

Figure 1

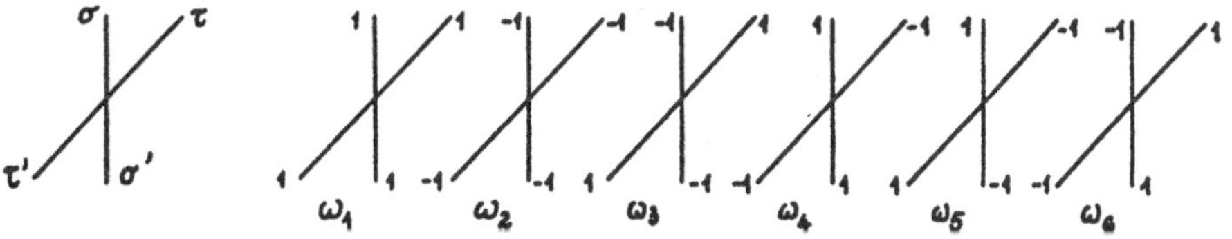

Figure 2: Ice-rule $\sigma+\tau=\sigma'+\tau'$, the 6 vertex configurations and their weights.

whole lattice with $\sigma'_{I,J} = \sigma_{I,J+1}$, $\tau'_{IJ} = \tau_{I-1J-1}$. Then the partition function is the sum over all such configurations of their statistical weights:

$$Z_{MN} = \sum_{\{C\}} \prod_{I,J} \omega_{C(I,J)}$$

2.2. Diagonal-to-diagonal transfer matrix

We take a slice of the lattice of figure 1 made up of a row of vertices as shown in figure 3, and define a $4^N \times 4^N$ matrix by its matrix element:

$$T_{N\,\cdots\sigma'_j\,\tau'_j\cdots}^{\;\cdots\sigma_j\,\tau_j\cdots} = \prod_{J=1}^{N} \omega_{r_J}$$

where $\sigma_J + \tau_J = \tau'_J + \sigma'_J (J=1,\ldots N)$ and ω_{r_J} is the weight of the vertex J with the configuration σ_J, τ_J, σ'_J and τ'_J. Then we have

$$Z_{MN} = Tr\,(T_N)^M$$

as in the usual row-to-row approach /12/. Permuting in and out labels of T_N, namely $(\sigma_J, \tau_J) \leftrightarrow (\sigma'_J, \tau'_J)$ we obtain the transpose of T_N which is represented in figure 4.

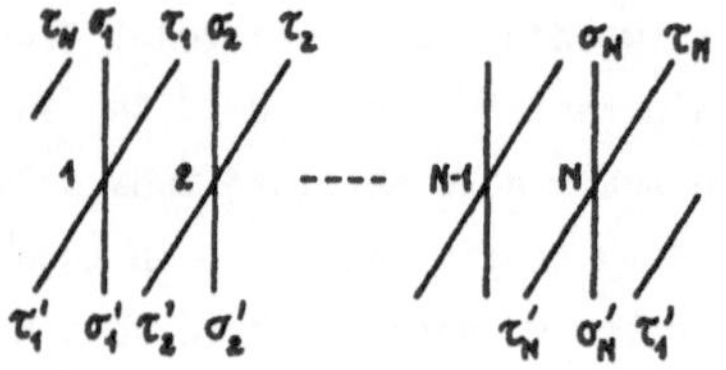

Figure 3. The matrix T_N

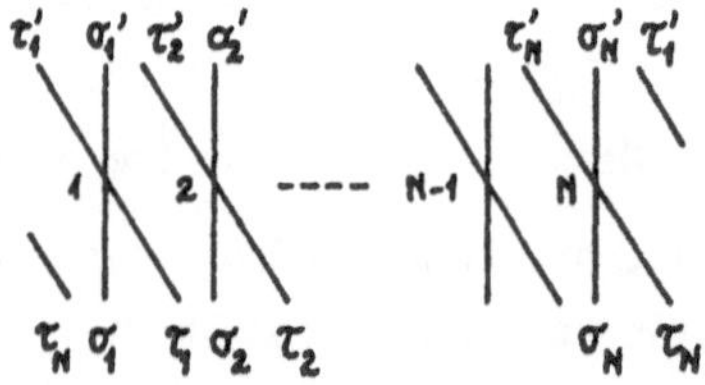

Figure 4. Transpose of T_N:T_N

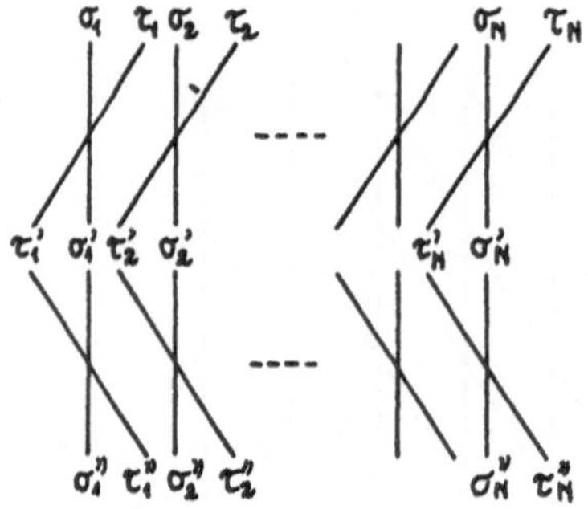

Figure 5 shows the simple structure of the product $T'_N T_N$, and how the inverse of T_N can be computed from a local product of vertex operators. It is thus straightforward to find the inverse relations of the partition function /13/.

Figure 5. The product $T'_N T_N$.

§3. Triangle relations and eigenvectors of transfer matrices

3.1. Parametrization. As noted earlier by E.Lieb, the number of down-arrows (or negative spin on an edge; i.e. either $\sigma = -1$ or $\tau = -1$(remains constant in a row. One treats them as "particles" and constructs the eigenstates of T_N as Bethe-ansatz wave functions. This has been done by R.Z. Bariev, as well as by E.H. Lieb earlier in the row-to-row formulation. We shall by- pass this conventional method and refer the reader to references /11/ and /12/. Instead we would like to recall that the integrability of T_N (or its row-to-row counterpart) hinges on a canonical parametrization discovered by R.J. Baxter in 1970 /5/ according to which the zero-field 6-vertex model weights are expressed by trigonometric/hyperbolic functions:

$$\omega_1 = \omega_2 = a = \sin(\mu-\theta)$$
(3.1.1) $$\omega_3 = \omega_4 = b = \sin\theta$$
$$\omega_5 = e^{i\theta}c \ , \quad \omega_6 = e^{-i\theta}c \ , \quad c = \sin\mu$$

Eliminating θ, we obtain the Lieb invariant $\Delta = (a^2 + b^2 - c^2)(2ab)^{-1} = -\cos\mu$. The presence of unimodular factory in ω_5 and ω_6 is immaterial to the partition function since it depends only on the product $\omega_5\omega_6 = c^2$. They do, however, add an extra symmetry to the problem in the sense that θ, the "spectral" parameter, as called by L.D. Faddeev and his school /14/ can be now defined modulo π. On the other hand θ labels commuting row-to-row transfer matrices as it is well known /5/.

This canonical parametrization realizes, in a sense, a uniformization of the following algebraic problem. Consider the simplest non-trivial lattice made up of three vertices, intersections of three straight lines. When we shift any line parallel to itself we obtain a new configuration geometrically inequivalent as shown in figure 6.

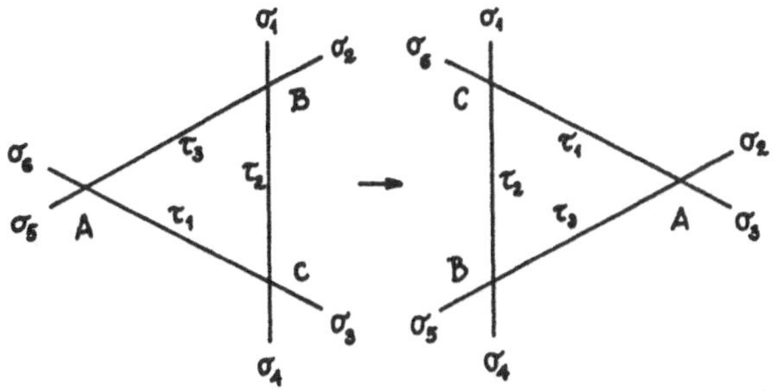

Figure 6: Horizontal shift of vertical line and spin configuration on edges.

Given any spin-configuration on the outer edges of the two diagrams, i.e. $\sigma_1,\dots,\sigma_6$, if the total statistical weight of the left diagram were to be equal to the total statistical weight of the right diagram whenever the inner spins τ_1,τ_2,τ_3 are chosen according to the ice-rule, then the vertices A,B,C would have the same Lieb invariant and their spectral parameters fulfill the relation:

(3.1.2) $$\Delta_A = \Delta_B = \Delta_C = \Delta \qquad\qquad \theta_C = \theta_A + \theta_B \pmod{\pi}$$

Baxter has proposed to identify effectively θ_A, θ_B, θ_C to the geometric straight line angles of the respective vertices with an appropriate North pole convention at vertices A, B and C. Equation (3.1.2) simply expresses that θ_C is the outer angle of an euclidean triangle is the sum of the two opposite inner angles. The angular relation has appeared earlier in the theory of two-dimensional factorizable, elastic S-matrices /15/:

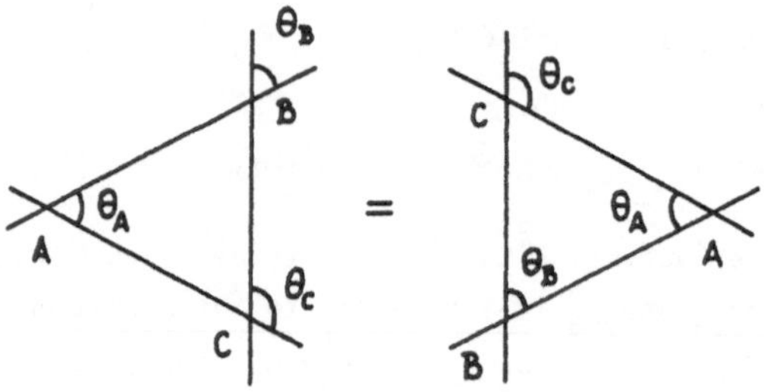

Figure 7: Angular relation

The immediate consequence is the following. Given a set of N lines intersecting mutually only pairwise, if one chooses on each edge a spin variable so that round a vertex the ice-rule is fulfilled, provided that each vertex has the same Lieb invariant and its angle θ is chosen to agree with the previous "triangle rule", then one is free to shift any line parallel to itself without changing the total statistical weight of the system, once the set of outer spins is fixed. /16/. We next show that this property will basically generate more or less graphically Bethe ansatz eigenfunctions of T_N.

3.2. First application of the triangle relations

As pointed out earlier the object on which one shauld focus our attention is a "down-arrow" or "particle" described by a spin: -1 on an edge. It traces a con-tintuous path through the lattice. The simplest extension of the triangle diagram of figure 6 is the addition of parallel vertical lines (see figure 8). We should understand that each side of figure 8 represents a sum over inner spins τ and τ' for any given outer spin configurations: $(\sigma,\sigma';\rho,\rho';\lambda,\lambda')$:

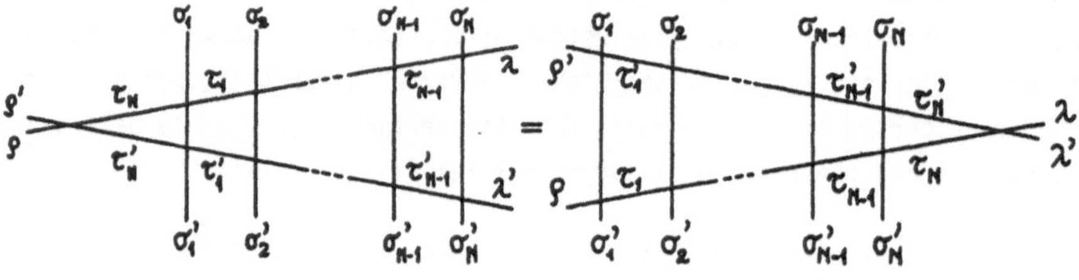

Figure 8:

Let us specialize to the following outer spin configuration of figure 9 where $\lambda=\lambda'=\sigma'_I=\sigma'_J = -1$ and every other spin equal to +1.

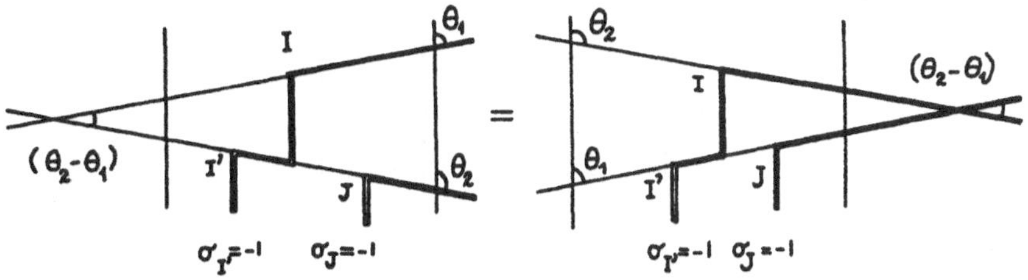

Figure 9: Paths of down-arrows are in heavy lines.

On each "almost" horizontal line, a "particle" comes from far right and appears on some vertical edge, say at vertex I. This is, as far as vertical edges are concerned, a creation of a particle on the $I^{\underline{th}}$ vertical edge. The statistical weight of such process is the product of the weights along this line from the first to the last vertex, it is equal to $b^N(\theta_1)c\,[a(\theta_1)/b(\theta_1)]^I$ where θ_1 is the angular parameter of the "almost" horizontal line. Thus the probability amplitude of finding a down-arrow at I is basically a plane wave with wave vector k_1 defined by $e^{ik_1} = a(\theta_1)/b(\theta_1)$.

What figure 9 tells us is two-fold. First the creation of two particles is commutative, since we can pull freely one line through the other. Second by computing the total statistical weight of two-down arrows appearing on the lowest row of vertical edges one finds that the probability amplitude for the existence of particles is precisely of Bethe-ansatz type (recall that one should resum and rearrange the terms so that particles appear in the sector I' < J since we have indistinguishable objects here.):

$$c^2\exp(ik_1I+ik_2J)\left\{\frac{b_2}{a_2}+\frac{a_1}{a_2}\frac{c^2}{a_2b_1-a_1b_2}\right\}+c^2\exp(ik_1J+ik_2I)\left\{\frac{a_2}{b_2}+\frac{b_2}{b_1}\frac{c^2}{a_1b_2-a_2b_1}\right\}$$

where $a(\theta_j) = a_j$ and $b(\theta_j) = b_j$, also $\exp ik_j = a_j/b_j$ as before. It is now straight forward to imagine the possibility of shifting mutally a set of "almost" horizontal lines defined by angles $\theta_1,\ldots,\theta_n$. Special configurations such that far-right spins $\lambda_1 = \lambda_2 = \ldots \lambda_n = -1$ reemerging on the lowest row (spin σ'_j) have statistical weights which yield in fact a Bethe-ansatz wave function for n-particle system on the lowest row of vertical edges.

This "graphical" construction of Bethe ansatz wave functions is extendable to the case of non parallel "vertical" lines. Each "almost" horizontal line is now labelled by a set of angles $\theta_j\theta'_j\ldots\theta_j^{(1)}$ subjected to the conditions

$$\theta_i-\theta_j = \theta'_i-\theta'_j = \ldots = \theta_i^{(1)}-\theta_j^{(1)} \qquad (\text{mod.}\,\pi)$$

This result has been obtained by Baxter /5/ long ago.

3.3. Second application of the triangle equations

We first deform slightly the triangle in figure 6 and add this time N vertices horizontally. The result is pictured in figure 10.

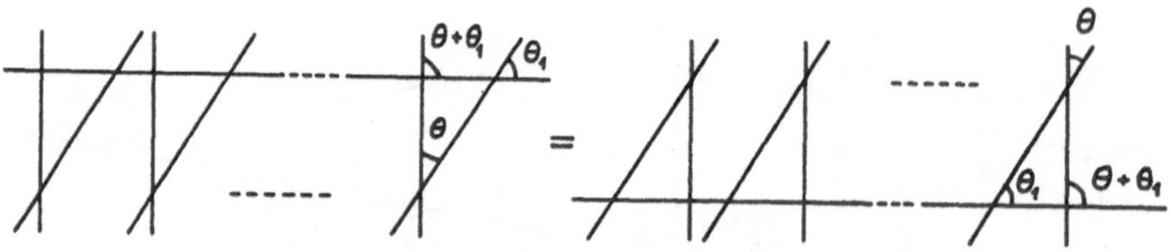

Figure 10:

Comparing with figure 3 we see that this is the vertical shift of a horizontal line of angular parameter θ_1 and $\theta + \theta_1$ through the diagonal-to-diagonal matrix $T_N = T_N(\theta,\mu)$. The horizontal line may be thought of a device to generated on this upper row of edges a down-arrow which will eventually propagate downward by performing a hopping to the left or staying on the same spot. Again by computing the statistical weight along the horizontal line one finds that the momentum of the plane wave of the down arrow is given by

$$\exp ik(\theta_1) = \frac{a(\theta+\theta_1)\,a(\theta_1)}{b(\theta+\theta_1)\,b(\theta_1)}$$

And by computing the statistical weight on both sides of figure 10, one sees that the plane wave exp ik_1J is an eigenstate of $T_N(\theta,\mu)$ with the eigenvalue $\Lambda_1 = a(\theta_1)/b(\theta_1)$. For more details see reference /17/.

A two-particle system can be generated by two "almost" horizontal lines. It is clear that, as explained in the previous paragraph, a proper choice of the angular parameters θ along the horizontal lines will ensure that they commute with each other, and yield the Bethe-ansatz wave function for two particles, see figure 11.

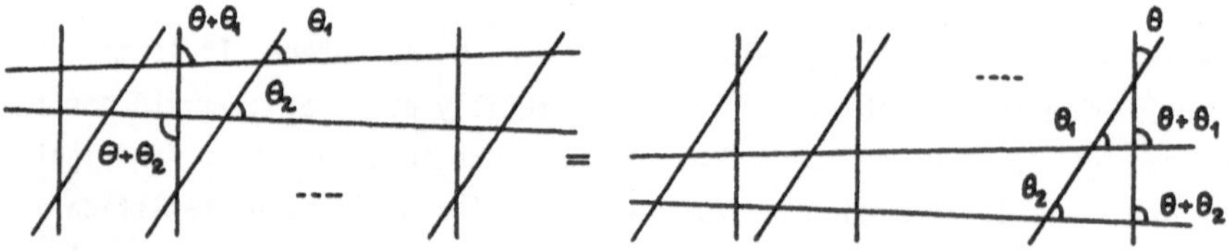

Figure 11

This construction may be extended to any number of horizontal lines. Thus we can directly, by calculating a global statistical weight, generate the Bethe-ansatz wave function of a system of n-down arrows susceptible of being an eigenstate of T_N. The corresponding eigenvalue is just equal to the product of single particle eigenvalues of equation (3.3.1). Periodic boundary conditions on the particle space variables determine the choice of the θ_j which is reduced to the calculation of a density of θ in the thermodynamic limit through an integral equation as it is usually the case.

We have sought here to present the highlights and main features of the method of R.Z. Bariev without tangling ourselves in the technical details which are available at length in references /11/ /17/ /18/. The crux of the matter resides basically in the triangle relations which once appropriately extended provide the correct construction of the Bethe-ansatz wave functions of the diagonal-to-diagonal transfer matrix, its eigenvalues and through the resolution of an integral equation governing the distribution of the eigenvalues in the thermodynamic limit the ultimate computation of the partition function. The method is attractive due to its simplicity which is best appreciated from the graphical interpretation and seems thereby to have an edge over the usual row-to-row approach.

§4. Staggered ice-rule on square lattice.

4.1. Simple two-fold staggering.

Two square interpenetrating sublattices A and B (of continuous and dotted lines) from our lattice as shown by figure 12, with the periodic boundary conditions described in paragraph 2.1. Geometrically there exist four types of vertices; two resulting from intersections of lines of the same type and two from intersections of lines of different types. We shall assume complete symmetry between the two types of lines and shall consider vertices which have the same Lieb invariant or equivalently the same μ. Hence we have the following distribution of spectral parameters in the vertices, see figure 13

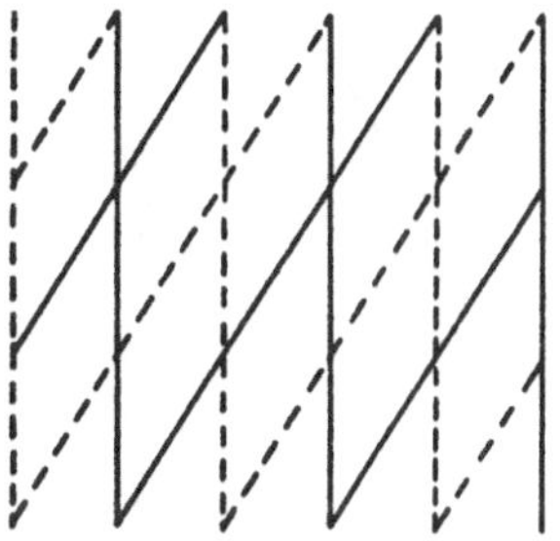

Figure 12.

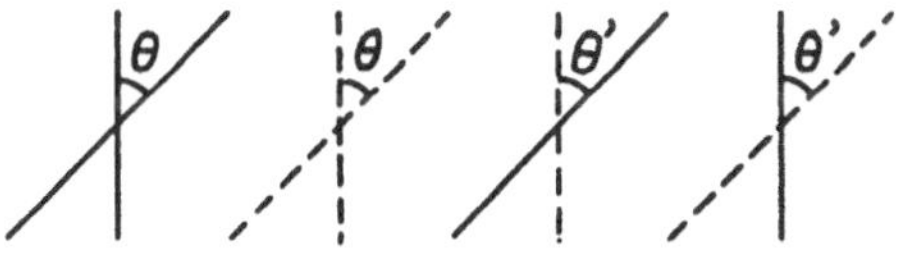

Figure 13

The partition function can be calculated from a transfer matrix which is the product of two diagonal-to-diagonal transfer matrices $T_N(\theta)$ and $T_N(\theta')$ of figure 14. Note that the partition function does not depend on the order with which the product of the matrices is carried out. This is an intrinsic symmetry of staggered vertex systems and is in fact a duality symmetry for an equivalent spin

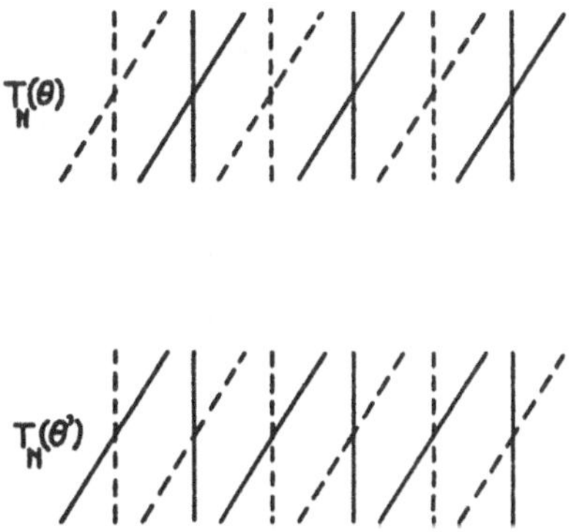

Figure 14.

model: Ashkin-teller or Potts model.

It is worthwhile mentioning that the inverse relation for the partition function of a staggered vertex system is quite transparent here, the so-called automorphy factor is just the product of automorphy factors of the separate systems (θ,μ) and (θ',μ) /13/.

4.2. Integrability condition

In section 3 we have shown that so long as we can shift vertically a line through a line of vertex defining a matrix $T_N(\theta)$. We would be able to construct Bethe-ansatz states that are eigenvectors of $T_N(\theta)$. The problem is now to determine under which condition(s) would these Bethe-ansatz states be also eigenvectors of $T_N(\theta')$. Put it differently, when could we shift a horizontal line (either continuous line or dotted line) through the second row of vertices defining $T_N(\theta')$?

In order to be able to go through the first row we should choose the following angular parameters for the horizontal line, see figure 15.

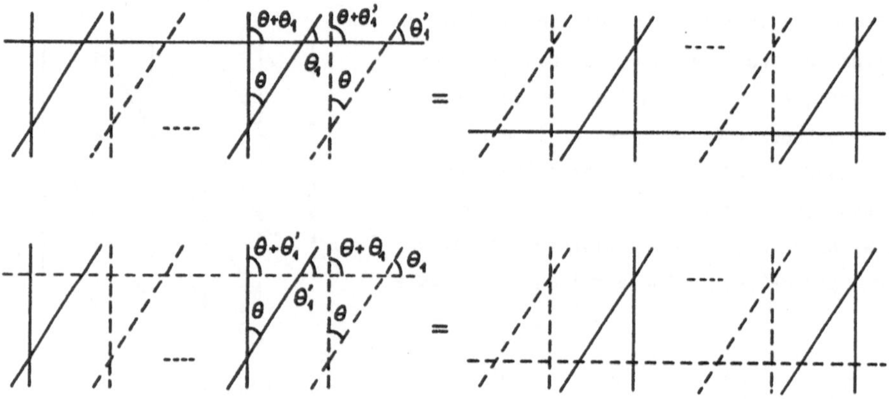

Figure 15: Shift of a line (continuous or dotted) through vertices (θ,μ).

Now we require that the lines cross the second row of vertices (figure 16).

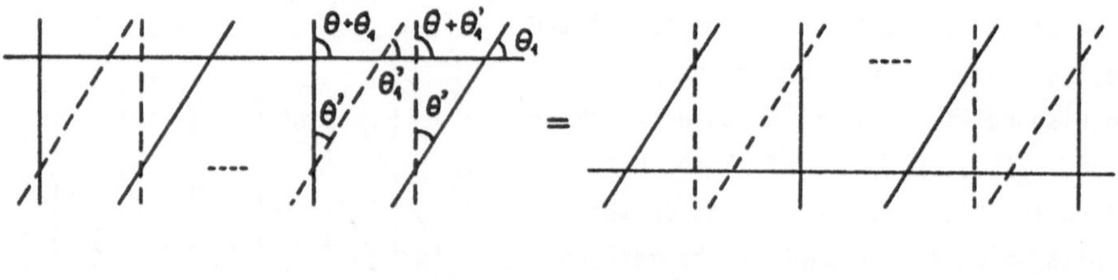

Figure 16: Shift of a line (continuous or dotted) through vertices (θ',μ).

The angles in the two triangles of figure 16 must fulfill then the relations

(4.2.1)
$$\theta + \theta_1 = \theta' + \theta_1' \qquad (\mathrm{mod.}\,\pi)$$
$$\theta + \theta_1' = \theta' + \theta_1 \qquad (\mathrm{mod.}\,\pi)$$

θ_1 and θ_1' are parameters of horizontal lines, they are eliminated from equations (4.2.1) and yield a relation between θ and θ, namely:

(4.2.2)
$$\theta' = \theta \pm \ell\frac{\pi}{2} \qquad (\mathrm{mod.}\,\pi)$$

The unconvinced reader can use the weights of equations (3.1.1) to see that the triangle equations can only be verified if we have (4.2.1). If we choose $\ell = 0$ we reobtained the non staggered vertex system. But for $|\ell| = 1$ we recover the case found by Baxter /10/. Consequently we are led to choose also for the horizontale line $\theta_1' = \theta_1 \pm \ell^{\pi}/2$ (modπ). These results would be best understood for an "isotropic" model where $\theta' = \mu - \theta$. This corresponds to the exchange of weights a and b when passing from one sublattice labelled by (θ,μ) to the other labelled by (θ',μ). In figure 17, we have plotted the solubility lines on a diagram (a/c, b/c)

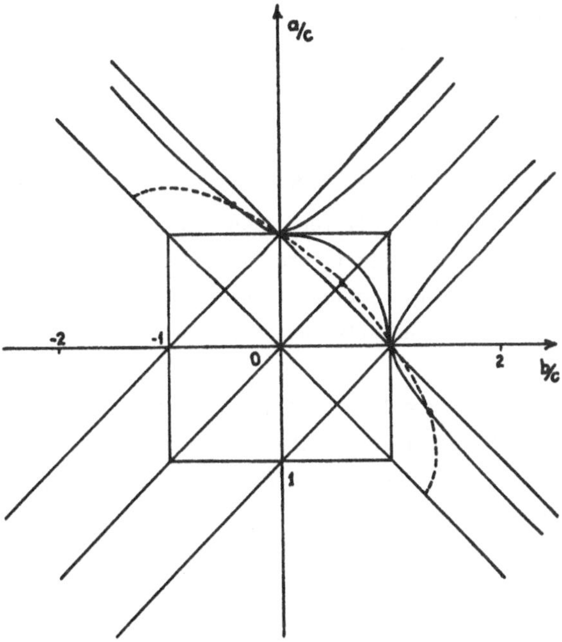

Figure 17.

The case $\ell = 0$ is represented by the line a = b whereas $|\ell| = 1$ implies the curve

(4.2.3)
$$(a^2 - b^2)^2 = c^2(a^2 + b^2)$$

This curve is symmetric with respect to the exchange a $\leftrightarrow$ b even in a, and b and of fourth order. The locus of Δ = constant are second order curves.

(4.2.4) $$a^2 + b^2 - c^2 = 2\Delta ab$$

they pass all through two fixed points (a/c = 1, b/c = 0) and (a/c = 0, b/c = 1). Degenerate cases are $|\Delta| = 1$ represented by a/c = $\mp$ b/c $\pm$ 1. Finally the free fermion conditions $\Delta = 0$ implies the unit circle.

4.3. Application to the q-state Potts model

The mapping of the q-state Potts model to a staggered six -vertex model has been done by H.N.V. Temperley and E.H. Lieb as well as by R.J. Baxter, S.B. Kelland and F.Y. Wu /19/. Let us first consider an isotropic Potts model on a square lattice with interaction parameter K between Potts-spins. The connection to the 6-vertex weights a,b,c is given by the relations

(4.3.5) $$b/a = \frac{e^K - 1}{\sqrt{q}} \qquad \Delta = -\frac{\sqrt{q}}{2}$$

The first soluble line a = b corresponds to the self-dual line:

$$e^K = 1 + \sqrt{q}$$

whereas the $|\ell| = 1$ line has now the equation

$$e^K = -1 + \sqrt{4-q}$$

found by Baxter /10/ and physical only for q < 4. It is the critical curve for the antiferromagnetic Potts model. In figure 17 an antiferromagnetic Potts model (K < 0) implies b/a < 0, the intersection of the curves of equations (4.2.4) are only real if q < 4 because Δ is negative and the intersections are in the area b/a < 0.

Now the Potts model with q = 1,2 is soluble, in fact these cases are the one-component Potts trivially solved and the Ising model. Looking back at figure 17, we can say that the lines $\Delta = 1/2, \sqrt{1/2}$ are also soluble lines for our staggered vertex model.

Finally let us quote the result for the anisotropic Potts model where the identification to the 6-vertex is made through:

$$\frac{e^K - 1}{\sqrt{q}} = \frac{\sin\theta}{\sin(\mu-\theta)} \qquad \frac{e^{K'} - 1}{\sqrt{q}} = \frac{\sin(\mu-\theta')}{\sin\theta'}$$

with $\Delta = -\sqrt{q}/2$ and $\theta' = \theta \pm \ell\pi/2$. Eliminating θ we have a relation between the coupling constant K and K':

(4.3.6a) $\qquad \ell = 0 \qquad (e^{K}-1)(e^{K'}-1) = q$

(4.3.6b) $\qquad |\ell| = 1 \qquad (e^{K}+1)(e^{K'}+1) = 4-q$

Both curves may be obtained also by the dual and automorphic properties of the partition function of the Potts model /20/.

4.4. Application to the Ashkin-Teller model

This model has been shown to be equivalent to two sets of Ising spins $\{S_i\}$ and $\{T_i\}$ on a site of a square lattice interacting with a statistical weight for neighboring pairs (i,j)

$$W_{ij} = \exp\{K S_i S_j + K' T_i T_j + K'' S_i S_j T_i T_j\}$$

A partial duality transformation on the T_j spin shows that the model is mapped into a staggered 8-vertex model /21/. However for K = K'' a subsequent weak-graph transformation on the staggered 8-vertex model reduces it to a staggered 6-vertex model such that the weights are distributed as follows (see figure 14)/22/:

	$\omega_1 = \omega_2 = a$	$\omega_3 = \omega_4 = b$	$c = \sqrt{\omega_5\omega_6}$
Matrix $T_N(\theta)$	$\sinh 2K$	$e^{-2K''}$	$\cosh 2K$
Matrix $T_N(\theta')$	$e^{-2K''}$	$\sinh 2K$	$\cosh 2K$

Vertices belonging to $T_N(\theta)$ and $T_N(\theta')$ have the same Lieb invariant:

$$\Delta = -\frac{\sinh 2K''}{\sinh 2K}$$

We plot now the solubility curves in the plane $\omega'' = \exp(-4k)$, $\omega = \exp(-2k-2k'')$ see figure 18.

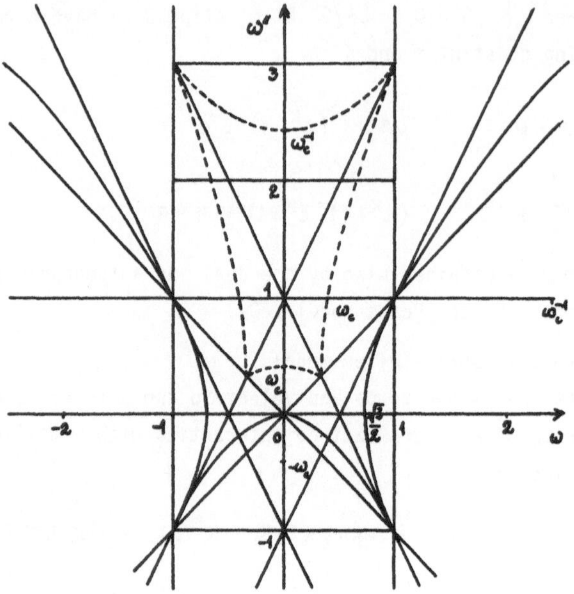

Figure 18: The symmetric Ashkin-Teller diagram.

The non staggering case $\ell = 0$ of equation (4.2.2) is simply the self-dual curve:

(4.4.7a) $$a = b \quad \text{or} \quad e^{-2K''} = \sinh 2K \quad \text{or} \quad 2\omega + \omega'' = 1$$

The other case with $|\ell| = 1$ is the curve of equation, also invariant under duality:

(4.4.7.b) $$(a^2 - b^2)^2 = c^2(a^2 + b^2) \quad \text{or} \quad \left[(1-\omega'')^2 - 4\omega^2\right]^2 = (1+\omega'')^2\left[(1-\omega'')^2 + 4\omega^2\right]$$

Upon comparison with figure 17, we could now add to these solubility curves the curves of equation deduced indirectly from the Potts model:

$$\Delta = -1/2\,,\, -1/\sqrt{2} \quad \text{with} \quad \Delta = \frac{\omega - \omega''/\omega}{1 - \omega''}$$

The free fermion condition is always $\Delta = 0$ corresponding to K'' = 0. And there exists two Ising cases for $K'' \to \infty$ or K = 0, for which our staggered six-vertex model takes precisely the form obtained by F.Y. Wu to represent the Ising model /6/.

In figure 14, the solubility line of equation (4.7.b) is beyond the duality envelop defined by the equation:

$$\omega'' = 2\omega - 1$$

We have also sketched the critical lines in dotted lines which are not known exactly /23/. We do not wish here to elaborate on them.

4.5. Generalisation to p-fold staggering on square lattice

The previous discussion has led us to consider the integrability conditions for the ice-rule on a square lattice from a purely geometric point of view, which may be applied to the following generalisation. We go back to the square lattice of figure 1. We allow each line (vertical or inclined) to carry a "color" labelled by $j = 1,\ldots, p$. Let θ_{ij} (resp. α_{ij}) be the angle between a vertical (resp. horizontal) line of color i and an inclined line of color j. We construct our p-fold staggered lattice of six-vertex as follows. We take the first row, which defines the first diagonal-to-diagonal transfer matrix $T_N^{(1)}$ as made up of vertices, intersections of lines bearing the same color so that the angles are $\theta_{11},\theta_{22},\ldots,\theta_{pp}$ and they are ordered horizontally in periodic pattern. The next row defining the diagonal-to-diagonal transfermatrix $T_N^{(2)}$ is associated with the periodic pattern of angles: $\theta_{12},\theta_{23},\ldots, \theta_{p-1,p,1}$. We construct in this way successively p transfer matrices, whose product is the transfer matrix of the staggered lattice with p-sublattices. Again we shall assume permutational symmetry among the p-sublattices we end up only with p independent angles:

$$\begin{array}{llllllll}
\text{Matrix } T_N^{(1)} & \theta_{11} = & \theta_{22} = & \theta_{33} = & \ldots & \theta_{p-1\,p-1} = & \theta_{pp} = & \theta \\
\text{Matrix } T_N^{(2)} & \theta_{12} = & \theta_{23} = & \theta_{34} = & \ldots & \theta_{p-1\,p} = & \theta_{p1} = & \theta' \\
\text{Matrix } T_N^{(3)} & \theta_{13} = & \theta_{24} = & \theta_{35} = & \ldots & \theta_{p-1\,1} = & \theta_{p2} = & \theta'' \\
 & \vdots & \vdots & \vdots & & \vdots & \vdots & \\
\text{Matrix } T_N^{(p-1)} & \theta_{1\,p-1} = & \theta_{2p} = & \theta_{31} = & \ldots & \theta_{p-1\,p-3} = & \theta_{p\,p-2} = & \theta^{(p-2)} \\
\text{Matrix } T_N^{(p)} & \theta_{1p} = & \theta_{21} = & \theta_{32} = & \ldots & \theta_{p-1\,p-2} = & \theta_{p\,p-1} = & \theta^{(p-1)}
\end{array}$$

A horizontal line carrying a color i will intersect an inclined line of color j with an angle α_{ij}. The assumption of color permutational symmetry forces the α_{ij} to be subjected to the same restrictions as for θ_{ij}, i.e:

$$\alpha_{11} = \alpha_{22} = \alpha_{33} = \ldots \quad \alpha_{p-1\,p-1} = \alpha_{pp} = \alpha$$

$$\begin{array}{cccccc} \alpha_{12} = & \alpha_{23} = & \alpha_{34} = & \cdots & \alpha_{p-1p} = & \alpha_{p1} = \alpha' \\ \vdots & \vdots & \vdots & & \vdots & \vdots \quad \vdots \\ \alpha_{1p} = & \alpha_{21} = & \alpha_{32} = & \cdots & \alpha_{p-1p-2} = & \alpha_{pp-1} = \alpha^{(p-1)} \end{array}$$

We use now the angular relation in a triangle (equation (3,1,2)) to determine the angles between vertical and horizontal lines along the first row of vertices. However, this determination can only be consistent with the angular relation in triangles belonging to the (p-1) rows below if we have

$$\begin{array}{l} \theta + \alpha = \theta' + \alpha' = \theta'' + \alpha'' = \cdots = \theta^{(p-1)} + \alpha^{(p-1)} \\ \theta + \alpha' = \theta' + \alpha'' = \theta'' + \alpha''' = \cdots = \theta^{(p-1)} + \alpha \\ \vdots \\ \theta + \alpha^{(p-2)} = \theta' + \alpha^{(p-1)} = \theta'' + \alpha = \cdots = \theta^{(p-1)} + \alpha^{(p-3)} \\ \theta + \alpha^{(p-1)} = \theta' + \alpha = \theta'' + \alpha' = \cdots = \theta^{(p-1)} + \alpha^{(p-2)} \end{array}$$

We eliminate all the horizontal line parameters $\alpha, \alpha', \ldots, \alpha^{(p-1)}$ to obtain constraints on $\theta, \theta^*, \ldots \theta^{(p-1)}$. They are of the types:

$$p\theta = p\theta' = \cdots = p\theta^{(p-1)} \quad (\text{mod.}\,\pi)$$

$$2\theta^{(j)} = \theta^{(j+1)} + \theta^{(j-1)} \quad (\text{mod.}\,\pi)$$

$$\text{for } j = 0,1,\ldots,p-1$$

Hence we have the solutions:

$$\theta^{(j)} = \theta \pm \ell^{(j)}\pi/p \quad (\text{mod.}\,\pi)$$

$$\text{with } 2\ell^{(j)} = \ell^{(j+1)} + \ell^{(j-1)} \quad (\text{mod.}\,p)$$

Figure 19 illustrates our conventions for p = 4.

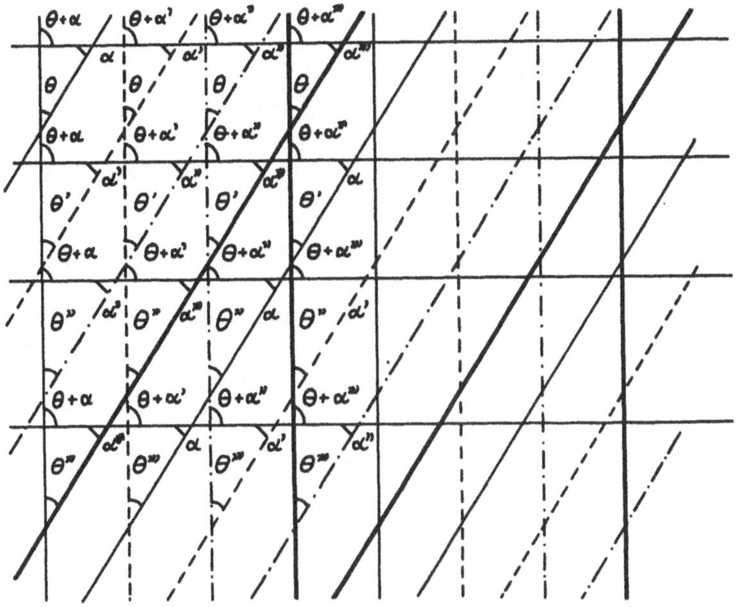

Figure 19: 4-fold staggered vertex system

Examples: For $p=3$ $\quad \theta' = \theta \pm \ell' \frac{\pi}{3}$ and $\theta'' = \theta \pm \ell'' \frac{\pi}{3}$ $\quad \ell' + \ell'' = 0 \ (\mathrm{mod.} 3)$

$\ell' = \ell'' = 0 \ (\mathrm{mod.} 3)$ $\quad \theta = \theta' = \theta''$ No staggering

$\ell' = -\ell'' = 1,2 \ (\mathrm{mod.} 3)$ one can take $\theta' = \theta + \frac{\pi}{3}$ and $\theta'' = \theta + 2\frac{\pi}{3}$

For $p=4$, $\theta' = \theta \pm \ell' \frac{\pi}{4}$, $\theta'' = \theta \pm \ell'' \frac{\pi}{4}$ and $\theta''' = \theta \pm \ell''' \frac{\pi}{4}$ $\quad (\mathrm{mod.} \pi)$

the constraints on ℓ', ℓ'' and ℓ''' are:

$$2\ell'' = \ell' + \ell''' \ (\mathrm{mod.} 4), \quad 0 = \ell' + \ell''' \ (\mathrm{mod.} 4), \quad 2\ell''' = \ell'' \ (\mathrm{mod.} 4), \quad 2\ell' = \ell'' \ (\mathrm{mod} 4)$$

There are many possible choices, namely,

if $\ell'' = 0$ then $\ell' = \ell''' = 0$ (non-staggered) or $\ell' = \ell''' = 2$

if $\ell'' = 2$ then $\ell' = 1$, $\ell''' = 3$ or $\ell' = 3$, $\ell''' = 1$

A non-trivial particular solution for general p is the choice $\ell^{(j)} = (j-1)$ for which we have the addition of π/p to each successive $\theta^{(j)}$, $j = 0,\dots,p-1$

$$\theta' = \theta + \pi/p \ , \quad \theta'' = \theta + 2\pi/p \ , \ \dots\dots \ , \theta^{(p-1)} = \theta + (p-1)\pi/p$$

The free energy per vertex of such a system, in the thermodynamic limit, is equal to:

$$F_p(\theta,\mu) = 1/p\left[f(\theta,\mu) + f(\theta+\pi/p,\mu) + \dots + f(\theta+(p-1)\pi/p,\mu)\right] \tag{4.5.1}$$

where $f(\theta,\mu)$ is the free energy per vertex of the square lattice six-vertex model with parameter θ and μ, already known /16/.

Equation (4.5.1) has a limit for $p \to \infty$:

$$\lim_{p\to\infty} F_p(\theta,\mu) = \int_\theta^{\theta+\pi} f(x,\mu)\,dx = F(\theta,\mu) \tag{4.5.2}$$

For example in the disorder state $|\Delta| < 1$ with $\theta < 2\mu$ after performing the x-integrations we obtain a new expression for $F(\theta,\mu)$ which seems to exhibit a different singularity structure from that of $f(\theta,\mu)$ but a detailed study of (4.5.3) is beyond the scope of this paper

$$F(\theta,\mu) = \text{Const.} + \int_{-\infty}^{\infty} \frac{\sinh(2\theta+\pi)x \ \sinh(\pi-\mu)x}{2x^2 \cosh\mu x}\,dx \tag{4.5.3}$$

To close this section we observe that for p = 2m we have an equivalence with an m-fold layered q-state Potts model (alternatively symmetric Ashkin-Teller model) for which soluble cases exist. However, as already observed in reference /10/ case must be taken in expressing the related free energy per site since periodic boundary conditions for vertices do not correspond to periodic boundary conditions for spin-systems.

§5. Conclusions

That field theories are directly connected with statistical mechanics is long ago known due to the work of J. Schwinger and then later systematized by K. Osterwalder and R. Schrader. The class of non-linear field theories which are integrable has considerably grown since the connection between the 8-vertex model solved by R.J. Baxter and a lattice version of the massive Thirring model has been established by A. Luther and M. Lüscher /24/. Recently we have also shown that staggered vertex systems have given rise to non-linear field theory models involving internal degrees of freedom which are precisely implemented by the sublattices of a staggered lattice /25/.

In the context of an integrable theory we could escape the ardous task of solving non-linear equations of motion and instead deal with the linear problem obtained by the inverse transform. Here we have shown that, indirectly, the associated linear problem is simply a uniformisation of the vertex system which leads further to a geometrisation of the solution, as already proposed by Baxter in 1978. The amusing thing is at least up to now only euclidean geometry is needed.

It is quite often believed that non-integrable systems become integrable at criticality /26/. This is true for example for the Potts model in both ferro and antiferromagnetic regimes. But conversely solubility needs not imply always a critical point as been shown recently /27/. Moreover, there is no criterion which could insure that a critical line is also a solubility line, this is the case of the presumed Ising-like critical lines in the symmetric Ashkin-Teller model.

To complete this study we have gathered in the appendix a review of the Bariev's model as well as some academic examples of the technique which include a generalization to multicomponent ice-rule /28/ on a staggered lattice, as well as ice-rule associated to "Solid-on-Solid" (S.O.S.) model /29/ now on a staggered lattices. Finally we make some comments related to models parametrized by elliptic functions.

REFERENCES

1. L. Onsager, Phys. Rev. 65 (1944) 117.

2. J. Ashkin & E. Teller, Phys. Rev. 64 (1943) 178.

3. R.B. Potts, Proc. Camb. Phil. Soc. 48 (1952) 106.

4. E.H. Lieb, Phys. Rev. 162 (1967) 162.
Phys. Rev. Lett. 18 (1967) 1046
Phys. Rev. Lett. 19 (1967) 108.

5. R.J. Baxter, Ann. Phys. (N.Y.) 70 (1972) 193.

6. F.Y. Wu and K.Y. Lin, Phys. Rev. B12 (1975) 419.

7. K.Y. Lin, J. Phys. A8 (1975) 1899.

8. K.Y. Lin and I.P. Wang, J. Phys. A10 (1977) 813.

9. R.J. Baxter, Stud. appl. Math. 50 (1971) 51.

10. R.J. Baxter, Proc. R. Soc. Lond. A383 (1982) 43.

11. R.Z. Bariev, Theor. Math. Phys. 49 (1982) 1021.

12. E.H. Lieb and F.Y. Wu, in "Phase Transitions and Critical Phenomena" (C. Domb and M.S. Green, eds.) Vol. 1 (1972) pp. 321-490. Academic Press, London.

13. R.J. Baxter, Jour. Stat. Phys. 22 (1982) 1.

14. L.A. Takhtadzhan and L.D. Faddeev, Russ. Math. Surveys 34 (1979) 11.

15. A.B. Zamolodchikov and Al.B. Zamolodchikov, Nucl. Phys. B133 (1978) 525.
M. Karowski, H.J. Thun, T.T. Truong and P.H. Weisz, Phys. Lett. 67B (1977) 321.

16. R.J. Baxter, Phil. Trans. Roy. Soc. (London) 289 (1978) 315.

17. T.T. Truong and K.D. Schotte, Nucl. Phys. B220 (FS 8) (1983) 77.

18. T.T. Truong, Physica 122A (1984) 603.

19. H.N.V. Temperley and E.H. Lieb,Proc. Roy. Soc. (London) A322 (1971) 251.
R.J. Baxter, S.B. Kelland and F.Y. Wu, J. Phys. A9 (1976) 397.

20. M.T. Jaeckel and J.M. Maillard, J. Phys. A15 (1982) 2241.
T.T. Truong, J. Phys. A17 (1984) L473.

21. C. Fan, Phys. Lett. 39A (1972) 136.
F.J. Wegner, J. Phys. C5 (1972) L131.

22. M. Kohmoto, M. den Nijs and L.P. Kadanoff, Phys. Rev. B24 (1981) 5229.

23. R.V. Ditzian, J.R. Banavar, G.S. Grest and L.P. Kadanoff, Phys. Rev. B22 (1980) 2542.

24. A. Luther, Phys. Rev. B14 (1976) 2153.
M. Lüscher, Nucl. Phys. B117 (1976) 475.

25. T.T. Truong and K.D. Schotte, Nucl. Phys. B230 (FS 10) (1984) 1.

26. R. Shankar, Phys. Lett. B102 (1981) 257.

27. I. Peschel and F. Rys, Phys. Lett. 91A (1982) 187.
M. Barber and F. Rys, "The critical lines of an asymmetric 8-vertex model". Preprint 1984. Australian National University.

28. I.V. Cherednik, Theor. Math. Phys. 43 (1980) 117, 356.
C.L. Schultz, Phys. Rev. Lett. 46 (1981) 629.
D. Babelon, H.J. de Vega and C.M. Viallet, Nucl. Phys. B150 (FS 8) (1981) 542.

29. G.E. Andrews, R.J. Baxter and P.J. Forrester, Jour. Stat. Phys. 35 (1984) 193.

ACKNOWLEDGMENTS

The author would like to thank Professors N. Sánchez and H. J. de Vega for their invitation to participate to the series of seminars on " Non-linear Equations in Field Theory ".

APPENDIX .

A complete survey of all soluble models on staggered lattices would be difficult to set up. In the text we have concentrated on models of vertex admitting always the same Lieb's invariant. It was R. Z. Bariev who first found that there exits a case for which Δ may be chosen unequal on different sublattices. We will review this case which is quite original and go on to other types of vertices.

1. The Bariev Model /11/. Here we show that the previous "geometrical" construction of Bethe-ansatz wave function still applies. We shall use the notations of §4.1 whereby the weights associated to figure 13 are now given by the table:

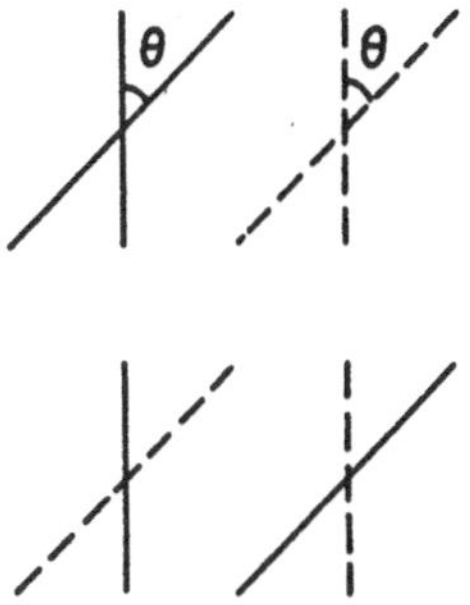

$\omega_1 = \omega_2 = a$	$\omega_3 = \omega_4 = b$	$\omega_5 = \omega_6 = c$	Δ
$\cos\theta$	$\sin\theta$	1	0
e^{γ}	$e^{-\gamma}$	0	$\cosh 2\gamma$

Again there is complete symmetry between continuous and dotted lines. Due to this choice of weights a down-arrow on a continuous line will never jump to a dotted line and vice-versa, thus once created on one sublattice it will stay there. Horizontal lines associated to them are defined by the auxiliary weights:

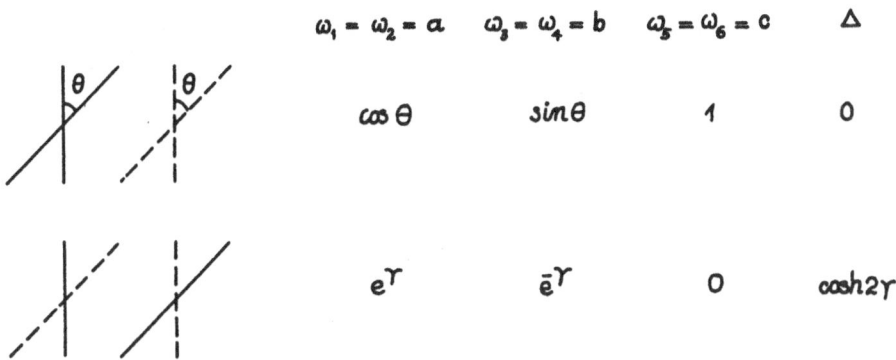

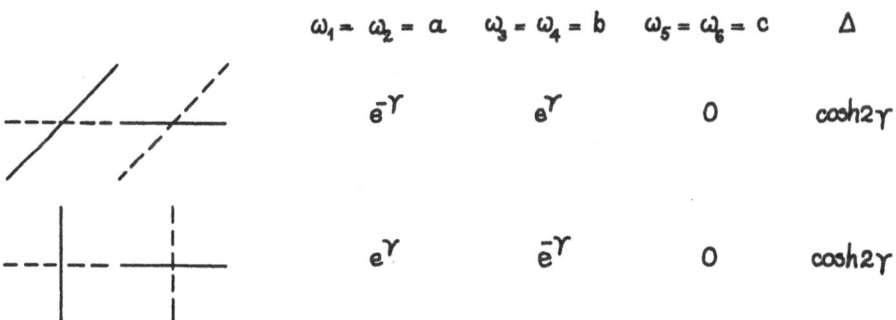

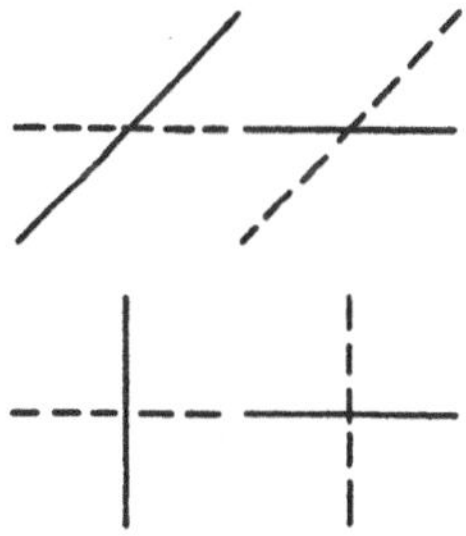

$\omega_1 = \omega_2 = a$	$\omega_3 = \omega_4 = b$	$\omega_5 = \omega_6 = c$	Δ
$e^{-\gamma}$	e^{γ}	0	$\cosh 2\gamma$
e^{γ}	$e^{-\gamma}$	0	$\cosh 2\gamma$

The other auxiliary weights for vertices involving lines of the same type are deduced from the usual triangle identities. Then it is possible to shift them through T_{Bariev}

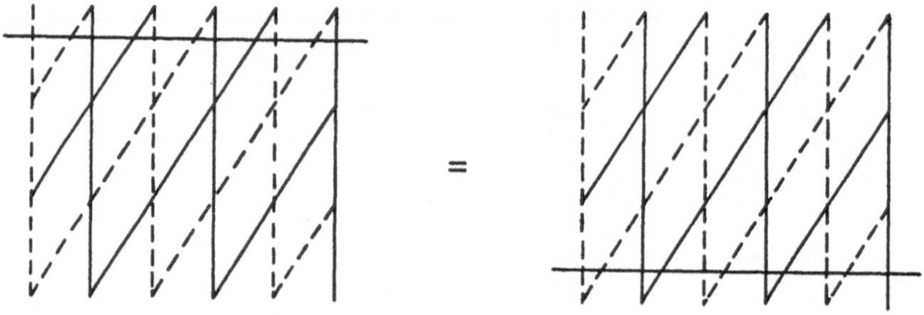

A similar diagram holds for a dotted horizontal line.

Horizontal lines of the same type (continuous or dotted) commute among themselves obviously as in section 3. Their resulting Bethe-ansatz wave function is simply a Slater determinant since we have a free fermion condition $\Delta = 0$. But horizontal lines of different type "create" distinguishable down arrows and therefore do not commute. To construct a Bethe-ansatz wave function for them which is an eigenstate of the diagonal-to-diagonal transfer matrix it is necessary to be able to shift a pair of horizontal lines of different type through the second row of vertices defining T_N'. To do this we have to introduce coexistence amplitudes:

$\mathcal{A}^{|\,\vdots}(\theta_1,\theta_2)$: amplitude for presence of down-arrow on dotted line with parameter θ_2 and down-arrow on continuous line with parameter θ_1.

so that we have the diagram:

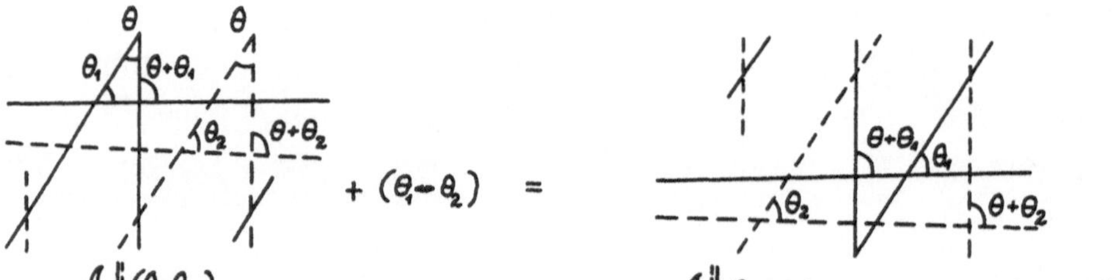

$\mathcal{A}^{|\,\vdots}(\theta_1,\theta_2)$ $\qquad$ $\mathcal{A}^{\vdots\,|}(\theta_2,\theta_1)$ $\quad + \ (\theta_1 \leftrightarrow \theta_2)$

There is another diagram obtained by permuting dotted and continuous lines. The linear relations between all the amplitudes $\mathcal{A}^{\vdots\vdots}(\theta_1\theta_2)$ etc.... define a 2-body S-matrix which turns out to be a factorizable elastic one /15/ having again a six-vertex model type of parametrization. Applying conventional techniques, this model is fully solved.

2. Multicomponent-generalisation of the ice-rule, fulfilling the triangle relations have been proposed and solved by many authors /28/. We consider now the possibility to have such vertex configurations on staggered lattice, and ask the question whether the method advocated in the text still applied. To simplify the discussion we restrict ourselves to 3-component systems. There exist basically two types of parametrization as shown below:

$$a = \sinh(\lambda-\theta) \quad b = \sinh\theta \quad c = \sinh\lambda$$

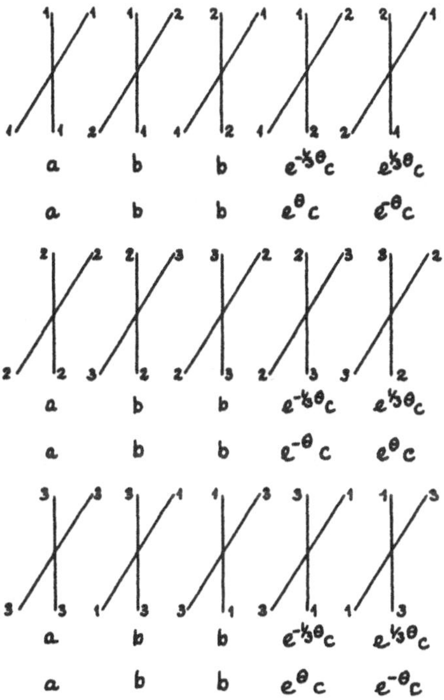

Hence the special parameter θ is defined modulo 3π or π. Here only horizontal lines creating the same type of particle, are commutative. To push through a set of lines of different types one needs amplitudes in analogy with the Bariev model. The net effect is that a "nested" Bethe ansatz wave function will then be eigenstate of the corresponding diagonal-to-diagonal transfer matrix. If one deals with a staggered vertex system then this "nested" Bethe ansatz would be eigenstate of two successive transfer matrices of parameters (θ,λ) and (θ',λ) provided that :

$$2\theta' = 2\theta + 3\pi \quad (\text{mod.}\, 3\pi)$$

or

$$2\theta' = 2\theta + \pi \quad (\text{mod.}\, \pi)$$

3. Solid-on-solid vertex systems: There is a well-known connection between the ice-rule and the variations of atom-heights on the lattice when the variations are restricted to be ± 1 when passing from one site to the neighboring site. Let ℓ be the height at some site then the 6 possible arrangements of heights round a dual site are:

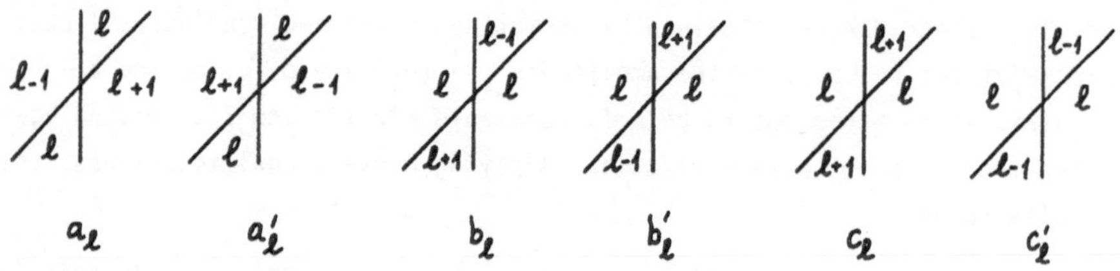

such that $a_\ell = a'_\ell = \sin(\mu+\theta).\sin\mu\ell$, $b_\ell = b'_\ell = \sin\theta\sqrt{\sin\mu(\ell-1)\,\sin\mu(\ell+1)}$ and $c_\ell = \sin\mu.\sin(\mu\ell-\theta)$, $c'_\ell = \sin\mu.\sin(\mu\ell+\theta)$. Again θ is defined modulo π. As shown in reference 17, a horizontal line can be served to define a down-arrow and the subsequent Bethe-ansatz for a many particle system. Hence on a staggered lattice we expect the relation (4.2.2) to hold again as solubility condition.

The generalisation to an elliptic parametrisation exists /29/ but then θ is only defined modulo 2 K(k) where the K(k) is the elliptic integral of the modulus k. There we cannot take into account the imaginary part of the period pallalelogram, everything else remains the same. To close this appendix note that a staggered eight-vertex model cannot be quite transformed into a staggered "elliptic" version of the SOS model, in this sense it is not clear how a staggered eight-vertex does have non trivial solutions.

FIELDS ON A RANDOM LATTICE II

RANDOM SURFACES : A SEARCH FOR A DISCRETE MODEL

M. Bander* and C. Itzykson
Service de Physique Théorique
CEN-Saclay, 91191 Gif sur Yvette, Cedex, France

I. PRESENTATION

It is our purpose to elaborate on certain concepts introduced in the study of flat random lattices [1][2] and apply them in the construction of models for random manifolds. Here we choose to discuss random (two dimensional) triangulated surfaces. We will use some of the methods and notations developped in reference [2].

In these days, when sophisticated differential geometry becomes a familiar tool in theoretical physics, it is nevertheless instructive to increase one's insight by appealing to discrete (and old fashioned) geometry. This is of interest, if only because it affords means to attack numerically non linear problems, classical and quantum mechanical. A reference work is Regge's [3] discretization of Euclidean relativistic gravity. An other landmark is Wilson's [4] lattice gauge theory, which enables one to extract numerical information, however incomplete, from chromodynamics. The existence of modern computer facilities is a strong incentive to pursue along these lines.

To return to our specific subject, it has been advocated by a number of authors, including Wallace [5], Polyakov [6], Parisi [7], Fröhlich [8] and others, that a theory of random surfaces might play a role in a diversity of situations, ranging from interfacial effects in condensed matter physics to string models and gauge theories in particle physics.

The notion of a surface is itself not such a straight-forward matter as it looks at first. One may wish to consider it as an abstract object endowed with internal properties like connectedness and its topological generalization (Betti numbers), metric, curvature of various other bundle structure that it can support. Or one may think of its imbeddings in larger manifolds, in particuler Euclidean spaces. In this case new concepts emerge, like topological invariants attached to its complementary, or its relation to three-dimensional manifolds it can bound. An important circumstance is when one deals with an interface in $\mathbb{R}^3$ between distinct material phases.

An other new aspect is the non trivial structure of boundary problems. There is a clear distinction between curves bounded by structureless sets of points and surfaces bounded by curves. Typical of such questions is the Plateau problem of minimal surfaces, or the study of Wilson loops in gauge theories.

*Permanent address : Physics Dpt., University of California, IRVINE, CA 92717.

A third aspect has to do with parametrization and correlatively the dynamical generation of surfaces. A curve may be viewed as the evolution of a point, while a surface is generated by a string with infinitely many degrees of freedom. String theories are notoriously difficult and have given rise to a large body of literature [9]. On the other hand, when viewed as an intrinsic object, one may want to think of the surface without prejudice about any coordinate choice, i.e. insist on reparametrization invariance, or general covariance of its "physical" properties.

It is perhaps a combination of all these aspects, which has prevented up to now to develop a simple reference model with easily computable properties (at least some of them). Of course a number of very ingenious attempts have been made, but a clear picture of their inter-relationships is still missing. The work of Polyakov [6] will be an inspiration for our presentation, together with some analytical and numerical results of Billoire, Gross, and Marinari [10][11].

We shall first present a down to earth model for random curves and recall on this example a poor man's definition of Hausdorff (or scaling) dimension. We shall then generalize the formulation to triangulated (piece-wise linear) surfaces and focus on the Euler characteristic and its relation to curvature. The latter can be described as a "frustration" preventing to identify locally the surface as a piece of Euclidean (metric) space. The generalization of this point of view to higher dimensional triangulated manifolds and the definition of the Euler class in terms of metric properties is not a trivial problem, as discussed by Cheeger, Müller and Schrader [12]. We shall briefly digress on this point.

We then turn to the discussion of (massless) free fields on an arbitrary "curved", triangulated, surface. Eventhough we seem at first to lack the analog of the Voronoï construction described in the flat case in [1],[2], we shall nevertheless overcome this difficulty using an embedding property as a guide. As a result we will obtain the geometrical elements of a "virtual" dual lattice, which lead naturally to the definition of differential and co-differential. This construction generalizes to higher dimensional manifolds.

The need to complete the definition of the corresponding path integrals by giving an a priori measure (the entropy problem) leads one naturally to study the two-dimensional conformal anomaly. We include a short "pedagogical" review of this question following Polyakov [6] and Fujikawa [13].

We are then in a position to complete the discretization of a Liouville type model for random surfaces. The model has some analogy with a two dimensional (neutral) Coulomb gas with unquantized charges. At the present stage we are unfortunately unable to present a convincing analytical discussion of this model even in the mean field case (large embedding dimension). Numerical simulations, which are not out of reach, might help to figure out whether its content is richer than the one studied in reference [10].

Eventhough our original contributions are somehow meager, we hope that this presentation might be useful to some readers and help stimulate further work on this subject.

II. RANDOM CURVES

An abstract model for a closed curve is a circle. An actual realization can be viewed as a mapping from the circle in the Euclidean space R^d. The parameter space, the circle, enables one to assign to each point of the curve a "time" or angle, and distinguishes possible multiple points on the image. We may be interested in properties specific to the image and independent of the particular mapping chosen, assumed at least to be continuous. Indeed a complete mathematical description requires the specification of the allowed mappings. A physicist's point of view might be more concrete, with the curve being thought as an idealization for a material object, be it for instance a (closed) long molecule, a defect line in an otherwise ordered medium, a particle trajectory in a dense material... In such instances, the infinitesimal (or short distance) structure might be of little relevance to the large scale properties under investigation. One should then be allowed, in certain circumstances to replace the continuous aspect by a discrete one (with a very fine mesh) without altering in a noticeable way the overall picture. This we do, for instance, by replacing the continuous parameter space (the circle) by a sequence of densely packed point labelled from 0 to N-1 (with N identified with 0), N very large, to which we assign N points in R^d : $\underset{\sim}{x}_o, \underset{\sim}{x}_1, \dots, \underset{\sim}{x}_{N-1}$ and we think of the curve as a linear interpollation between $\underset{\sim}{x}_i$ and $\underset{\sim}{x}_{i+1}$.

To define a statistical model on these "curves" requires two ingredients which are really not independent. The first is to give a (relative) statistical weight to each curve e^{-S} where S is a dimensionless action, or an energy devided by kT. The second, is to figure out a mean to distinguish and count the curves (this is the entropy). As the curves are imbedded in $\mathbb{R}^d$ which has a metric structure, these prescriptions are of course required to respect Euclidean invariance. The arbitrariness is further reduced by a locality requirement (short range interactions). The most stringent form assumes S to be a sum of contributions, each from a successive pair of neighboring points

$$S = \sum_{i=0}^{N-1} L(\ell_{i,i+1}) \qquad \ell_{i,i+1} = |\underset{\sim}{x}_i - \underset{\sim}{x}_{i+1}| \tag{1}$$

The a priori measure can be taken as the product measure over all but one $\underset{\sim}{x}_i$. We indicate this by $\pi' \, d^d \underset{\sim}{x}_i$, so the overall measure reads

$$d\mu = \pi' \, d^d x_i \, e^S \tag{2}$$

If one introduces $\underset{\sim}{y}_1 = \underset{\sim}{x}_1 - \underset{\sim}{x}_o, \dots, \underset{\sim}{y}_{N-1} = \underset{\sim}{x}_{N-1} - \underset{\sim}{x}_{N-2}$, the measure factorises into a product except for the term $L(\ell_{N-1,0})$. We then add $\underset{\sim}{y}_N$ in such a way that $\sum_{i=1}^{N} y_p = 0$.

Standard choices for $L(\ell)$ are $\alpha \frac{\ell}{a}$ or $\alpha \frac{\ell^2}{a^2}$ where a is an arbitrary unit of length. The first choice is attractive since S acquires the geometrical meaning of being the total length of the curve up to a factor. In any case we assume all moments of the measure $d^d\underset{\sim}{y}\ e^{-L(|\underset{\sim}{y}|)}$ to be finite. Random curves such that this would not be fulfilled would fall in a very different class.

A universal property of such models is then the following. Choose the arbitrary position of the center of the center of mass to be the origin

$$\frac{1}{N}\sum_{0}^{N-1} \underset{\sim}{x}_i = 0 \tag{3}$$

Relative to this origin, the mean square radius R_N^2 is

$$\begin{aligned} R_N^2 &= \langle\frac{1}{N}\sum_{0}^{N-1} \underset{\sim}{x}_i{}^2\rangle = \langle\underset{\sim}{x}_o^2\rangle \\ &= \langle(\frac{1}{N}\sum_{1}^{N} p\underset{\sim}{y}_p)^2\rangle \end{aligned} \tag{4}$$

For convenience set

$$\langle\underset{\sim}{y}_i^2\rangle = a_N^2 \tag{5}$$

Then it follows from $\sum_{1}^{N} \underset{\sim}{y}_p = 0$, that $\langle\underset{\sim}{y}_i \cdot \underset{\sim}{y}_j\rangle = -\frac{a_N^2}{N-1}$ $(i \neq j)$, and therefore

$$R_N^2 = \frac{a_N^2}{12}(N+1) \tag{6}$$

To keep the average extent finite in the limit $N \to \infty$, requires therefore to adjust the bare parameters (in the Lagrangian) in such a way that $a_N \sim \frac{1}{\sqrt{N}}$. This is typical of a Brownian curve, and make it look more like a two dimensional manifold than an ordinary regular curve. For think of a fixed manifold of dimension δ, imbedded in a space of dimension d, and of finite extent. Approximate the manifold by a set of N points in some regular fashion. The distances between neighbors will obviously scale as $a_N \sim \frac{1}{N^{1/\delta}}$. This motivates a more rigorous definition of the Hausdorff dimension (not given here) which leads to $\delta_H = 2$ for a Brownian curve. What lies at the heart of the matter here, is that, apart from the overall constraint of being closed, the curve was constructed from independent increments. Their mean square add, leading to the above conclusion. Details of the short range structure are immaterial, as is the dimension of the imbedding space. All these are lumped together in a_N^2.

We cannot expect such simple properties when discussing higher dimensional manifolds. As far as surfaces are concerned, extreme suggestions have been made with the Hausdorff dimension being 4 or infinity, pointing to the fact that either one were not using the same concept, or one were discussing utterly different models.

III. PIECEWISE LINEAR TRIANGULATED SURFACES

Some of the preceding ideas do however generalize to higher dimensions. To be specific, we discuss here two dimensional (compact orientable) manifolds of fixed topology. First one picks a compact abstract model (or parameter space) which fixes the topology. We choose it to be orientable. It is characterized by its genus, or number of handles, g, related to the Euler characteristics χ through $\chi = 2-2g$. Then we discretize and triangulate it, for instance introducing a metric and using a Voronoi construction. This triangulation introduces the notion of nearest neighbor pairs (or links) and elementary 2-simplices or triangles. If $N_o \equiv N$ is the number of sites, N_1 of links, N_2 of simplices, then $N_o-N_1+N_2 = \chi$. Note that (specific to a triangulation) $2N_1 = 3N_2$ since each link belongs to two triangles and each triangle has three sides, thus

$$\begin{aligned} N_o &= N \\ N_1 &= 3(N-\chi) \\ N_2 &= 2(N-\chi) \end{aligned} \tag{1}$$

A somehow troublesome question is the one of inequivalent triangulations, even for given $N_o=N$, that is as abstract (unlabelled) graphs. We shall assume that one is selected for a growing sequence of N's. There is also the possibility to sum for each N over all possible inequivalent choices adding further entropy to the system [8][16].

We then map the N points in $\mathbb{R}^d$ and interpolate linearly between the images of neighbors. This reproduces in $\mathbb{R}^d$ a set of points, links (linear segments), planar faces (triangles), which may of course have numerous self intersections which we do not take into account. For these sets we want to introduce a measure with the same requirements as in the previous section.

The metric on $\mathbb{R}^d$ induces on the image surface, lengths and areas for the links and triangles as well as angles. The triangles are Euclidean, so that their internal angles add to π. For each triangle we can therefore split unity into $1 = \frac{1}{\pi}(\theta_1+\theta_2+\theta_3)$ with each θ between 0 and π. Summing over triangles gives N_2. But we can rearrange the sum by collecting all θ's pertaining to a vertex, then summing over vertices. Call θ_i the sum at each vertex. Then $N_2 = \sum_i \frac{\theta_i}{\pi}$. From (1) this is $2N-2\chi = -2\chi + 2\sum_i 1$. Identification leads to

$$\chi = \sum_i \left(1 - \frac{\theta_i}{2\pi}\right) \tag{2}$$

a classical formula in terms of deficit angles $2\pi - \theta_i$. This is the discrete form of the Gauss-Bonnet formula. When the deficit angle vanishes at a vertex the corresponding triangles fit in flat space. So we have the identification : deficit angle ⟷ curvature ⟷ frustration from planar situation. The identifiation with curvature can be easily understood on the example of a smooth surface like a sphere. Let its radius be r and therefore its curvature $\frac{1}{r^2} \equiv R$. For a spherical triangle with inner angles

α_1, α_2, α_3 we have the well known relation that the area A is given by $RA = (\alpha_1+\alpha_2+\alpha_3-\pi)$. The total amount that a tangent vector has rotated in one circumnavigation (rounding vertices) is $\theta = \sum_1^3 (\pi-\alpha_i) = 2\pi - RA$ so that there is a total angular deficit of matter $2\pi - \theta = RA$. The limiting value of the right hand side as the spherical area shrinks to zero and R scales as $\frac{1}{A}$ is the above angular defect. On the triangulated piecewise linear surface, curvature is entirely concentrated at the vertices. Incidentally with our normalization the continuum version of the Gauss Bonnet formula reads $\chi = \frac{1}{2\pi}\int dA\, R$, i.e. 2 for a sphere. For surfaces, curvature is a scalar concept and of course (2) relates geometry and topology. One observes a disymmetry between positive and negative "deficit", since, at it is defined, θ_i is positive so $1 - \frac{\theta_i}{2\pi}$ runs from 1 to $-\infty$ (more precisely between 1 and $-(q_i/2-1)$ if q_i triangles meet at the point i). Angular-wise we cannot have more than a 2π deficit but of course we can have as much as we want of extra matter.

In higher dimensions, for piece-wise linear triangulated compact manifolds (dimension D) expression (2) generalizes in two ways. The notion of deficit angle extend easily and represents the frustration in being able to paste together in $\mathbb{R}^D$ all D-simplices incident on a D-2 simplex of "volume" a. But now the "direction" of the (D-2) simplex matters. So curvature is no more a scalar but a tensor. Regge [3] has shown that if we label $\{\alpha\}$ the D-2 simplices, the Euclidean action for Einstein's gravity in discretized form is proportional to

$$S_E = \sum_\alpha a_\alpha \left(1 - \frac{\theta_\alpha}{2\pi}\right) \tag{3}$$

a rather remarkable formula. We shall not elaborate further on this point, refering to some recent work dealing with its derivation [14] and possible modification [15].

An other generalization is to ask for an expression of the Euler characteristics χ in terms of curvature, when D is larger than 2 (and even, otherwise it vanishes) which would generalize (2). To get as simple and explicit a formula in terms of deficit angles is not easy, as discussed by Cheeger, Müller and Schrader [12]. A partial and unsatisfactory answer is as follows. We want to dissect the number of p dimensional simplices occuring in $\chi = \sum_{p=0}^{D} (-1)^p N_p$ in contributions from its vertices. This is obtained by noticing that the exterior normal of a p simplex sweeps the entire S_{p-1} sphere if we round off corners, in such a way that we can assign to each vertex i the corresponding fraction $\varphi_i^{(p)}$ (normalized angle) of S_{p-1}. The sphere S_0 has two points, so for a link we assign $\frac{1}{2}$ to each end point, and for a point itself $\varphi=1$. In the sum for χ we collect for each vertex i and each p the contributions from the different p simplices incident on i and call it again $\varphi_i^{(p)}$, thus getting [12]

$$\chi = \sum_i \left(\sum_{p=0}^{D} (-1)^p \varphi_i^{(p)}\right) \tag{4}$$

While (4) reproduces (2) in the case D=2, as it is easy to see, it has serious drawback otherwise. First one fails to see that χ vanishes if D is odd. Also, if regions of the manifold are flat, their contribution is not seen to vanish, as one could expect. Some improvements can however be made which don't seem to have the elegance of the corresponding continuum formula, nor of (2). Of course the Euler characteristics is not the only topological invariant of higher dimensional manifolds.

IV. FREE FIELDS

Given a triangulated surface embedded in flat space $\mathbb{R}^d$, lengths, areas and angles being defined, we have a metric on the abstract structure. We want to write the action for free fields, and the corresponding classical equations and quantum mechanical path integrals. We have in mind generalizing the expressions given in [1] and [2]. We start with a massless scalar field defined on vertices and look for a natural quadratic form which approximates the "kinetic" term on a continuous surface, i.e.

$$S_{\text{cont.}} = \frac{1}{2}\int d^2\alpha \sqrt{g}\, g^{ab}\, \partial_a\varphi\, \partial_b\varphi \tag{1}$$

Here α stands for the parameters, g^{ab} is the inverse of the metric giving the length square $ds^2 = g_{ab}\, d\alpha^a d\alpha^b$, and $g = \det g_{ab} > 0$. To implement our program, take on each (flat) triangle with vertices $\underset{\sim}{x}_1$, $\underset{\sim}{x}_2$, $\underset{\sim}{x}_3$ in $\mathbb{R}^d$ barycentric coordinates

$$\underset{\sim}{x}(\alpha) = \alpha^1 \underset{\sim}{x}_1 + \alpha^2 \underset{\sim}{x}_2 + \alpha^3 \underset{\sim}{x}_3 \qquad \alpha_i \geq 0 \qquad \Sigma\alpha_i = 1 \tag{2}$$

and extend the field linearly through

$$\varphi(\underset{\sim}{x}(\alpha)) = \alpha^1\varphi_1 + \alpha^2\varphi_2 + \alpha^3\varphi_3 \tag{3}$$

given its values at the three vertices. This is the natural harmonic extension inside the triangle. We then apply (1), with the metric inherited on each triangle. It follows that

$$S_{\text{discrete}} = \sum_{(ijk)} \frac{1}{8} \frac{[\varphi_i(\underset{\sim}{x}_j-\underset{\sim}{x}_k)+\varphi_j(\underset{\sim}{x}_k-\underset{\sim}{x}_i)+\varphi_k(\underset{\sim}{x}_i-\underset{\sim}{x}_j)]^2}{\ell_{ijk}} \tag{4}$$

Here the sum is over the triangles and ℓ_{ijk} is the area of the corresponding triangle i.e. in terms of length edges (which satisfy the triangle inequalities)

$$\ell^2_{ijk} = \frac{1}{16}(\ell_{ij}+\ell_{jk}+\ell_{ki})(\ell_{ij}+\ell_{jk}-\ell_{ki})(\ell_{ij}-\ell_{jk}+\ell_{ki})(-\ell_{ij}+\ell_{jk}+\ell_{ki}) \tag{5}$$

Translation of the surface as a whole (or rotation), or shift of φ_i by a constant, does not affect (4) as it should. It is also possible to give to (4) a form analogous to the one for a flat random lattice, in spite of the fact that a dual lattice is missing. Very much as one can define $\ell_i \equiv 1$, ℓ_{ij}, ℓ_{ijk} for the direct lattice, we are going to introduce corresponding quantities $\sigma_{ijk} = 1$, σ_{ij}, σ_i related to a virtual dual lattice in a natural way starting from (4). Pick one of the triangles, call it (123) with interior angles θ_1, θ_2, θ_3 between 0 and π. Then it is an easy exercise in elementary geometry to show that the corresponding contribution to (4) can be written

$$S_{123} = \frac{1}{4}\,[\text{cotg}\ \theta_1(\varphi_2-\varphi_3)^2 + \text{cotg}\ \theta_2(\varphi_3-\varphi_1)^2 + \text{cotg}\ \theta_3(\varphi_1-\varphi_2)^2]$$

as a sum of squares with algebraic dimensionless coefficients. Nevertheless the wole expression is of course positive. It is easily seen that if R denotes the radius of the circumscribed circle to the triangle then

$$\frac{1}{4}\,\text{cotg}\ \theta_i = \frac{R\cos\theta_i}{2\ell_{23}}$$

The quantity $R\cos\theta_i$ is the (algebraic) distance from the center 0 of the circumscribed triangle to the edge (jk) and it is of the sign of $\frac{\pi}{2} - \theta_i$, i.e. positive or negative according to wether i and 0 are on the same side or not of the chord jk. Two and only two triangles say (123) and (1'23) share a given link (23). Define then σ_{23} as the length (algebraic) of the virtual link dual to (23) as

$$\sigma_{23} = \sigma_{23}^{+} + \sigma_{23}^{-} = R\cos\theta_1 + R'\cos\theta_1' \qquad (6)$$

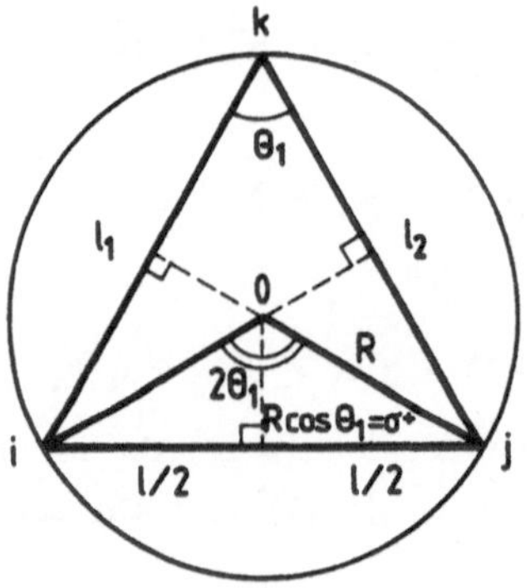

Figure 1

It is easy to show that from the lengths ℓ_{ij} alone one can construct the coefficients $\frac{\sigma_{ij}}{\ell_{ij}}$. Each one is a sum of two terms pertaining to adjacent triangles. one of them, say a_{ij}, is given as follows

$$a = \frac{\sigma^{+}}{\ell} = \frac{\cos\theta}{2\sin\theta} = \frac{1}{2}\ \frac{\ell_1^2 + \ell_2^2 - \ell^2}{\sqrt{[(\ell_1+\ell_2)^2-\ell^2][\ell^2-(\ell_1-\ell_2)^2]}}$$

provided of course that triangular inequalities are satisfied. To get $\frac{\sigma_{ij}}{\ell_{ij}}$ just add two such values for adjacent triangles.

The definition (6) agrees with the one given in the random flat case corresponding to a Voronoï construction, where thanks to its definition, the quantity σ_{23} was positive in this case, and equal to the length of the edge common to two adjacent Voronoï cells. The fact that σ_{23} was then positive arose from the circumstance that a circle circumscribed to a triangle contained no other point of the triangulation. But in the general case considered here, the quantities $\sigma_{ij} = \sigma_{ji}$ can unfortunately be negative. With this definition we can rewrite the positive action (4), by collecting the contribution of edges, as

$$S_{discrete} = \frac{1}{2} \sum_{(ij)} \frac{\sigma_{ij}}{\ell_{ij}} (\varphi_i - \varphi_j)^2 \tag{7}$$

The positivity of S (actually non negativity) entails certain inequalities, like

$$\sum_{j(i)} \frac{\sigma_{ij}}{\sigma_{ij}} \geq 0 \tag{8}$$

Anyhow, we have now σ_{ij} defined for each virtual dual link and of course $\sigma_{ijk} = 1$. It remains to define σ_i's as areas of virtual dual cells, with the constraint that $\sum_i \sigma_i = \sum_{(ijk)} \ell_{ijk}$ = total area. A natural definition is

$$\sigma_i = \frac{1}{4} \sum_{j(i)} \sigma_{ij} \, \ell_{ij} \tag{9}$$

and one verifies of course that the above requirements are satisfied. Again (9) reduces to t'e flat space definition of the area of a cell in a Voronoï construction. But here σ_i is in general algebraic. At least inequality (8) goes in the right direction to show that if the ℓ_{ij}'s are not too different, σ_i is likely to be positive.

For the time being we shall assume that no σ is zero. A small deformation of the surface would likely restore this condition.

We can repeat the construction of the operators d and d^* defined in [2] and analogous to gradient (and curl) and divergence on the surface. First apart from scalar fields φ_i, we introduce vector fields $\varphi_{ij} = -\varphi_{ij}$ associated to links, and antisymmetric tensor fields (pseudo scalars) $\varphi_{ijk} = (-1)^P \varphi_{p_i p_j p_k}$ associated to triangles. Then we set

$$\begin{aligned}
\varphi_i &\to (d\varphi)_{ij} = \frac{\varphi_i - \varphi_j}{\ell_{ij}} \qquad\qquad (\ell_i \equiv 1) \\
\varphi_{ij} &\to (d\varphi)_{ijk} = \frac{\ell_{ij}\varphi_{ij} + \ell_{jk}\varphi_{jk} + \ell_{ki}\varphi_{ki}}{\ell_{ijk}} \\
\varphi_{ijk} &\to 0
\end{aligned} \tag{10}$$

$$\begin{aligned}
\varphi_i &\to 0 \\
\varphi_{ij} &\to (d^* \varphi)_i = \frac{1}{\sigma_i} \sum_{j(i)} \sigma_{ij} \, \varphi_{ij} \\
\varphi_{ijk} &\to (d^* \varphi)_{ij} = \frac{1}{\sigma_{ij}} \sum_{k(ij)} \varphi_{ijk} \qquad\qquad (\sigma_{ijk} \equiv 1)
\end{aligned}$$

These operations lead to a natural definition of Laplacians. Thus for instance for scalars

$$(-\Delta\varphi)_i = (d^*d\varphi)_i = \frac{1}{\sigma_i}\sum_{j(i)} \frac{\sigma_{ij}}{\ell_{ij}}(\varphi_i-\varphi_j) \tag{11}$$

which enables one to "integrate by parts" in (7) with the result

$$S_{\text{discrete}} = \frac{1}{2}\sum_i \sigma_i \varphi_i \; (-\Delta\varphi)_i \tag{12}$$

Of course Δ has dimension $(\text{length})^{-2}$ while S is dimensionless, so is φ as it should in two dimensions. Equation (4) indicates that on a compact surface $S_{\text{disc.}} > 0$ as soon as two adjacent φ_i's are unequal ; it follows that the only harmonic functions are constants, and $-\Delta$ is a non negative operator with respect to the (possible indefinite) square norm $\sum_i \sigma_i \varphi_i^2$.

It is interesting to note that for pseudoscalar fields assigned to (oriented) triangles one can write a formula analogous to (7) namely

$$\frac{1}{2}\sum_{(ijk)} 1_{ijk}\,\varphi_{ijk}\,(dd^*\varphi)_{ijk} = \frac{1}{2}\sum_{(ij)} \frac{\ell_{ij}}{\sigma_{ij}}(\varphi_{ijk}-\varphi_{jik'})^2 \tag{13}$$

where (ijk) and (jik') are the two triangles adjacent to the link ij with compatible orientations. Typically ℓ's and σ's appear interchanged as compared to (7). Assuming all σ's positive, then again an harmonic pseudo scalar is a "constant" namely a constant multiple of the orientation η_{ijk} $(=\pm 1)$. Furthermore for vector fields φ_{ij}, harmonicity is equivalent to being both divergenceless and curl-less

$$[(dd^*+d^*d)\varphi]_{ij} = 0 \leftrightarrow \left\{(d\varphi)_{ijk} = 0,\ (d^*\varphi)_i = 0\right\} \tag{14}$$

Indeed if φ_{ij} is harmonic the $\psi_i = (d^*\varphi)_i$ satisfies $d^*d\psi = d^*dd^*\varphi = -d^{*2}d\varphi = 0$ so ψ is constant and $\sum_i \sigma_i\psi_i = \psi\times\text{Area} = \sum_i \sigma_i (d^*\varphi)_i = \sum_i\sum_{j(i)} \sigma_{ij}\varphi_{ij} = 0$, hence $\psi=0$. So $d^*d\varphi = 0$ and $dd^*(d\varphi) = 0$, hence $d\varphi$ is harmonic, therefore a constant multiple $\tilde{\psi}$ of the orientation : $\tilde{\psi}\times\text{Area} = \sum_{(ijk)} \eta_{ijk}\,\ell_{ijk}(d\varphi)_{ijk} = \sum_{(ijk)} \eta_{ijk}(1_{ij}\varphi_{ij}+1_{jk}\varphi_{jk}+\ell_{ki}\varphi_{ki})$ $= 0$. So (14) is fully justified. Incidentally (10) also shows that $d^*\eta = 0$. It is easy to show that the usual counting of harmonic vector fields still holds in this discrete context. Indeed there are N_1 linearly independent φ_{ij}'s and from (14) the number of conditions is $(N_0-1)+(N_2-1)$ (since $\sum_{ijk} \ell_{ijk}\,\eta_{ijk}\,(d\varphi)_{ijk} = 0$ and $\sum_i \sigma_i (d^*\varphi)_i = 0$). We then come to the conclusion that there are $N_1-(N_0+N_2-2) = 2-\chi = 2g$ linearly independent solutions to (14).

For the time being, we continue to assume the σ's to be positive and promote free field theory from classical to quantum mechanical by constructing the corresponding path integral. This requires defining properly the a priori measure (on fields). Let us ignore for the moment the zero mode problem (for scalar fields $\varphi_i = \text{cst}$) or else introduce a mass term $\frac{m^2}{2}\sum_i \sigma_i\,\varphi_i^2$ in the action, with m having the dimension of an inverse length.

The a priori measure is not the product of Lebesque measures of the φ_i's, or else we would miss the two dimensional "dilatation anomaly" of the continuous case. Since we are interested in comparing results for different surfaces this difference matters. The point is that we want the free field path integral to be related to the determinant of $-\Delta$ in the subspace orthogonal to its zero eigenvalue mode, call it $\det'(-\Delta)$ or equivalently to the product of all its non zero (hence positive) eigenvalues E_n, $1 \leq n \leq N-1$, $E_o = 0$. Let $\psi^{(n)}$ denote the corresponding eigenfunctions of $-\Delta$ and expand the field φ in eigenmodes as

$$\varphi_i = \sum_n c_n \psi_i^{(n)} \tag{15}$$

Then

$$S_{disc} = \frac{1}{2} \sum_n E_n \varphi_n^2 \tag{16}$$

$$Z = \frac{1}{\left(\prod_1^{n-1} E_n\right)^{1/2}} = \frac{1}{(\det'-\Delta)^{-1/2}} = \sqrt{2\pi}\int \prod_0^{N-1} \frac{dc_n}{\sqrt{2\pi}} \delta(c_o) \quad e^{-S_{disc}} \tag{17}$$

The ψ's are normalized through $\delta^{n_1 n_2} = \sum_i \sigma_i \psi_i^{(n_1)} \psi_i^{(n_2)}$. Hence $\prod_o^{N-1} dc_n = (\prod_i \sigma_i)^{1/2} \prod_i d\varphi_i$. If A is the total area then $\psi_i^{(o)} = \frac{1}{\sqrt{A}}$ and $c_o = \frac{1}{\sqrt{A}} \sum_i \sigma_i \varphi_i$. Thus

$$Z = \sqrt{2\pi} \int \prod_1 \frac{d\varphi_i}{\sqrt{2\pi}} (\prod_i \sigma_i)^{1/2} e^{-S_{discr.}(\varphi)} \delta\left(\frac{1}{\sqrt{A}} \sum_i \sigma_i \varphi_i\right) \tag{18}$$

and this can readily be extended to include an external source. In flat space the extra factor plays of course no role, but it is seen to restore sensitivity to scale transformations, if only in the crudest sense, when we dilate all lengths by a constant factor.

In the best of all worlds we would like to estimate (18), a finite integral. Since we are unable to do so in a manageable way, we turn now to a continuum evaluation of it.

V. CONFORMAL ANOMALY

In a continuous theory the action (IV-1) is conformally invariant. This is best appreciated if one recalls that *locally* any two dimensional metric can be written (isothermal coordinates)

$$g_{ab} = \rho^2 \delta_{ab} \qquad \sqrt{g} = \rho^2 \tag{1}$$

in a coordinate patch. As an example on a sphere of radius r by projection on a plane ρ_1, ρ_2 we can write

$$ds^2 = \rho^2 d\alpha^2 \qquad \rho = \frac{1}{1 + \frac{\alpha^2}{4r^2}} \tag{2}$$

this being valid on all the sphere except one point. Then form (1)

$$S_{cont} = \frac{1}{2}\int d^2\alpha \sum_{a=1,2} \partial_a \varphi\, \partial_a \varphi \tag{3}$$

which makes obvious the independence over the local scale parameter ρ.On the other hand the Laplacian is

$$\Delta\varphi = \frac{1}{\rho^2}\sum_{a=1}^{2} \partial_a^2 \varphi \tag{4}$$

and does depend on ρ. The free field path integral is required to be some renormalized verion of the determinant of $-\Delta$ to the power $-1/2$ with the zero eigenvalue omitted. The ultraviolet difficulty is related to the continuum infinite number of modes. We then consider the variation of Z under a local variation of scale

$$\delta\, \ell n\, Z = \delta\, \ell n \int \pi' \frac{d\varphi(m)}{\sqrt{2\pi}}\, e^{-\frac{1}{2}\Sigma E_n \varphi^2_{(n)}} \tag{5}$$

Again the field has been expanded in eigenmodes of the Laplacian with amplitudes $\varphi_{(n)}$ so that the action $\Sigma E_n \varphi^2_{(n)}$ is invariant under changes of ρ. As ρ varies however, for a fixed field , both the $\varphi_{(n)}$ and the E_n (eigenvalues of $-\Delta$)vary. And of course π' is to indicate that the zero mode is omitted. Call the corresponding amplitude $\varphi_{(o)}$ ($E_o = 0$). Therefore

$$\delta\, \ell n\, Z = \left< \left(\sum_1^\infty \frac{\partial\delta\varphi_{(n)}}{\partial\varphi_{(n)}} - \frac{\partial\delta\varphi_{(o)}}{\partial\varphi_{(o)}} \right) \right> \tag{6}$$

arising from the Jacobian of the transformation from $\varphi_{(n)} + \delta\varphi_{(n)}$ to $\varphi_{(n)}$. Since the field is invariant

$$0 = \delta\varphi = \sum_0^\infty \delta\varphi_{(n)}\, \psi^{(n)} + \varphi_{(n)}\, \delta\psi^{(n)} \tag{7}$$

and

$$\frac{\partial\delta\varphi_{(n)}}{\partial\varphi_{(n)}} = -\int d^2\alpha\, \rho^2\, \psi^{(n)}\, \delta\psi^{(n)} = +\int d^2\alpha\, \rho\delta\rho\, {\psi^{(n)}}^2 \tag{8}$$

The quantity in (6) does not require to be computed in the mean, since it is $\varphi_{(n)}$ independent. Therefore

$$\delta\, \ell n\, Z = \int d^2\alpha\, \rho\delta\rho \sum_0^\infty {\psi^{(n)}}^2 - \frac{1}{2A}\,\delta A \tag{9}$$

The second term is an effect of the zero mode subtraction. The first one is ultraviolet infinite, due to the infinity of modes ,and requires a further subtraction.

It is regularized in the most natural way, as

$$\sum_0^\infty \psi^{(n)^2}(\alpha) \underset{s\to 0}{\to} \sum_0^\infty \psi^{(n)^2}(\alpha)\ e^{-sE_n} = U(s;\alpha,\alpha) \tag{10}$$

where

$$U(s;\alpha,\beta) = U(s;\beta,\alpha) = \sum_0^\infty \psi^{(n)}(\alpha)\ \psi^{(n)}(\beta)\ e^{-sE_n} \tag{11}$$

$$\left(\frac{\partial}{\partial s} - \Delta\right) U(s;\alpha,\beta) = 0 \qquad s > 0$$

$$\lim_{s\to 0} U(s;\alpha,\beta) = \delta_{\text{inv.}}(\alpha,\beta) \tag{12}$$

The invariant Dirac distribution satisfies

$$\varphi(\alpha) = \int d^2\beta\ \rho^2(\beta)\ \delta_{\text{inv.}}(\alpha,\beta)\ \varphi(\beta) \tag{13}$$

Obviously $\delta_{\text{inv.}}(\alpha,\beta) = \sum_0^\infty \psi^{(n)}(\alpha)\ \psi^{(n)}(\beta)$ is singular when $\beta \to \alpha$.

We consider therefore

$$\delta\ \ell n\ Z_{\text{reg}}(s) = \int d^2\alpha\ \rho^2\ \frac{\delta\rho}{\rho}\ U(s;\alpha,\alpha) - \frac{1}{2A}\ \delta A \tag{14}$$

and ask for the small s behaviour. We expect terms of order $\frac{1}{s}$ and finite ones as $s\to 0$. In flat space

$$U_{\text{flat}}(s;\alpha,\beta) = \frac{e^{-\frac{1}{4s}(\alpha-\beta)^2}}{4\pi s} \tag{15}$$

In that case $U_{\text{flat}}(s;\alpha,\alpha) = \frac{1}{4\pi s}$ independently of α. In curved space, and for dimensional reasons, we expect $U_{\text{curved}}(s;\alpha,\alpha) = \frac{1}{4\pi s}\ (1 + a\ R_\alpha s + \ldots)$ where R_α is the curvature at α. If this is so, it is sufficient to do the calculation for a sphere with the curvature R equal to r^{-2}. Then to first order in R, using (2), we find in isothermal coordinates

$$U_{\text{sphere}}(s;\alpha,0) = \frac{e^{-\frac{\alpha^2}{4s}}}{4\pi s}\left[1 + \frac{Rs}{3}\left(1 + \frac{1}{4}\ \frac{\alpha^2}{s} + \frac{1}{8}\ \frac{\alpha^4}{s^2}\right) + \ldots\right] \tag{16}$$

and therefore

$$\begin{aligned} U_{\text{curved}}(s;\alpha,\alpha) &= \frac{1}{4\pi s}\left[1 + \frac{R_\alpha s}{3} + \ldots\right] && \text{(a)} \\ &= \frac{1}{4\pi s} - \frac{1}{12\pi}\ \Delta_{\text{curv.}}\ \ell n\ \rho && \text{(b)} \end{aligned} \tag{17}$$

The second expression is easily checked in the case of the sphere once again. Consequently, we end up with

$$\delta \ \ln Z_{reg} = \left(\frac{1}{8\pi s} - \frac{1}{2A}\right) \delta A + \frac{1}{12\pi} \int d^2\alpha \ \rho^2 (\delta \ \ln \rho)(-\Delta_{curv} \ \ln \rho)$$

$$= \delta\left[\frac{A}{8\pi s} - \frac{1}{2} \ln \frac{A}{A_o} + \frac{1}{24\pi} \int d^2\alpha \ \rho^2 \ \ln \rho \ (-\Delta_{curv}) \ \ln \rho\right] \tag{18}$$

We note that (17a) could be slightly improved to read for small s and in the vicinity of a point α

$$U_{curved}(s;\alpha,\beta) = \frac{e^{-d^2_{\alpha\beta}/4s}}{4\pi s} \left[1 + R_\alpha \frac{s}{3} + R_\alpha \frac{d^2_{\alpha\beta}}{12} + \dots\right]$$

ignoring derivatives of the curvature. Here $d_{\alpha\beta}$ is the geodesic distance between α and β. We also observe that the factor $\frac{1}{4\pi s}$ in the flat case can be interpreted as the inverse area of a circle of radius $r_{eff} = 2\sqrt{s}$ as is reasonable in a Brownian motion interpretation of U as a probability density. Then on a curved surface a circle of same radius (assumed to be small) has an area $(\approx 2\pi R^{-1}(1-\cos\theta) \approx 2\pi R^{-1}\left(\frac{\theta^2}{2} - \frac{\theta^4}{24}\right)$ with $\theta = R^{+1/2}\, 2\sqrt{s}$, i.e. $4\pi s\left(1 - \frac{Rs}{3}\right)$. The inverse of this area expanded in powers of $Rs \ll 1$ yields precisely (17a), a neat way of understanding this expression.

When discussing (18) we first realize a shortcoming : as was emphasized ρ can only be defined in a defined in a coordinate patch, which means sensitivity to boundary conditions, if we want to interpret $\ln Z_{reg} - A\left(\frac{1}{s} - \frac{1}{s}\right)$ as a $\ln Z_{renormalized}$. This is exmplified in the case of a sphere with ρ given by (29).While it is tempting to relate $\int d^2\alpha \ \rho^2 \ \ln \rho(-\Delta_{curved} \ \ln \rho) = \int d^2\alpha \ \ln \rho - \Delta_{flat} \ \ln \rho$ to $\int d^2\alpha \sum_{a=1}^{2} (\partial_a \ \ln \rho)^2$

the first integral is finite and the second is not. Note that with our convention

$$-\Delta_{curved} \ \ln \rho = R \tag{20}$$

showing why for a compact surface with non vanishing Euler characteristic

$$\chi = \frac{1}{2\pi} \int d^2\alpha \ \rho^2 \ R \tag{21}$$

$\ln \rho$ cannot be defined everywhere as a non singular function. In (21) the interpretation is that $d^2\alpha \ \rho^2$ is an invariant element of area, and several coordinate patches may be needed to compute the integral. Of course this is nothing but the Gauss Bonnet theorem. We do not have these difficulties, on a torus, $\chi=0$, and we therefore assume such a simplification from now on. These subtleties do not play any role in a variation of $\delta \ \ln Z_{ren.}$ when $\delta\rho$ is contained in a coordinate patch. We emphasize again that the non local term $\ln \frac{A}{A_o}$ is due to the fact that we consider compact surfaces.

To proceed further, we define $\tilde{G}_{curved}(\alpha,\beta)$ as a subtracted propagator -again because of the zero mode problem-as

$$\tilde{G}(\alpha,\beta) = \sum_{1}^{\infty} \frac{1}{E_n} \psi^{(n)}(\alpha)\psi^{(n)}(\beta) \quad ; \quad -\Delta_{curv} \tilde{G}(\alpha,\beta) = \delta_{cur}(\alpha,\beta) - \frac{1}{A} \tag{22}$$

Then for $\chi=0$

$$\ell n\ \rho(\alpha) = \int d^2\beta\ \rho^2(\beta)\ \tilde{G}(\alpha,\beta)\ R(\beta) \tag{23}$$

and the r.h.s. is insensitive to the addition to $\tilde{G}(\alpha,\beta)$ of an arbitrary constant. We therefore have

$$\chi = 0 \quad \ell n\ Z_{ren} = \frac{A}{s_0} - \frac{1}{2}\ell n \frac{A}{A_0} + \frac{1}{24\pi}\iint d^2\alpha\ d^2\beta\ \rho^2(\alpha)\ \rho^2(\beta)\ R(\alpha)\ \tilde{G}(\alpha,\beta)\ R(\beta) \tag{24}$$

up to an arbitrary additive constant. Each elementary piece of the surface is endowed with a curvature charge

$$dq(\alpha) = d^2\alpha\ \rho^2(\alpha)\ R(\alpha) \tag{25}$$

The total charge is zero

$$\int dq(\alpha) = 0 \tag{26}$$

and $\tilde{G}(\alpha,\beta)$ has a (classical two-dimensional) long range character of a Coulomb potential. For all its aesthetic appeal, equation (24) is not that great a simplification. Z was given in terms of the product of eigenvalues and it requires as much effort to compute $\tilde{G}$. On the other hand given a metric in the form (1), it is easy to find the variations of $\ell n\ Z_{ren}$. by simple integrations.

We have tried to find analogs of (18) and (24) in the discrete situation of the previous sections but failed to obtain "nice" expressions. This is perhaps not unrelated to localization problems on such a random lattice. However inspired by (24), the so called "Liouville action", we may obtain a natural discretization for it. First the analog of $\tilde{G}$ is the solution of the equation

$$\sum_{k(i)} \frac{\sigma_{ik}}{\ell_{ik}} \left(\tilde{G}_{(i,j)} - \tilde{G}_{(k,j)}\right) = \delta_{ij} - \sigma_i \tag{27}$$

To make things more symetric, we observe that, adding a small mass term to the Laplacian shifts all eigenvalues but a constant positive amount. The lowest eigenmode with positive eigenvalue is orthogonal to the curvature for $\chi=0$(call it the neutral case), hence the limit $m^2 \to 0$ is well defined. We can replace $\tilde{G}$ by G such that

$$\sum_{k(i)} \frac{\sigma_{ik}}{\ell_{ik}} [G(i,j;m^2)-G(k,k;m^2)] + m^2\sigma_i\ G(i,j;m^2) = \delta_{ij} \tag{28}$$

and write in discretized form

$$S_{\text{Liouville}} = \lim_{m^2 \to 0} \frac{1}{2} \sum_{i,j} q_i G(i,j;m^2)\, q_j$$

$$q_i = \left(1 - \frac{\theta_i}{2\pi}\right) \quad ; \quad \sum_i q_i = 0 \tag{29}$$

The limit $m^2 \to 0$ has to be taken when the double sum is first computed. We can relax here the condition that the σ's be positive. If we define

$$\psi_i = \sum_{m^2 \to 0\; j} G(i,j;m^2)\, q_j \tag{30}$$

Then

$$S_{\text{Liouville}} = \frac{1}{2} \sum_i \sigma_i\, \psi_i\, (-\Delta\psi)_i \tag{31}$$

In the neutral case, (30) has a limit when $m^2 \to 0$, ψ remains finite, hence $S_{\text{Liouville}}$ is positive.

VI. RANDOM SURFACES

As we discussed in the introduction there is not at the moment a consensus on good candidates for models of random surfaces (the most random ones !). Here we limit ourselves to repeat the analysis of the simplest one [10][11], and suggest some "natural" modification.

Take a torus ($\chi=0$). Choose a triangulation. Then map it in $\mathbb{R}^d$ as explained above. The simplest statistical weight is

$$d\mu = \pi' \, d^d x_i \, e^{-\beta_o S_o} \tag{1}$$

$$S_o = \frac{1}{a^2} A = \frac{1}{a^2} \sum_{(ijk)} \ell_{ijk} \,, \quad a \text{ : arbitrary length unit.} \tag{2}$$

This looks like the direct analog of the Brownian case and we may wish to estimate what happens when the number points gets large and the parameter points dense on the torus. This is was has been done numerically and analytically (for large d) by Billoire Gross and Marinari [10]. Let us briefly recapitulate what happens following Duplantier [11]. Use for instance the triangulation shown on Figure 2, with $n=n_1=n_2$, hence $N_o=n^2$, $N_1=3n^2$, $N_2=2n^2$. The base space has translational invariance, and it is assumed unbroken in the statistical model. Set for simplicity $f_{ijk} = \beta_o \frac{\ell_{ijk}}{a^2}$. Then from dilatation covariance we have

$$Z = \int \pi' \, d^d x_i \, e^{-\beta_o S_o} = \lambda^{d(n^2-1)} \int \pi' \, d^d x_i \, e^{-\lambda^2 \beta_o S_o}$$

$$\langle \beta_o S_o \rangle = (n^2-1)\, \frac{d}{2} \approx N\, \frac{d}{2} \qquad \langle f \rangle = \frac{d}{4} \tag{2}$$

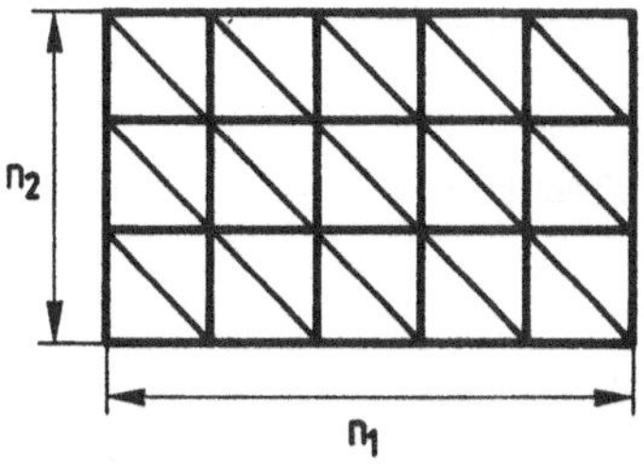

Figure 2

This suggests that the limit $d \to \infty$ is a good place to look for a valid saddle point approximation, to which one then proceeds. One rewrites

$$Z = \int \pi' \, d^d x_i \int \pi \frac{d\lambda_{ij}}{2\pi i} \, d\ell_{ij} \; e^{\sum\limits_{(i)} \lambda_{ij}[(x_i - x_j)^2 - \ell_{ij}^2] - \beta_o S_o\{\ell_{ij}\}} \tag{3}$$

with λ_{ij} integrated over the imaginary axis (anticipating that the saddle point value will then be real) and assumes a (base space) translational invariant saddle point with $\lambda_{ij}^o \equiv \lambda$, $\ell_{ij}^o = \ell$. This leads to

$$Z \to Z_o = \text{cst} \quad e^{-S_{eff}(\lambda,\ell)}$$

$$S_{eff} = N\left[\frac{d}{2} \ln \lambda - 3\lambda \, \ell^2 + 2f(\ell,\ell,\ell)\right] \tag{4}$$

where $f(\ell,\ell,\ell) = \beta \frac{\sqrt{3}}{4} \frac{\ell^2}{a^2}$ and the elementary area is effectively the one for an equilateral triangle. The variational equations to be satisfied by λ and ℓ reduce to

$$\lambda = \frac{d}{6\ell^2} \qquad f = \beta_o \frac{\sqrt{3}}{4} \frac{\ell^2}{a^2} = \frac{d}{4} \tag{5}$$

The second equation has of course to agree with (2). Therefore $S_{eff} = \text{cst} + \frac{Nd}{2} \ln \beta_o$ and $\beta \frac{dS_{eff}}{d} = \frac{Nd}{2} = \langle S_o \rangle$ as it should. Equipped with these values of λ and ℓ from (5) we see from (3) that the statistical weight reduces to d-uncoupled gaussian (free field) models (in flat space) for any question pertaining to the x_i's. In particular it is straightforward to see that $\langle (x_i - x_j)^2 \rangle = \ell^2$ and that

$$\begin{aligned} R_N^2 &= \frac{1}{N} \langle \sum x_i^2 \rangle \\ &= \frac{1}{\pi} \frac{\sqrt{3}}{4} \ell^2 \ln N = \frac{1}{\pi} \langle \text{elementary area} \rangle \ln N \end{aligned} \tag{6}$$

The $\ln N$ term results from the infrared behavior of the 2d massless propagator very much as in section III the linear behavior in N arose from a similar infrared singularity of the one-dimensional propagator. The numerical factor on the r.h.s. of (6) is specific to triangulations. It behaves like d/4 for large d and is corrected by a factor (1 + 2/d + ...) to first order in 1/d. This is in fair agreement with the observations made in [10] as shown in the following table giving $R_N^2 \frac{4\pi}{\sqrt{3}\,\ell^2} \cdot \frac{1}{\ell n\, N}$

d	First order correction $1 + \frac{2}{d}$	Numerical data[10]
3	1.667	1.73 ± 0.04
4	1.500	1.43 ± 0.03
6	1.333	1.27 ± 0.01
12	1.167	1.16 ± 0.03

It is unlikely that inequivalent triangulations of the torus would lead to a qualitatively different result or that going to a surface with a different topology would change the $\ell n\, N$ behavior also observed in numerical computation. This means that the image surface is a strongly *collapsed* object in the mean.

Our discussion of the last section suggests to modify equations (1), (2) by taking as probability weight (forgetting about the $\ell n\, A/A_o$ term presumably irrelevant as compared to S_o)

$$d\mu = \prod_i{}' d^d x_i \; e^{-\beta_o S_o - \beta_1 S_{Liouville}} \tag{7}$$

as a slight generalization of the above model -where $S_{Liouville}$ has been written in (V.29) or (V.31).

A rough way to estimate the size of the Liouville term would be, in the same type of simulations that have been performed before, to compute $\langle q_i^2 \rangle$. Or else one could try to estimate it from the Gaussian model emerging from (3). We believe that it is of order 1. Hence the Liouville term could become relevant even when $d \to \infty$ provided we take β_1 of order d. This would then require to find different saddle points.

To see what is involved let us look at the Gaussian model implied by (3). Since the sum of angles of a given triangle is π and each of them has equal mean value it is indeed very reasonable that their average value is $\frac{\pi}{3}$. Then look at figure 3 which involves the points A, B, with A second neighbor to B, and compute $(N \to \infty)$

$$\frac{\ell'^2}{\ell^2} = \frac{1}{(2\pi)^2} \int_{-\pi}^{+\pi}\int_{-\pi}^{+\pi} dk_1 dk_2 dk_3 \, \frac{3 - \cos(k_1-k_2) - \cos(k_2-k_3) - \cos(k_3-k_1)}{3 - \cos k_1 - \cos k_2 - \cos k_3} \, \delta(k_1+k_2+k_3) \tag{8}$$

$$= \frac{9}{2} - \frac{6\sqrt{3}}{\pi} = 1.192$$

The points ABC fit in a plane and either O is in this plane or else OABC fits in a 3d linear subspace. If the picture were flat and regular (as on the regular triangulated plane) then $\frac{\ell'^2}{\ell^2} = 3$. Apparently this is not the case showing the importance of fluctuations. Most likely $\langle q_i^2 \rangle$ does *not* vanish with $\frac{1}{d}$.

We can also rewrite (7) using a Lagrange multiplier

$$d\mu = \frac{\pi' \, d^d x_i \; \pi'_i \, d\psi_i \; e^{-\beta_o S_o - \beta_i S_{\text{free field}}(\psi) + i\sum_i \psi_i q_i}}{\int \pi'_i \, d\psi_i \; e^{-\beta_1 S_{\text{free field}}(\psi)}} \tag{9}$$

where $S_{\text{free field}}(\psi)$ is given by (IV.4) or (IV.12).

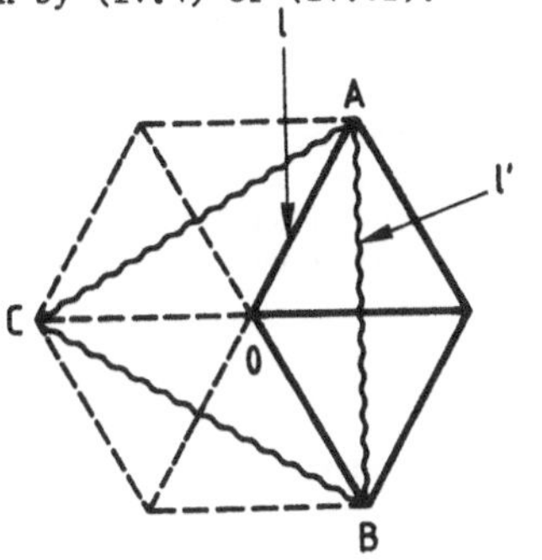

Figure 3

It is perhaps easier numerically to deal with (9) than with (7). At any rate they provide a non trivial discretization of the Liouville theory and it is perhaps worthwhile to investigate their relevance to a model of random surfaces. Of course this is only a very first step towards proposing a consistent scheme to study quantized gravity in a discretized version, which is one of the motivations to get a better understanding of random geometry.

REFERENCE

[1] N.M. Christ, R. Friedberg, T.D. Lee, Nucl. Phys. B202, 89 (1982), B210 (FS6) 310, 337 (198).
T.D. Lee "Discrete mechanics", to be published in the proceedings of the International School of Subnuclear Physics, Erice, 1983.

[2] C. Itzykson "Fields on a random Lattice", to be published in the proceedings "Progress in Gauge fields Theory", Cargèse 1983.

[3] T. Regge, Nuovo Cimento 19, 551 (1961).

[4] K. Wilson, Phys. Rev. D10, 2445 (1974).

[5] D.J. Wallace in "Recent advances in field theory and statistical mechanics" , J.B. Zuber and R. Stora, eds, Les Houches 1982, North Holland Amsterdam (1984).

[6] A.M. Polyakov, Phys. Lett. B103, 1207 (1981).

[7] G. Parisi, Phys. Lett. B81, 957 (1979).

[8] J. Fröhlich "The statistical mechanics of surfaces" presented at the Sitges meeting (1984).

[9] "String theories", ed. by Jacob, North Holland , Amsterdam (1974)

[10] A. Billoire, D.J. Gross, E. Marinari, Phys. Lett. 139B, 239 (1984)
D.J. Gross, Phys. Lett. 138B, 185 (1984).

[11] B. Duplantier , Phys. Lett. 141B, 239 (1984).

[12] J. Cheeger, W. Müller, R. Schrader, Comm. Math. Phys. 92, 405 (1982).

[13] K. Fujikawa, Phys. Rev. D21, 2848 (1980), D23, 2262 (1981).

[14] R. Friedberg, T.D. Lee "Derivation of Regge's action from Einstein's Theory of General Relativity" Columbia preprint CU-T.P-281.
C. Feinberg, R. Friedberg, T.D. Lee, M.C. Ren, "Lattice gravity near the continum limit" - Columbia preprint CU-T.P-281.

[15] H.W. Hamber, R.M. Williams, "Higher derivative quantum gravity on a simplicial lattice" IAS preprint (1984).

[16] F. David, "Planar Diagrams, two dimensional lattice gravity and surface models", Saclay preprint SPhT/84/25 (1984).

PRODUCTS OF RANDOM MATRICES AND ONE DIMENSIONAL DISORDERED SYSTEMS

Bernard Derrida
Service de Physique Théorique
CEN-Saclay,
91191 Gif sur Yvette, Cedex, France

ABSTRACT

Several examples of disordered one dimensional models are discussed : a spin glass chain, an Ising chain in a random field, the diffusion on a random chain, a Schrödinger equation with a random potential. For each problem, one can develop analytic methods to expand the Lyapounov exponent.

INTRODUCTION

A lot of problems in the physics of one dimensional disordered systems can be reduced to the study of products of random matrices [1,2,3] : spin models with random interactions (spin glass), random field models, Schrödinger equation with a random potential, diffusion on a random chain, etc... One usually needs to calculate the Lyapounov exponent γ of a product of independent random matrices M_n which are randomly distributed according to a probability distribution which depends on the problem that one considers

$$\gamma = \lim_{N\to\infty} \frac{1}{N} \log \left[\mathrm{tr} \left(\prod_{n=1}^{N} M_n \right) \right] \tag{1}$$

For an Ising chain with random interactions J_i

$$\mathcal{H} = - \sum_i J_i \, \sigma_i \sigma_{i+1} - h \sum_i \sigma_i \tag{2}$$

the random matrices M_n are given by

$$M_n = \begin{pmatrix} e^{J_n + h} & e^{-J_n + h} \\ e^{-J_n - h} & e^{J_n - h} \end{pmatrix} \tag{3}$$

and the Lyapounov exponent gives the free energy of the chain.

Similarly for an Ising chain in a random field

$$\mathcal{H} = -J \sum_i \sigma_i \sigma_{i+1} - \sum_i h_i \sigma_i \tag{4}$$

the random matrices have the following form

$$M_n = \begin{pmatrix} e^{J+h_n} & e^{-J+h_n} \\ e^{-J-h_n} & e^{J-h_n} \end{pmatrix} = e^{J+h_n} \begin{pmatrix} 1 & e^{-2J} \\ e^{-2J-2h_n} & e^{-2h_n} \end{pmatrix} \tag{5}$$

and the Lyapounov exponent gives again the free energy.

For a one dimensional Schrödinger equation with a random potential λV_n

$$\psi_{n+1} + \psi_{n-1} + \lambda V_n \psi_n = E \psi_n \tag{6}$$

the random matrices M_n are given by

$$M_n = \begin{pmatrix} E-\lambda V_n & -1 \\ 1 & 0 \end{pmatrix}$$

and the Lyapounov exponent is related to the localisation length and to the density of states by the Thouless formula [4].

The main difficulty with products of random matrices is that there does not exist any general analytic method to calculate the Lyapounov exponents. Only two approaches are available. Either one can calculate numerically the Lyapounov exponent. Or one can try to make analytic expansions around well understood situations : weak disorder expansions, expansions around the situation where the matrices M_n commute, etc...

In this paper I shall present several analytic expansions which can be done for the models defined above.

I. A SPIN GLASS CHAIN IN A UNIFORM FIELD [5-7]

Let us consider an Ising chain described by the hamiltonian (2) with random interactions J_i distributed according to a given distribution $\rho(J_i)$. It is possible to find analytically the Lyapounov exponent γ in the limit of zero temperature for any value of the uniform field h if the distribution $\rho(J_i)$ is the sum of two delta functions [5]

$$\rho(J_i) = \frac{1}{2}\left[\delta(J_i-J) + \delta(J_i+J)\right] \tag{7}$$

and only for small h for arbitrary distributions $\rho(J_i)$ which behave near $J_i=0$ as

$$\rho(J_i) \sim |J_i|^K \tag{8}$$

To explain the idea which allows to solve the problem at 0 temperature, let us take the case of the distribution (7). Then we have to calculate the Lyapounov exponent of a product of matrices M_n which can take two possible forms :

$$M_n = \begin{pmatrix} z^{1+\alpha} & z^{-1+\alpha} \\ z^{-1-\alpha} & z^{1-\alpha} \end{pmatrix} \quad \text{or} \quad M_n = \begin{pmatrix} z^{-1+\alpha} & z^{1+\alpha} \\ z^{1-\alpha} & z^{-1-\alpha} \end{pmatrix} \tag{9}$$

where $z = e^J$ and $\alpha = \frac{h}{J}$. The 0 temperature limit corresponds to $z \to \infty$.

The main simplification which occurs in this limit $z \to \infty$ is the following : when one performs the product of two matrices A and B, each element of the product AB is the sum of products of elements of A and B. In the limit $z \to \infty$, one can replace this sum by its largest term. If we start with an arbitrary vector V_o and we define V_n by

$$V_n = \prod_{i=1}^{n} M_i \ V_o \tag{10}$$

Then if we write

$$V_n = \begin{pmatrix} z^{a_n} \\ z^{b_n} \end{pmatrix} \tag{11}$$

one finds that for the random matrices given by (9), the problem reduces in the limit $z \to \infty$ to the following recursion for the a_n and b_n

$$\begin{cases} a_{n+1} = \max(a_n+1+\alpha \ , \ b_n-1+\alpha) \\ b_{n+1} = \max(a_n-1-\alpha \ , \ b_n+1-\alpha) \end{cases} \quad \text{with probability } \frac{1}{2}$$

and

$$\begin{cases} a_{n+1} = \max(a_n-1+\alpha \ , \ b_n+1+\alpha) \\ b_{n+1} = \max(a_n+1-\alpha \ , \ b_n-1-\alpha) \end{cases} \quad \text{with probability } \frac{1}{2} \tag{12}$$

In the limit $n \to \infty$, the difference a_n-b_n has a stationnary probability distribution which can be calculated exactly [5]. Then one finds that for the distribution (7) of the bonds that the Lyapounov exponent behaves like

$$\gamma = \frac{E}{J} \log z + \text{cste} \qquad \text{in the limit } z \to \infty$$

where E is the ground state energy per spin and is given by

$$E = \frac{J(n^2+3n)+2h(n+1)}{(n+1)(n+2)} \quad \text{when} \quad \frac{2}{n+1} < \alpha < \frac{2}{n} \tag{13}$$

So n is just the integer part of $2/\alpha = 2J/h$.

For more general distributions $\rho(J_i)$ which have a behaviour given by (8), one can find γ at zero temperature for small values of the field h [6,7] and from the expression of γ one can show that the zero temperature magnetization m of a one dimensional spin glass chain is given for small h by

$$m = h^{\frac{K+1}{K+3}} C_K \tag{14}$$

where C_K is a constant which can be calculated [6] from the knowledge of $\rho(J_i)$. It is interesting to notice the non analytic behaviour of the magnetization (for a gaussian distribution of the J_i, $m \sim h^{1/3}$). In this first example, the main simplification was that sums of products of elements could be replaced by the largest term in the sum.

In the next example, we shall consider a case where the matrices M_n almost commute.

II. AN ISING CHAIN IN A RANDOM FIELD [8]

For the Ising chain described by the hamiltonian (4), we need to calculate the Lyapounov exponent $\gamma(\varepsilon)$ of the following product (see (5))

$$\gamma(\varepsilon) = \lim_{N\to\infty} \frac{1}{N} \log \operatorname{tr} \prod_{i=1}^{N} \begin{pmatrix} 1 & \varepsilon \\ \varepsilon z_i & z_i \end{pmatrix} \tag{15}$$

where $\varepsilon = e^{-2J}$ and the z_i are random positive numbers $z_i = e^{-2h_i}$ whose distribution is known when the distribution of random field is known.

For $\varepsilon = 0$ the matrices commute and the problem is therefore very easy

$$\gamma(0) = \max(0, \overline{\log z}) \tag{16}$$

A natural question is to try and expand $\gamma(\varepsilon)$ around $\varepsilon=0$. Let us see how such an expansion can be done. To simplify the discussion let us assume that the distribution of the z_i is such that

$$\overline{\log z} < 0 \tag{17}$$

If we define V_n by

$$V_n = \begin{pmatrix} a_n \\ b_n \end{pmatrix} = \begin{pmatrix} 1 & \varepsilon \\ \varepsilon z_{n-1} & z_{n-1} \end{pmatrix} \begin{pmatrix} a_{n-1} \\ b_{n-1} \end{pmatrix} \tag{18}$$

Then one can use several equivalent definitions to calculate $\gamma(\varepsilon)$

$$\gamma(\varepsilon) = \lim_{N\to\infty} \frac{1}{N} \log \left(\frac{||V_n||}{||V_o||} \right) = \lim_{N\to\infty} \left(\frac{1}{N} \sum_{i=1}^{N} \log R_i \right) \tag{19}$$

where the R_i are defined by

$$R_i = \frac{a_{i+1}}{a_i} \tag{20}$$

It is easy to see from (18) that the R_i satisfy the following recursion relative

$$R_{i+1} = 1+z_i + z_i(\varepsilon^2-1)/R_i \tag{21}$$

To expand $\gamma(\varepsilon)$, one can assume that the R_i can be expanded in the following form

$$\log R_i = A_i\, \varepsilon^2 + B_i\, \varepsilon^4 + \ldots \tag{22}$$

Then (21) gives recursion relations for A_i, B_i etc...

$$A_{i+1} = z_i + z_i A_i \tag{23}$$

$$B_{i+1} + \frac{A_{i+1}^2}{2} = -z_i A_i + z_i B_i - z_i \frac{A_i^2}{2} \tag{24}$$

and the expression of $\gamma(\varepsilon)$ is then given by

$$\gamma(\varepsilon) = \varepsilon^2 \overline{A} + \varepsilon^4 \overline{B} + \dots \tag{25}$$

When one calculates $\overline{A}$, one finds that $\overline{A}$ is finite only if

$$\overline{z} < 1 \tag{26}$$

and is given by

$$\overline{A} = \frac{\overline{z}}{1-\overline{z}} \tag{27}$$

Similarly, when one calculates $\overline{B}$, one finds a more restrictive condition

$$\overline{z^2} < 1 \tag{28}$$

for $\overline{B}$ to be finite and one finds

$$\overline{B} = -\frac{1}{2} \frac{(1+\overline{z})^2 \, \overline{z^2} + 2(\overline{z})^2 (1-\overline{z^2})}{(1-\overline{z})^2 \, (1-\overline{z^2})} \tag{29}$$

One expects that the coefficient of ε^{2p} in the expansion of $\gamma(\varepsilon)$ will be finite only if

$$\overline{z^p} < 1 \tag{30}$$

One can of course ask the question of the behaviour of $\gamma(\varepsilon)$ when $\varepsilon \to 0$ if one of the conditions (26), (28), (30) is not satisfied. For example if the distribution of the z_i is such that

$$\overline{\log z} < 0 \quad \text{but} \quad \overline{z} > 1 \tag{31}$$

Then one can show [8] that $\gamma(\varepsilon)$ has a non analytic behaviour when $\varepsilon \to 0$

$$\gamma(\varepsilon) \sim C\, \varepsilon^{2\alpha} \tag{32}$$

where α is given by the positive solution of

$$\overline{z^\alpha} = 1 \tag{33}$$

Except in a few special cases, the exact calculation of the constant C is a very hard problem [9].

A physical interpretation can be given to the non analytic behaviour (32). The expansion around $\varepsilon = 0$ is a low temperature expansion around the ferromagnetic configuration. When $\gamma(\varepsilon) \sim \varepsilon^2$, the expansion is given by the contribution of configurations with finite clusters of spins flipped. When $\gamma(\varepsilon) \sim \varepsilon^{2\alpha}$, the contribution of configurations with an infinite number of spins flipped is dominant.

This random field Ising is mathematically very similar the problem of the diffusion on a random chain. Consider a particle which diffuses on a chain according to the following Master equation [10,11]

$$\frac{dP_i}{dt} = W_{i,i+1}\, P_{i+1} + W_{i,i-1}\, P_{i-1} - (W_{i+1,i} + W_{i-1,i})\, P_i \tag{34}$$

where P_i denotes the probability of finding the particle at time t on site i and the hopping rates W_{ij} are randomly distributed according to a given probability distribution. When one considers the Laplace transforms $Q_i(\omega)$ of the P_i

$$Q_i(\omega) = \int_0^\infty P_i(r)\, e^{-\omega t}\, dt \tag{35}$$

One can try to expand in the limit $\omega \to 0$ the Lyapounov exponent $\gamma(\omega)$ defined by

$$\gamma(\omega) = \lim_{N\to\infty} \frac{1}{N} \log \left(\frac{Q_N(\omega)}{Q_o(\omega)}\right) \tag{36}$$

where we assume that Q_o and Q_1 have been chosen arbitrarily and that the Q_n for $n \geq 0$ are calculated using (34).

In the expansion, the term linear in ω gives the velocity whereas the coefficient of ω^2 is related to the diffusion constant. As for the random field problem, the condition for the coefficient of ω^2 to be finite (i.e. for the diffusion constant to exist) is stronger than the condition for the coëfficient of ω to be finite (i.e. for the velocity to exist)[11].

We have seen with this second example that the expansion around the situation where the matrices M_n commute may be singular. We shall see in the next example that the weak disorder expansion of the Lyapounov exponent may also become singular.

III. THE ONE DIMENSIONAL SCHRODINGER EQUATION IN A RANDOM POTENTIAL [12,13]

Let us consider now the discretized 1d Schrodinger equation (6) with a random potential λV_n where the V_n are the distributed according to a given distribution $\rho(V_n)$. Like in the previous example, one can introduce the ratios R_n defined by

$$R_n = \frac{\psi_n}{\psi_{n-1}} \tag{37}$$

The R_n satisfy the following recursion

$$R_{n+1} = E - \lambda\, V_n - \frac{1}{R_n} \tag{38}$$

and a possible definition of γ is

$$\gamma = \lim_{N\to\infty} \frac{1}{N} \sum_{n=1}^{N} \log R_n \tag{39}$$

A way of expanding γ around $\lambda = 0$ is to assume that the R_n can be expanded in the following way

$$R_n = A_n \exp(\lambda B_n + \lambda^2 C_n + \ldots) \tag{40}$$

Then the recursion (38) gives recursion relations for the A_n, B_n, C_n etc...

$$\begin{aligned} A_{n+1} &= E - \frac{1}{A_n} \\ A_{n+1} B_{n+1} &= -V_n + B_n/A_n \\ A_{n+1}\left(C_{n+1} + \frac{1}{2} B_{n+1}^2\right) &= \left(C_n - \frac{1}{2} B_n^2\right) / A_n \end{aligned} \tag{41}$$

For any complex value of the energy E which does not belong to the spectrum of the pure system i.e.

$$E \neq 2 \cos q \quad \text{with} \quad q \text{ real} \tag{42}$$

the A_n converge to A the root (with largest modulus) of

$$A = E - 1/A \tag{43}$$

and the expansion of γ is given by

$$\gamma = \log A + \lambda \overline{B_n} + \lambda^2 \overline{C_n} + \ldots \tag{44}$$

From (41) one can calculate $\overline{B}$, $\overline{C}$ etc ... and for a distribution of V_n with zero mean $\langle V_n \rangle = 0$, one finds

$$\begin{aligned} \gamma = \log A &- \frac{1}{2} \lambda^2 \frac{A^2}{(A^2-1)^2} \langle V^2 \rangle \\ &- \frac{1}{2} \lambda^3 \frac{A^3}{(A^2-1)^3} \langle V^3 \rangle \\ &- \frac{1}{4} \lambda^4 \frac{A^4}{(A^2-1)^4} \langle V^4 \rangle \\ &- \frac{1}{2} \lambda^4 \frac{3+2A^2}{A^4-1} \frac{1}{(A^2-1)^4} \langle V^2 \rangle^2 + O(\lambda^5) \end{aligned} \tag{45}$$

This expansion is valid as soon as condition (42) is satisfied. One sees in this expansion that if one tries to approach energies like $E = \pm 2$ (i.e. $A \to \pm 1$) or $E = 0$ (i.e. $A \to \pm i$) the expansion (45) becomes singular. Moreover one should expect that if the expansion (45) was pushed further, the denominator $(A^{2n}-1)^{-1}$ would appear at order λ^{2n} and therefore at all energies of the form $E = \cos 2\pi\alpha$ with α rational, the expansion (45) would become singular.

To expand γ for energies which belong to the spectrum ($E = 2\cos q$ for q real), the easiest way is to work with the stationnary probability $P(R,E,\lambda)$ of the R_n. From the recursion (38) on the R_n, one can show that $P(R,E,\lambda)$ satisfies the following integral equation

$$P(R,E,\lambda) = \int \rho(V)\, dV \frac{1}{(E-R-\lambda V)^2} P\left(\frac{1}{E-R-\lambda V}, E, \lambda\right) \tag{46}$$

and the Lyapounov exponent γ is given by

$$\gamma = \int P(R,E,\lambda)\, dR \log |R| \tag{47}$$

The band edge (E=2)

One can show [12] that in the limit $E \to 2$ and $\lambda \to 0$, the distribution $P(R,E,\lambda)$ takes a scaling form

$$P(R,E,\lambda) = F\left(\frac{R-1}{\lambda^{2/3}}, \frac{E-2}{\lambda^{4/3}}\right) \tag{48}$$

and that it satisfies a differential equation which can be solved.

For E=2, one finds [12] that for $\lambda \to 0$

$$\gamma \sim (6\lambda^2 \langle V^2\rangle)^{1/3} \frac{\sqrt{\pi}}{2\Gamma(\frac{1}{6})} = .2893\ldots(\lambda^2\langle V^2\rangle)^{1/3} \tag{49}$$

and for $E \to 2$, γ has the following scaling form

$$\gamma \sim (\lambda^2\langle V^2\rangle)^{1/3} f\left(\frac{E-2}{(\lambda^2\langle V^2\rangle)^{2/3}}\right) \tag{50}$$

where f is a function which can be calculated.

The band centre (E=0)

One can also calculate the Lyapounov exponent γ in the limit $E \to 0$ and $\lambda \to 0$. Again in that limit, one cannot expect the expansion (45) to be valid since the term of order λ^4 diverges.

One can again solve the integral equation (46) in that limit and one finds that the Lyapounov exponent γ takes again a scaling form

$$\gamma \sim \lambda^2\langle V^2\rangle f\left(\frac{E}{\lambda^2\langle V^2\rangle}\right) \tag{51}$$

where the scaling function f can be calculated [11,14] for example $f(0) = \left(\Gamma(\frac{3}{4})/\Gamma(\frac{1}{4})\right)^2$

and $f(\infty) = 1/8$. One sees that for E=0 and $\lambda \to 0$, one has

$$\frac{\gamma}{\lambda^2 \langle V^2 \rangle} \to f(0) = \left(\frac{\Gamma(\frac{3}{4})}{\Gamma(\frac{1}{4})} \right)^2 \simeq .1154 \tag{52}$$

which is in agreement with the result of a Monte Carlo calculation [15] but disagrees with what could be expected from formula (45) (which would give $\gamma/\lambda^2 \langle V^2 \rangle \to 1/8$).

The limit $E \to 1$ has also been calculated recently [12,16] and again, the expansion (45) becomes singular and the result contains a scaling function of $(E-1)/\lambda^2 \langle V^2 \rangle$.

Let me just mention here that the expansion (45) can also be done for quasiperiodic potentials, for example

$$V_n = \cos(kn + \varphi) \tag{53}$$

where $k/2\pi$ is irrational.

One finds that up to second order in λ^2

$$\gamma = \log A - \frac{\lambda^2}{4} \frac{A^2+1}{A^2-1} \; \frac{A^2}{A^4-2A^2 \cos k+1} \tag{54}$$

It is interesting to compare (54) with (45) because in the limit $E \to 2 \cos q$ (which should be done carefully as discussed above) formula (45) gives a real part to γ indicating that a weak disorder is enough to produce a finite localisation length whereas formula (54) shows, at least at order λ^2, that the quasiperiodic potential (53) does not localise for λ small.

ACKNOWLEDGEMENTS

I would like to thank E. Gardner, H.J. Hilhorst, C. Itzykson, Y. Pomeau and J. Vannimenus with whom the works that I presented here were done. I am also grateful to J. Avron, J. Lacroix, D. Mukamel, J.L. Pichard and H. Sompolinsky for discussions.

REFERENCES

[1] F.J. Dyson, Phys. Rev. 92, 1331 (1953)

[2] H. Schmidt, Phys. Rev. 105, 425 (1957)

[3] S. Alexander, J. Bernasconi, W.R. Schneider and R. Orbach, Rev. Mod. Phys. 53, 175 (1981)

[4] D.J. Thouless, J. Phys. C5, 77 (1972)

[5] B. Derrida, Y. Pomeau and J. Vannimenus, J. Phys. C11, 4749 (1978)

[6] E. Gardner and B. Derrida, preprint 1984

[7] H.H. Chen and S.K. Ma, J. Stat. Phys. 29, 717 (1982)

[8] B. Derrida and H.J. Hilhorst, J. Phys. A16, 2641 (1983)

[9] C. de Calan, J.M. Luck, T.M. Nieuwenhuizen and D. Petritis, to appear in J. Phys. A

[10] B. Derrida and Y. Pomeau, Phys. Rev. Lett. 48, 627 (1982)

[11] B. Derrida, J. Stat. Phys. 31, 433 (1983)

[12] B. Derrida and E. Gardner, J. Physique (Paris) 45, 1283 (1984)

[13] E. Gardner, C. Itzykson and B. Derrida, J. Phys. A17, 1093 (1984)

[14] M. Kappus and F. Wegner, Z. Phys. B45, 15 (1981)

[15] G. Czycholl, B. Kramer, A. Mackinnon, Z. Phys. B43, 5 (1981)

[16] C.J. Lambert, Phys. Rev. B29, 1091 (1984)

EXACT DISORDER SOLUTIONS

Paul Ruján
Institute für Festkörperforschung der KFA, Jülich and
Institute for Theoretical Physics, Eötvös University
Budapest.

The topic of my talk is the discussion of order and disorder trajectories in lattice systems with competing interactions. My main goal is to show that although the mathematical formulation of the problem is extremly simple, the resulting body of physical informations is surprisingly rich. I take the opportunity to give a rather pedagogical presentation of the topic and also to comment on the history of this subject. The seminar is organized as following :

1. Mathematical formulation
2. The scheme of dimensionality reduction
3. General physical properties on and near disorder trajectories
4. Examples: a, Decoupling lines for quantum-spin Hamiltonians at T=0; b, Crystal growth formalism and disorder lines for Ising and Potts models on a triangular lattice; c, Phase diagram of anisotropic three-dimensional closed packed lattices
5. Further applications and conclusions.

1. Definitions and mathematical formulation : Order and disorder trajectories are trajectories /subspaces/ in the parameter space of lattice systems with competing interactions. In such systems one has not only the usual /thermal/ competition between energy and entropy possibly leading to a phase transition-but also an additional competition between two or more T=0 ground states with different periodicities. Schematically I shall represent a phase diagram in terms of a temperature like parameter τ and competition ratio denoted by κ. The lattice systems under consideration may be classical statistical systems with short range interactions, quantum spin Hamiltonians at T=0 or lattice gauge systems with mixed actions. A disorder /order/ trajectory is a $\tau_D(\kappa)$ line in this phase diagram with well defined properties /see section 3/, lying entirely into a disordered /paramagnetic of fluid/ phase or, respectively, into on ordered phase For some models the partition function and some correlation functions can be determined exactly along these trajectories and this is the point of view I would like to put forward here. A given model with competing interactions may have or may have not an exact solution along some trajectory on the (τ,κ)-space. According to my experience, however, is much easier to inverse the problem and to construct that class of models which does have an

exact solution/I mean here ground state properties of Hamiltonians or transfer matrices/. To make this point clear let me present the simplest possible example. Consider a one-dimensional Schrödinger--equation:

$$H\Psi_n = \left[-\frac{1}{2}\partial_x^2 + U(x)\right]\Psi_n = E_n\Psi_n \tag{1}$$

The inverse problem in this case is to /re/construct the Hamiltonian operator from the known $\{\Psi_n\}$ set. This question was solved long ago by the french mathematician Darboux[1]. Assume that one has a discrete non-degenerate nodeless ground state

$$\Psi_o(x) = e^{-\phi(x)} \quad , \tag{2}$$

where $\phi(x)$ is some given function and Ψ_o is normalizable. An elementary calculation shows that (2) is the ground state if $U(x) = E_o - \frac{1}{2}\phi'' + \frac{1}{2}\phi'^2$ thus

$$\bar{H} \equiv -\frac{1}{2}\partial_x^2 + \frac{1}{2}\phi'^2 - \frac{1}{2}\phi'' - E_o = \frac{1}{2}AA^{\dagger} \tag{3}$$

is a semi-positive definite operator with $\bar{E}_o = 0$ and $\overline{\Psi_o}(x) = \Psi_o(x)$, $\hat{A} = \partial_x - \phi'$.

This factorization was known already by Schrödinger[2] himself and this "ground-state representation" is very useful[3] in constructing a supersymmetric quantum mechanics[4]. What is important for our pourpose, however, is the observation that the Schrödinger-equation in purely imaginary time can be considered as a Liouville time-evolution operator under the action of which any Ψ_t relaxes to the unique stationary state Ψ_o.

This idea can be generalized to any transfer matrix corresponding to classical statistical models with short range interactions. Since in the spin-basis the elements of the transfer matrix are Boltzmann-weights the matrix is non-negative and according to the Frobenius-theorem the eigenvector corresponding to the largest eigenvalue is nodeless:

$$\hat{T}\Psi_o = \lambda_o\Psi_o \quad , \quad (\Psi_o)_i > 0 \tag{4}$$

Assuming that Ψ_o is known the similarity transformation

$$P = \frac{1}{\lambda_o} \begin{bmatrix} \ddots & & \\ & (\Psi_o^{-1})_i & \\ & & \ddots \end{bmatrix} \begin{bmatrix} & & \\ & T & \\ & & \end{bmatrix} \begin{bmatrix} \ddots & & \\ & (\Psi_o)_i & \\ & & \ddots \end{bmatrix} \tag{5}$$

defines a stochastic P matrix. Again, this operator can be interpreted as a time-evolution operator leaving the ground state Ψ_o invariant by construction.

2. Dimensionality reduction

We are now in the position to present the general construction scheme. Our choice of Ψ_o $\underline{\alpha}$ will be such as that

$$(\Psi_o, \Psi_o) \equiv Z_{\underline{\alpha}} \tag{6}$$

where $Z_{\underline{\alpha}}$ is the partition function of some fictitious /I shall call it underlying/ system. $\underline{\alpha}$ is some free parameter-set characterizing the underlying system. Any equal-time operator has then the ground-state expectation value:

$$< \hat{O} > = (\Psi_o, \hat{O}\Psi_o) \tag{7}$$

Since the partition function (6) represents a "row" spin-system its dimensionality is less than the dimensionality of the original model represented by the transfer matrix T. This dimensionality reduction may be of two kinds: $d \to d-1$ or $d \to 0$ dimensionality reduction. The scheme of the dimensionality reduction is shown below

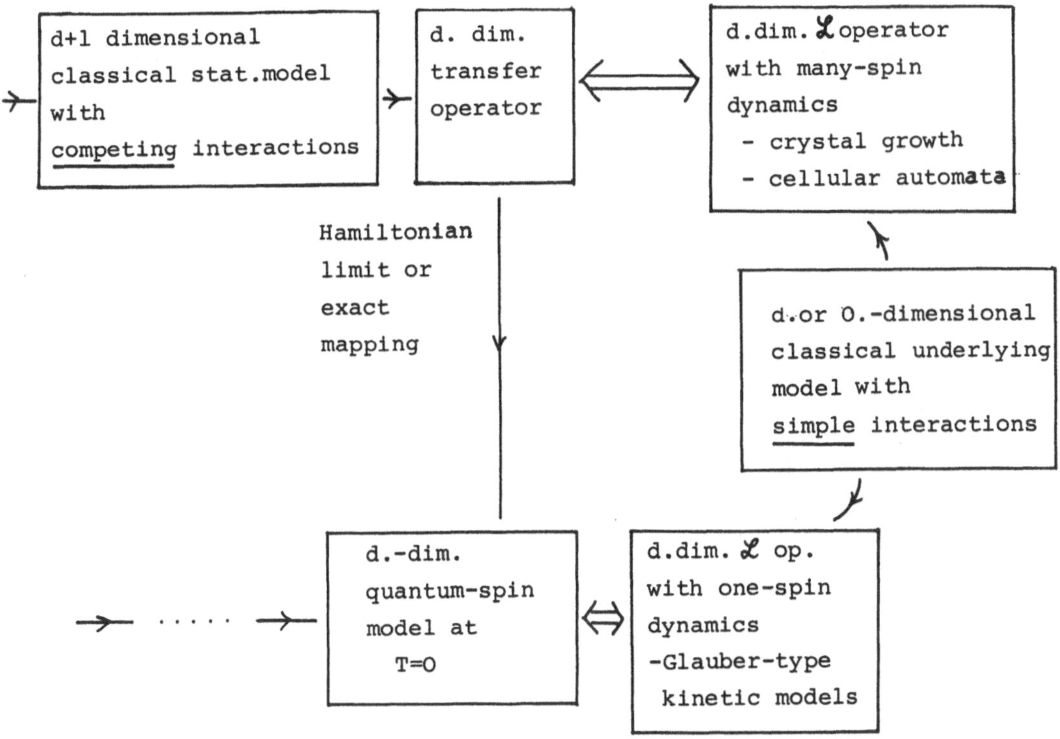

/ $\mathcal{L}$ -stands for a Liouville-type time evolution operator /

In this scheme quite different parts of the statistical and quantum physics are present, is thus not unexpected to find many independent and parallel discoveries of the disorder trajectory phenomena.

I should mention first the extensive work of Stephenson[5] /1964,1966,1970/ who solved exactly one - and two - dimensional Ising models with competing interactions and observed the presence of disorder points. At this disorder points some correlation functions simplified spectacularly and also the pattern of correlations changed. Such behaviour was reported also in other exactly soluble free-fermion models[6], although the dimensionality reduction was not yet realized and exploited in full. This first decade was followed by very interesting work on an apparently quite different field, namely the field of crystal growth. Wellberry and Galbraith[7] /1973,75/ and many others initiated the study of stochastic crystal growth models, some of them being amenable to exact solutions. Enting[8] and Verhagen[9] deserve the merit to have made the connections between these exactly soluble crystal growth models and disorder phenomena in Ising spin systems with competing interactions.

Independently, Peschel and Emery[10] realized the lower part of the dimensional-reduction scheme in the context of the two-dimensional axial next-to-nearest-neighbour Ising /ANNNI/ model and their work helped significantly to clarify the phase diagram of the 2D ANNNI-model. Myself I simplified and generalized the Peschel and Emery method[11,12] and later using the crystal growth representation I did a rather complete classification of disorder lines in two-dimensional models[13]. Very recently interest on cellular automata[14] models /which are equivalent to crystal growth models/ have again led to the rediscovery of disorder type solution[15], also in three-dimensional models[16].

Among the most interesting applications of these results stand the calculation of the generating function of oriented lattice animals in two- and three dimensions by Dhar[17].

3. Physical properties near and on disorder trajectories.

3a. Phase diagrams

One of the main informations one obtains from the presence of such disorder /order/ trajectories pertain to the degeneracy of the ground state. If the ground state is not degenerate the trajectory $\tau_D(\kappa)$ should be on the paramagnetic phase. If the ground state is degenerate one has an order line and $\tau_o(\kappa)$ lies on an ordered phase /see Section 4c/.

3b. Change in the pattern of correlations

From exactly soluble models one knows that along the axis of competition

$$\langle s_o s_R \rangle_{\parallel} \underset{R \to \infty}{\sim} R^{-\alpha} \exp(-R/\xi_{\parallel}) \qquad \text{if} \qquad \tau > \tau_D(\kappa) \tag{8}$$

but

$$\langle s_o s_R \rangle_{\parallel} \underset{R \to \infty}{\sim} R^{-\alpha} \cos q(\kappa,\tau) \exp(-R/\xi_{\parallel}) \text{ if } \tau < \tau_D(\kappa) \tag{9}$$

Here α is the Ornstein-Zernicke exponent, $q(\kappa,\tau)$ is in general a continuous function of (κ,τ) . For this reason one calls this region oscillatory or incommensurate disordered /ordered/ phase. Note that sometimes the disorder line is defined by the $\tau(\kappa)$ trajectory where the maximum of the structure factor

$$S(\underline{k}_{\parallel}) = \sum_{\underline{R}} \langle s_{\underline{O}}\, s_{\underline{R}} \rangle_{\parallel}\, e^{i\underline{k}_{\parallel}\underline{R}} \tag{10}$$

moves from $\underline{k}_{\parallel} = 0$ to some $\underline{k}_{\parallel} \neq 0$ value. This trajectory is different from our definition, based on the $R\to\infty$ behaviour of the spin-spin correlations.

3c. Singular behaviour along the disorder trajectory

$$\xi_{\parallel} \sim \left(\tau-\tau_D^c(\kappa)\right)^{-\nu_{\parallel}} \qquad \nu_{\parallel} = \nu \quad \text{in the underlying model} \tag{11}$$

$$\xi_{\perp} \sim \left(\tau-\tau_D^c(\kappa)\right)^{-\nu_{\perp}} \qquad \nu_{\perp} = \nu Z \quad - \text{"} - \tag{12}$$

τ_D^c : critical point in the underlying model

Z : critical dynamic exponent /Ruján[11],1982/

Using hyperscaling Domany[16] /1984/ predicted

$$Z = (2+\alpha)\,\nu \quad \text{in the underlying model} \tag{13}$$

Since in general $\nu_{\parallel} \neq \nu_{\perp}$ one has anisotropic scaling at $\tau_D^c(\kappa)$ and it is tempting to identify it as a Lifshitz-point /multicritical point at the common border of a disordered, an ordered and a incommensurate phase/.

3d. Exact solutions, analytic continuations

As already explained if $\tau = \tau_D(\kappa)$ one is able to calculate Z, $\langle s_O s_R \rangle$ as long as (Ψ_O,Ψ_O) , $(\Psi_O,\hat{s}_O\hat{s}_R\Psi_O)$ can be calculated exactly. These solutions can be analytically continued in the parameter set $\underline{\alpha}$ until $Z_{\underline{\alpha}} = (\Psi_O,\Psi_O)$ hits the Lee-Yang singularity edge /this happens for oriented lattice animals - see Dhar [17] /. A different possibility is to use the matrix inversion relations [18] as recently done for

2D Potts models[19] . In two-dimensions the fully integrable models do exhibit disorder solutions. It would be of interest to known which are the three-dimensional models with "disorder" solutions satisfying also the tetrahedron-equations.

4. Examples

4a. Decoupling /order/ lines in a general 1/2-spin Heisenberg model at T=0

Consider the Hamiltonian:

$$H = - \sum_{<\underline{r},\underline{r}'>} \Big\{ \underbrace{(1+\gamma)\sigma^x_{\underline{r}} \sigma^x_{\underline{r}'} + (1-\gamma)\sigma^y_{\underline{r}} \sigma^y_{\underline{r}'}}_{\text{"kinetic energy"}} + \underbrace{V[\sigma^z]}_{\text{"potential energy"}} \Big\} \qquad (14)$$

and the following Ansatz for Ψ_o :

$$|\Psi_o\rangle = \exp\left(\frac{\alpha}{2} \Sigma\sigma^z_{\underline{r}}\right)|0_+\rangle \; ; \qquad \sigma^x_{\underline{r}} \, |0_\pm\rangle = \pm \, |0_\pm\rangle \; \forall \underline{r} \qquad (15)$$

where $\sigma^x, \sigma^y, \sigma^z$ are the usual Pauli matrices.

Then

$$H \, |\Psi_o\rangle = -e^{\frac{\alpha}{2}\Sigma\sigma^z_{\underline{r}}} \sum_{<\underline{r},\underline{r}'>} \Big\{ \underbrace{\Big[a+b\,\Sigma\,\sigma^z_{\underline{\alpha}} + c\Sigma\sigma^z_{\underline{\alpha}}\,\sigma^z_{\underline{\beta}} + \ldots\Big]}_{\text{local terms involving } \sigma^z_{\underline{r}} \text{ and its nn } \sigma^z\text{ s}} + V[\sigma^z]\Big\}|0_+\rangle$$

In one-dimension

$$H = - \sum_j \left\{ J \left[(1+\gamma)\sigma^x_j \, \sigma^x_{j+1} + (1-\gamma)\sigma^y_j \, \sigma^y_{j+1} + \Delta\sigma^z_j \, {}^z_{j+1} \right] + h(\sigma^z_j + \sigma^z_{j+1}) \right\} \qquad (16)$$

decouples /has Ψ_o of form (15) / if

$$\gamma^2 + \left(\frac{h}{J}\right)^2 = (\Delta - 1)^2$$

$$\Delta < 1 - \gamma \qquad /\alpha \text{ is real}/ \tag{17}$$

Also because of the structure of (14) $|\Psi_o\rangle$ is degenerate with

$$\overline{|\Psi_o\rangle} = e^{\frac{\alpha}{2}\Sigma\sigma_r^z} |0_-\rangle \tag{18}$$

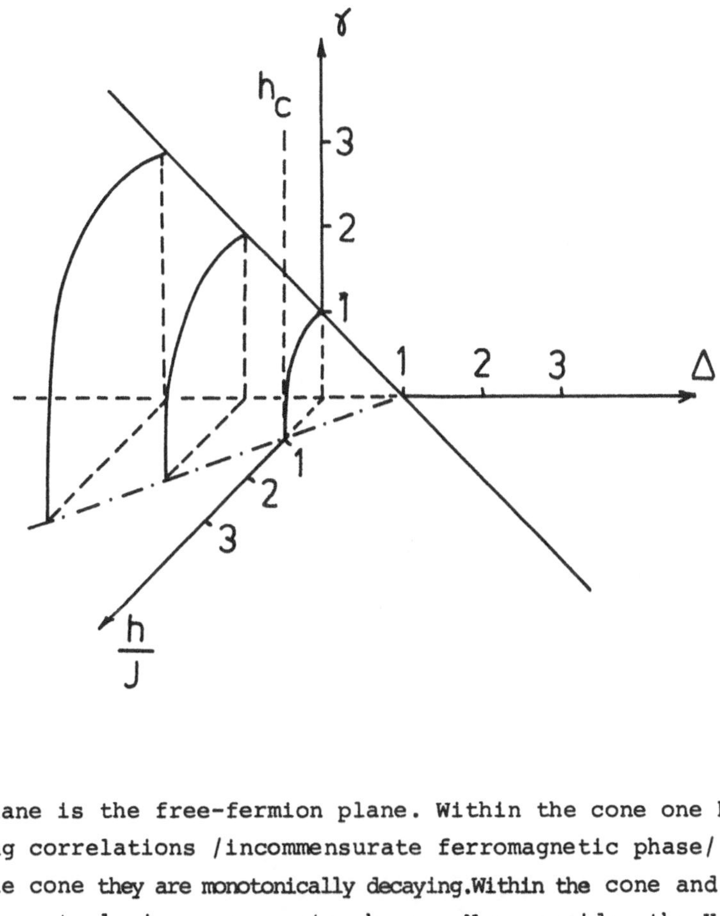

The Δ=0 plane is the free-fermion plane. Within the cone one has oscillating correlations /incommensurate ferromagnetic phase/ while outside the cone they are monotonically decaying. Within the cone and at γ=0 one expects a truly incommensurate phase. Now consider the Hamiltonian (16) with Δ=0 but $\gamma_j = p\,\delta(\gamma_j-\gamma_o)+(1-p)\,\delta(\gamma_j+\gamma_o)$ $h_j = q\,\delta(h_j-h_o)+(1-q)\delta(h_j+h_o)$ quenched probability distributions. Obviously if $\gamma_o^2 + h_o^2 = 1$ Eq. (17) is satisfied and one may calculate exactly the ground state properties of this random Hamiltonian on the one recovers the original Ansatz-layer. Moreover , the correlations between even-spins/t=1 row/ are the same as between odd-spins /t=0 row/ . Proceeding further with the construction of the crystal using the "detailed balance" rule (21) one gets a "two-dimensional" crystal where the spin-spin correlation on t=constant equal time rows are the same and are those of the original Ansatz-layer. For the form (21) one gets the disorder solution on the antiferromagnetic Δ Ising model [5] . If one includes also an external field one has the following phase diagram

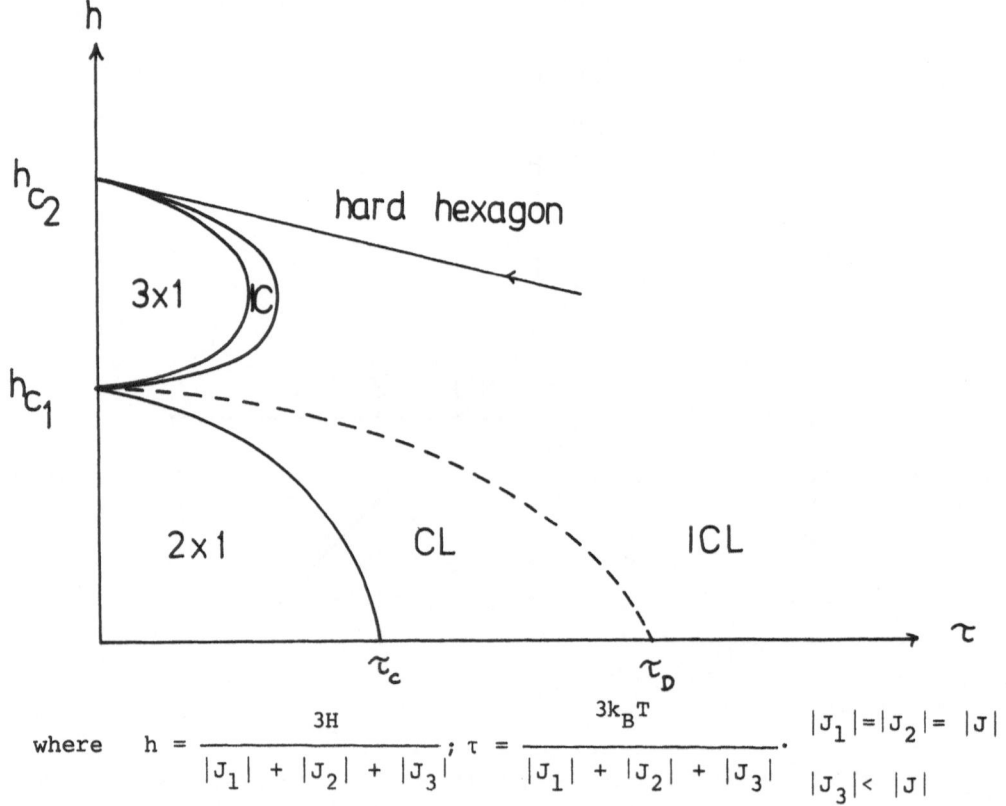

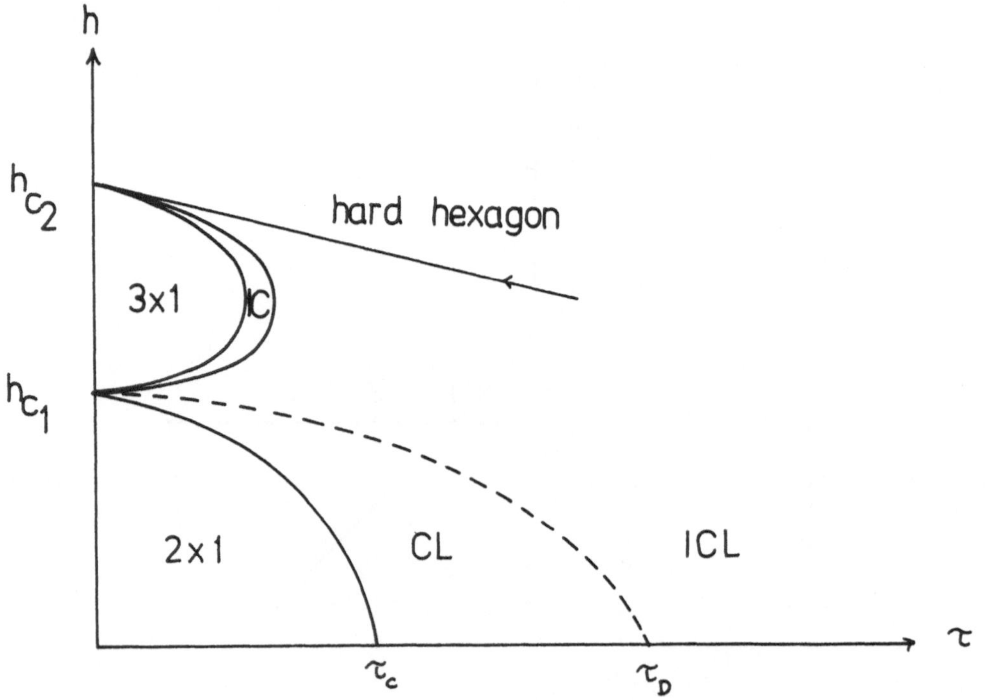

where $h = \dfrac{3H}{|J_1| + |J_2| + |J_3|}$; $\tau = \dfrac{3k_BT}{|J_1| + |J_2| + |J_3|}$. $|J_1| = |J_2| = |J|$, $|J_3| < |J|$

One has a 2x1 ordered and a 3x1 ordered phase as well as an incommensurate IC phase and a ferromagnetically disordered CL = commensurate-liquid phase where the correlatins are of the form

$$\langle s_0 s_R \rangle \underset{R\to\infty}{\sim} (-1)^R \; R^{-1/2} \; e^{-R/\xi} \qquad \text{in CL}$$

order line. The procedure can be generalized to higher dimensions. An example of a 3+1-dimensional Z_2 gauge lattice theory on a disorder line has been presented by Ruján and Patkós [12].

4b. Stochastic crystal growth alias cellular automata

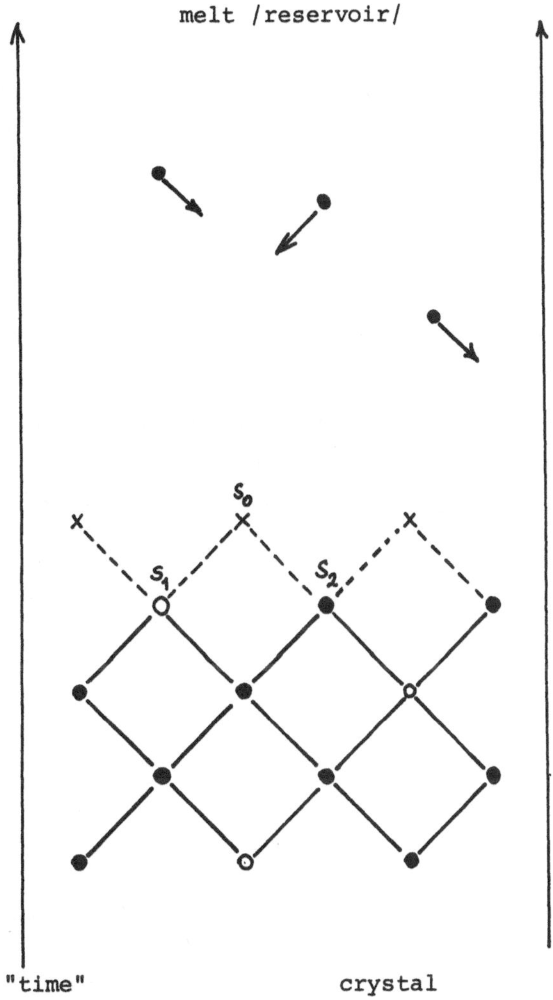

The probability to occupy s_o (s_o=+1) depends only on precedessor spins $s_1, s_2, \ldots$(o) but not on same generation spins (x)

One has a stochastic /Markov/ process:

$$P_t(\{s_i, \bar{s}_i\}) = \left\{\prod_i p\,(s_i \mid \bar{s}_{i,1}, \bar{s}_{i,2})\right\} P_{t-1}(\{\bar{s}_i\}) \tag{20}$$

new generation spin — old generations spin — new generation

Although the dynamic properties of such systems are very interesting[14,20] we treat here only equilibrium properties. Consider the Ansatz-layer /a row of Ising spins/

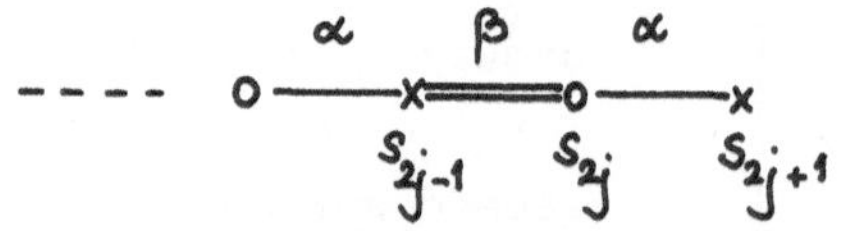

$$-\beta E = \Sigma\left(\alpha\, s_{2j}\, s_{2j+1} + \beta\, s_{2j-1}\, s_{2j}\right)$$

Summing out exactly/iteration/ the even or odd spins one obtains a new row of spins with half of spins left and with an effective coupling $K_3(\alpha,\beta) = \frac{1}{2}\ln\frac{\cosh(\alpha+\beta)}{\cosh(\alpha-\beta)}$. Thus the probability distributions of even and dd spins are identical. Imagine now the inverse operation: start with the odd spins distributed according to the effective coupling K_3 and decorate

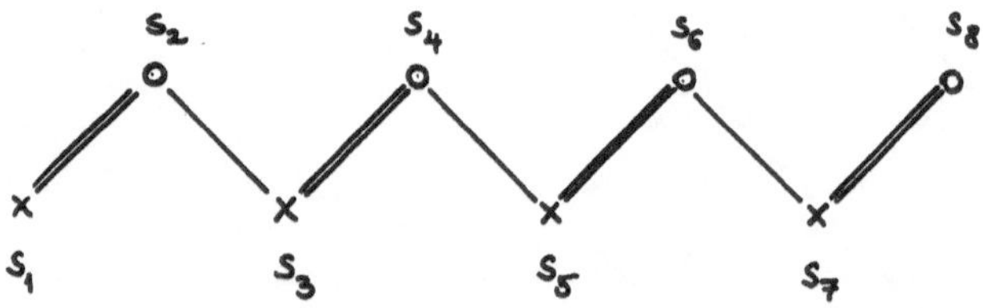

the chain in the manner shown above. If the decoration is made such as

$$p(s_{2j} \mid s_{2j-1}, s_{2j+1}) = e^{\alpha s_{2j} s_{2j+1} + \beta s_{2j-1} s_{2j}} \,/\, 2\cosh(\alpha s_{2j+1} + \beta s_{2j-1}) \tag{21}$$

divided by a disorder line from the disordered ICL = incommensurate liquid phase characterized by modulated correlations:

$$\langle s_o s_R \rangle \underset{R\to\infty}{\sim} \cos\left(\underline{q}\,(h,\tau)\,\underline{R}\right) \quad R^{-1/2}\, e^{-R/\xi} \quad \text{in ICL}$$

On the disorder line the correlations are one-dimensional

$$\langle s_o s_R \rangle = e^{-R/\xi} \quad \forall\, R \; !$$

Note that the slope of the IC-boundary at the h_{c2} point corresponds to the critical activity of the exactly soluble hard hexagon model.

The procedure is easily generalized to q-state spin models. For the isotropic Potts model on a Δ-lattice one obtains a T_D $q \neq 0$ disorder point for $0<q<2$ $\left(T_D(q=2) = 0\ ;\ T_D(q=0) = \infty\right)$ which may indicate that the critical point $T_c < T_D$ moved also to $T_c > 0$ temperatures.

4c. Possible phase diagram of closed packed three-dimensional lattices

The simplest generalization of the Δ-lattice to three dimensions consists of closed packed lattices. Consider hexagonal planes which are shifted on each other as shown below by the O, □ and x lattice points. Calling the planes denoted by o, □ , and x as A,B,C one may form closed packed 3D lattices as regular sequences of A,B,C planes /ABAB.... = closed packed hexagonal, ABCABC ... = fcc lattice/ . For our models it is not important to have regular sequences, only that the nn. coupling between planes K_o should be different than the intraplane nn. coupling K_1 . If one choses as Ansatz a 2D honeycomb lattice the same calculation as in two-dimensions leads to the trajectory [8,16]

$$K_1 = -\frac{1}{4}\ln\frac{\cosh 3K_o}{\cosh K_o} \qquad (22)$$

This trajectory is not strictly speaking a disorder line since if $K_o < K_c^{hex}$ the ground state is unique /disorder/ while for $K_o > K_c^{hex}$ it is /at least/ twice degenerate. The phase diagram of such a closed packed system is schematically shown below

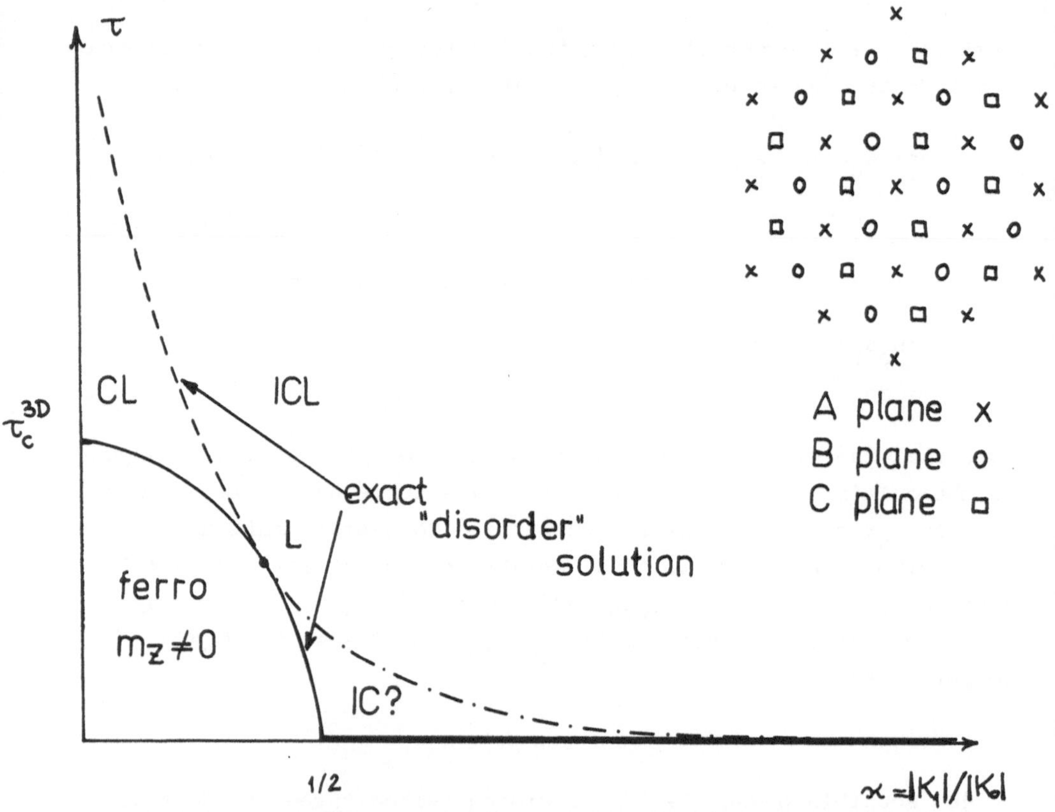

One expects a ferromagnetic ordered phase where the magnetization points perpendicular to the hexagonal planes. If $\kappa=0$ one expects an usual 3D-type Ising transition between the disordered and the ordered phase τ_c^{3D}. At $\tau=0$ one may show that for $\kappa>1/2$ the minimal energy is obtained for $m_z=0$ configurations while for $\kappa\to\infty$ one has decoupled antiferromagnetic Δ planes whose critical point is at $\tau_c^{\Delta}=0$. When the underlying model /here the 2D honeycomb Ising model/ undergoes a phase transition $K_o = K_c^{hex}$ the disorder line ends up probably on a Lifshitz-type multicritical point whith anisotropic scaling and for $K_o > K_c^{hex}$ it probably follows the ferromagnetic phase boundary. We think that the line $\tau=0$, $\kappa>\frac{1}{2}$ is a critical line, possibly with varying critical exponents. The possibility of a truly incommensurate phase as indicated in the figure is not ruled out by RG-arguments and seems quite plausible. Note that the presence of the (22) trajectory rules out the possibility of a "chiral" transition [21] between τ_c^{3D} and L. I think that these models deserve further work on view of their possible experimental realization in intercalated conpounds.

5. Conclusions

In conclusion we have seen that sometimes quite simple mathematics may lead to a rich body of physical informations. I think that new interesting results can be obtained through different analytic continuation of these calculations. It would be also useful to use more sophisticated Ansatze as well as to consider the dynamic poperties of systems with competing interactions.

References

1. G. Darboux: Comptes rendus, 94, p. 1456 (1982).
2. E. Schrödinger: Proc.Roy.Irish Acad. A46, 9 /1940/
3. E. Gozzi : Phys.Lett. 129B, 432 /1983/
   M.N.Nieto: Los Alamos preprint LA-UR-84-1457 /1984/
4. E. Witten: Nucl. Phys. B185, 513 /1981/
5. J. Stephenson: J.Math.Phys. 5, 1009 /1964/, ibid 7, 1123 /1966/; ibid 11 413, 420 /1970/; Can.J.Phys. 48, 2118 /1970/ ; Phys Rev. B1, 4405 /1970/
6. M.B. Geilikman: Sov.Phys. JETP 39 , 570 /1974/
   E.Fradkin and T.P. Eggarter, Phys.Rev. A14, 495 /1976/
   E. Barouch and B. McCoy, Phys.Rev A3, 786, /1971/
7. T.R. Welberry and R. Galbraith, J.Appl. Crystallogr. 6, 87 /1973/, ibid 8 636 /1982/
8. I.G. Enting; J.Phys. C10, 1379, 1023 /1977/ , ibid 11, 55, 2001 /1978/
9. A.M.Verhagen: J.Stat. Phys. 15, 219 /1976/
10. I. Peschel and V.J.Emery: Z.Phys. B43, 241 /1981/
11. P. Ruján: J.Stat.Phys. 29, 231, 247 /1982/
12. P. Ruján and A. Patkós: Phys.Lett. B129, 437, /1983/
13. P. Ruján: J.Stat.Phys. 34, 615 /1984/
14. For a review see S. Wolfram: Rev.Mod.Phys. 55, 601 /1983/
15. E. Domany and W. Kinzel: Phys.Rev.Lett. 53, 311 /1984/
16. E. Domany: Phys.Rev. Lett. 52, 871 /1984/
17. D. Dhar: Phys.Rev. Lett. 49 , 959 /1982/ , ibid 51 , 853 /1983/
18. Yu. G. Stroganov; Phys.Lett. A74, 116 /1979/
    R.J. Baxter; J. Stat. Phys. 28, 1 /1982/
19. M.T. Jaekel and J.M. Maillard: J.Phys. A17 2079 /1984/

20. P. Grassberger, F.Krause and T. von der Twer: J.Phys. A17 L 105 /1984/

21. D. Huse and M.E. Fisher: Phys.Rev.Lett. 49, 793 /1982/; Phys.Rev. B29, 239 /1984/

THE COULOMB GAS SYSTEMS : SOME RIGOROUS RESULTS +

F. NICOLO *

Centre de Physique Théorique x

C.N.R.S. Luminy Case 907

F-13288 MARSEILLE CEDEX 9

1 - INTRODUCTION

A Coulomb system is a system of identical spinless particles of charges $\pm$ e interacting via the Coulomb potential
It will be therefore a very rough approximation of a pure system of particles (for instance ions and electrons) interacting through the e.m. interactions.
The most relevant approximations are :

1) Relativistic effects are neglected
2) Particles with opposite charges are equal on any other respect
3) The spin dependent interactions are neglected
4) Interaction with the radiation field is neglected.

Therefore the models we want to discuss are a caricature of these models but nevertheless they present interesting features and have a very rich structure.
The interaction is a two-body interaction : the potential is

$$V(x) = \begin{cases} \frac{1}{2}|x| & d = 1 \\ \frac{1}{2\pi}\ln\left(\frac{1}{|x|}\right) & d = 2 \\ \frac{1}{4\pi}\frac{1}{|x|} & d = 3 \end{cases} \tag{1}$$

The first problem to face is to give a meaning to this system as a statistical mechanics system : this essentially means we have to control :
a) the ultraviolet problem (stability problem) (short distance)
b) the infrared problem (thermodynamic limit) (long-distance)

+ Talks given at C.P.T. Marseille, 1984.

* Università di Roma

x Laboratoire propre du C.N.R.S.

The ultraviolet problem

To control the stability problem means that we want the model to be such to prevent collapse of infinite particles in finite regions. Otherwise the model is not suited to be a reasonable approximation of systems when collapse is not present.

If, vice-versa, in our model some form of collpase is produced we want to control it and have a physical explanation of it. This is exactly what happens in a certain range of the temperature in the $d=2$ case, as will be discussed in detail later.

There are different non equivalent definitions of stability all amounting to give a meaning to the partition function of the system in the grand canonical ensemble.

H-stability : a system is said H-stable if for any positive integer N, calling $H^{(N)}$ the interaction energy of N particles, for any configuration the following inequality holds

$$H^{(N)}(q_1,\dots,q_N) > -\text{ const. } N \qquad (2)$$

the constant being uniform in N.

It is very easy to prove that for $d>2$ a classical Coulomb gas is never H-stable.

(The analogous of H-stability is proved in the quantum domains provided one of the two particle species is fermionic. (Dyson–Lenard ; Lieb) [1]).

Nevertheless there is another notion of stability, equally basic for statistical mechanics which is the notion of Ξ-stability, we are going to define now :

Let's call Ξ the grand canonical partition function for a Coulomb system at a certain "temperature" β and activity z . Then we will say that the system is Ξ-stable if

$$\exp - Cst|\Lambda| \leq \Xi_\Lambda(\beta, z) \leq \exp Cst|\Lambda| \tag{3}$$

The situation is therefore the following :

Classical case

Classical Coulomb gas are not H-stable for $d \geq 2$.

Classical Coulomb gas are Ξ-stable (in some interval of temperature and in some appropriate meaning) in d=2.

Classical Coulomb gas are not Ξ-stable in d=3.

Quantum case

d=3 ; if the negative charged particles are fermions, H-stability is satisfied (Dyson-Lenard) ;

if all the particles are bosons, there is no H-stability.

Of course an ultraviolet cutoff implies H-stability and also Ξ-stability.

In the second part of my talk I shall be interested mainly on the problem of the Ξ-stability for the classical Coulomb gas in d=2.

For this aspect the d=2 case is the most interesting one, as only in this case it is possible either having or not having Ξ-stability, depending on the value of β.

The reason of that is that the Coulomb potential, for d=2, is logarithmic. This implies that the Gibbs factor diverges, due to the absence of H-stability as a power (for some configurations). This divergence can be compensated by the entropy factor (which essentially counts the number of configurations of a certain type) provided β is not too large providing therefore the u.v. stability.

This mechanism vice-versa cannot be produced in d=3 as in this case the Coulomb potential as $r \longrightarrow 0$ diverges as $1/r$ producing a Gibbs factor diverging logarithmically and therefore the entropy cannot compensate its divergence.

Once the complexity of the d=2 case is clear ; let's remind which are the known results :

Ξ -stability has been proved by Frohlich [2] for $\alpha^2 \equiv \beta e^2 < 4\pi$.

Subsequently this result has been extended in a couple of papers by (Benfatto, Gallavotti and Nicolo) and by (F. Nicolo) in the interval of α^2 : $[4\pi, 8\pi]$ in the following sense. Neglecting the infrared problem or if one prefers in a finite volume or in the Yakawa gas case, the u.v. stability of the Coulomb gas is still true up to $8\pi-\epsilon$ provided we modify the partition function in an appropriate way subtracting to it some (constant) counterterms.

The physical interpretation of this phenomenon is the following : between 4π and 8π there is an infinite sequence of thresholds given by the following expression :

$$\alpha^2_{2n} = 8\pi\left(1-\frac{1}{2n}\right) \tag{4}$$

which corresponds to partial collapse of the Coulomb (or Yakawa) gas in the following sense :

for $\alpha^2 \in [4\pi, 6\pi) \equiv [\alpha^2{}_2, \alpha^2{}_4)$ the gas particles tend to collapse forming microscopic dipoles, therefore in this temperature interval it can be thought as a gas of free charges in a sea of microscopic dipoles, of infinite density ;

for $\alpha^2 \in [6\pi, \frac{20\pi}{3}) = [\alpha^2{}_4, \alpha^2{}_6)$ the particles tend to form also microscopic quadripoles, the partition function describing therefore a gas of charged particles in a sea of dipoles and quadripoles. It is therefore obvious how to extend this argument to every interval $[\alpha^2{}_{2n}, \alpha^2{}_{2(n+1)})$.

As $n \longrightarrow \infty$ in (4) $\alpha^2{}_{2n} \longrightarrow 8\pi$ and therefore above this value all the possible multipoles are formed, we are in a situation of complete collapse, the partition function diverges and we have not any more a statistic description of our collapsed system.

These infinite sequences of partial collapses are interesting for many reasons :

a) In their own as describing a short distance phenomenon of the d=2 Coulomb (Yukawa) gas.

b) As a renormalization phenomenon in the associated Sine-Gordon field theory.

c) As these thresholds appear also to play an interesting role in the infrared problem.

Let's therefore, before discussing a), b), c) in some details, discuss some generalities of the infrared problem.

The infrared problem

The basic results concerning the thermodynamic limit are due to Lieb and Lebowitz [4], which essentially proved in a beautiful paper the existence of thermodynamic functions, although the Coulomb potential decreases so slowly as $r \to \infty$. This is essentially due to the screening properties of these systems (quantum case). (Neutrality assumption is required).

In the classical case the more interesting results are the following :

a) Debye screening

This result is true for classical Coulomb systems in d=2,3 and has been proved in a series of paper by Brydges and Brydges and Federbush (see also Imbrie) [5].

This result has been proved for $\beta <<$ and activity $Z <<$ (dilute gas). The screening manifests itself in the fact that the correlation functions have exponential cluster properties.

The Debye screening has the effect that everything works as if the Coulomb interaction decays exponentially as $r \to \infty$. "therefore the effect of the Debye screening is that of producing a mass", in field theory language, in the propagator.

This situation describes a "phase" of the d=2 Coulomb gas which is called plasma phase, when the particles form a plasma and tend to screen one each other.

If we go to low temperature (β>>) there is a β_c finite above which the Coulomb gas is present in different phase : the dipole phase.

b) The Koesterlitz-Thouless phase

The dipole phase is characterized by the fact that the particles tend to form the "neutral multipoles" among which neutral dipoles may be expected the dominant configurations of the gas.

The Coulomb gas in this region of β and Z is therefore similar to a dipole gas.

It has been proved by Frohlich and Spencer [6] that dipole gas have correlations with power law decay i.e. Debye screening breaks down at low temperatures. This is proved to be true also for the Coulomb gas.

Therefore in this phase, often called the Koesterlitz-Thouless phase, we have a power decay for the correlation functions.

The existence of this phase has been proved for β>> but the common wisdom says that the temperature at which this transition takes place corresponds to $\alpha^2 = 8\pi$.

Now that a dipole phase exists is due to the fact that the Gibbs factor for a dipole of length l tends to 0 as $l \longrightarrow \infty$ which means that if β is large enough the average value of l is finite and therefore a relevant weight have the "finite size dipole" configurations.

This is not possible in d=3 which again implies that, $\forall$ β, all neutral multipoles are unstable.

Let's now shortly recall how the existence of the Koesterlitz-Thouless transition is proved by Frohlich and Spencer in [6].

The proof of the K.T. transition is based on the following facts :

a) The order parameter for the Coulomb gas will be the average of the fractional charge density $\langle e^{i\alpha\Phi(0)}\rangle$ (in the Sine Gordon formalism)
The existence of different phases will mean that

$\langle e^{i\alpha\Phi(0)}\rangle \neq 0 \quad (\alpha < 1)$ in the plasma phase

$$< e^{i\alpha \Phi(0)} > = 0$$ in the dipole phase

b) It is proved that for d=2 at any β there is the following lower bound

$$< e^{i\alpha(\Phi(x) - \Phi(0))} > \geqslant O(|x|^{-\alpha^2 \beta / 2\pi})$$

(Park-Frohlich) [7].

c) For small β we have the exponential decay of truncated correlation functions (Brydges) [5] therefore :

$$< e^{i\alpha(\Phi(x) - \Phi(0))} > \xrightarrow[|x| \to \infty]{} 0$$ exponentially

d) b) +c) imply that

$$< e^{i\alpha(\Phi(x) - \Phi(0))} > \xrightarrow[|x| \to \infty]{} < e^{i\alpha \Phi(0)} >^2 \neq 0$$

e) For β large Frohlich and Spencer prove that

$$< e^{i\alpha(\Phi(x) - \Phi(0))} > \leq \text{Cst} \; (1 + |x|)^{-\alpha^2 \beta / 2\pi}$$

f) e) implies that $< e^{i\alpha \Phi(0)} > = 0$ for β>>.

The next result I want to discuss has been conjectured and recently proved by G. Gallavotti, J. Lebowitz and myself [8].

To describe it in detail we have to introduce the general formalism with which we are going to study the Coulomb gas in d=2. What can we say at this moment is the following.

Although 8π is, in a sense we have to specify, the starting temperature for the K. T. phase the interval $[4\pi, 8\pi]$ seems to represent a region of the phase space where the Coulomb gas is in a sort of mixing phase between the pure plasma and the pure multipole one ; this means the following.

We can identify in this interval an infinite sequence of thresholds

$$\alpha_{2n}^2 = 8\pi\left(1 - \frac{1}{2n}\right) \tag{5}$$

(exactly the same as for the ultraviolet problem !) which have the following properties :

For $\alpha^2 > 8\pi$ the gas (neutral) is formed by all possible multipoles (of finite size). When $\alpha^2 < \alpha^2_{2n}$ then only the multipoles made by $\leq 2n$ particles are still stable while all the remaining ones are broken.
For $\alpha^2 < \alpha^2_2 = 4\pi$ there are no stable multipoles and we are in the pure plasma phase.

What I am going to discuss in the following is the Ξ -stability in the interval $[4\pi, 8\pi)$ (ultraviolet problem) and the existence and properties of this mixed phase region in the infrared problem.

To do that we have first of all introduced the Siegert transformation to transform this problem in a field theory problem.

2 - THE SINE GORDON TRANSFORMATION

Let's exploit the connexion between the Coulomb gas and the Sine Gordon theory.

Let's define a gaussian random field $\varphi(\xi)$, formally, as a field with covariance $C(\xi,\eta)$ and let's consider the following partition function :

$$Z_\Lambda(\beta,\lambda) = \int d\mu(\varphi) \; e^{2\lambda \int_\Lambda :\cos\alpha\varphi(\xi): \, d^2\xi} \tag{6}$$

where $\alpha^2 = \beta e^2$, $d\mu$ is the gaussian measure associated to the covariance C, λ is the "coupling constant" of the theory.

Expanding Z_λ in power of λ we get :

$$Z_\Lambda(\beta,\lambda) = \sum_{n=0}^{\infty} \frac{\lambda^n}{n!} \sum_{\sigma_1,\sigma_n} \int_{\Lambda^n} d\xi_1 \ldots d\xi_n \int :e^{i\alpha\sigma_1\varphi(\xi_1)}: \ldots :e^{i\alpha\sigma_n\varphi(\xi_n)}: d\mu(\varphi) \tag{7}$$

$$\text{where} \quad :e^{i\alpha\varphi(\xi)}: \; \equiv \sum_{k=0}^{\infty} \frac{(i\alpha)^k}{k!} :\varphi^k(\xi): \; = \frac{e^{i\alpha\varphi(\xi)}}{\langle e^{i\alpha\varphi(\xi)} \rangle} \tag{8}$$

It is easy to realize that :

$$\langle :e^{i\alpha\sigma_1\varphi(\xi_1)}: \ldots :e^{i\alpha\sigma_n\varphi(\xi_n)}: \rangle = e^{-\alpha \sum_{i<j} \sigma_i \sigma_j C(\xi_i,\xi_j)} \tag{9}$$

and therefore

$$Z_\Lambda(\beta,\lambda) = \sum_{n=0}^{\infty} \frac{\lambda^n}{n!} \sum_{\sigma_1\ldots\sigma_n} \int_{\Lambda^n} d\xi_1 \ldots d\xi_n \, e^{-\alpha \sum_{i<j} \sigma_i \sigma_j C(\xi_i,\xi_j)} \tag{10}$$

which is the partition function for a charged gas (of charges $\pm e$) whose potential is:

$$V(\xi_i, \xi_j) = C(\xi_i, \xi_j) \qquad (11)$$

therefore if C is $(-\Delta)^{-1}$ it follows that as

$$(-\Delta)^{-1}(\xi, \eta) = \frac{1}{2\pi} \ln \frac{1}{|\xi - \eta|} \qquad (12)$$

is just the Coulomb interaction.

This Sine-Gordon field theory describes the Coulomb gas.

This is still formal from the side of the field theory as $(-\Delta)^{-1}$ is not a well defined covariance. In fact $\frac{1}{2\pi} \ln \frac{1}{|\xi-\eta|}$ diverges both when $|\xi-\eta| \to 0$ and $|\xi-\eta| \to \infty$ and we need, therefore, to introduce some regularizations. At this point it is also clear that we need a regularization to control the short distance behaviour and another one to control the long distance behaviour.

This can be obtained in different ways a clear and instructive one will be the following : substitute $1/p^2$, associated to $(\widetilde{\Delta^{-1}})$ with

$$\hat{C}^{(-Q,N)}(p) = \frac{1}{p^2 + \gamma^{-2Q}} - \frac{1}{p^2 + \gamma^{2N}} \quad , \gamma > 1 \qquad (13)$$

which clearly reproduces $1/p^2$ when $Q \to \infty$ and $N \to \infty$.

Let's make the following remark :

$$C^{(-Q,N)}_{(\xi,\eta)} \sim e^{-\gamma^{-Q}|\xi-\eta|} \qquad \text{as } |\xi-\eta| \gg \tag{14}$$

$$C^{(-Q,N)}_{(0)} = \frac{1}{2\pi} \ln(\gamma^{2N}\gamma^{2Q})$$

Which implies that γ^{-Q} plays the role of a mass making the long distance behaviour decaying exponentially [some dimensional constants have been set=1. In other words $[\gamma^{Q}] = [L^{+1}]$, $[\gamma^{N}] = [L^{-1}]$].
while γ^{2N} plays the role of a cutoff stabilizing the u.v. behaviour.

The Coulomb gas with the right short distance properties and long distance properties are obtained when we remove both the cutoffs but one should accept that:

If one wants to study the short distance behaviour of the gas (ultraviolet problem). it should be enough to remove the ultraviolet cutoff. If one wants to study the long distance behaviour it would be enough to remove the infrared cutoff.

Therefore to study the ultraviolet problem we shall keep γ^{-Q} fixed (for instance γ=1) and we shall study the following partition function

$$Z^{(N)}_{\Lambda}(\beta,\lambda) = \int d\mu(\varphi^{[\leq N]})\, e^{2\lambda \int_{\Lambda} :\cos\alpha\, \varphi^{[\leq N]}_{(\xi)}:\, d^2\xi} \tag{15}$$

where $\varphi^{[\leq N]}$ is the Gaussian field with covariance

$$C^{(N)}(\xi,\eta) = \frac{1}{(2\pi)^2} \int d^2p\, e^{i p.(\xi-\eta)} \left(\frac{1}{p^2+1} - \frac{1}{p^2+\gamma^{2(N+1)}}\right) \tag{16}$$

Vice-versa if we want to study the thermodynamic limit we will keep fixed the ultraviolet cutoff γ^{N} (=1 for instance) and we will perform the limit γ^{-Q}

3 - THE ULTRAVIOLET STABILITY FOR $\alpha^2 \in [4\pi, 8\pi]$

The problem is reduced to prove that:

$$e^{-Cst|\Lambda|} \leq Z_\Lambda^{(N)} \leq e^{Cst|\Lambda|} \tag{17}$$

with Cst N-independent ; or, if this is not true, how one has to modify $Z_\Lambda^{(N)}$ to make this statement true. This is therefore a typical problem of field theory and will be discussed using the renormalization group (R. G.) methods.

Before doing that we want to discuss from a statistical mechanic point of view, via heuristic arguments, the appearance of the thresholds $\alpha^2{}_n$ corresponding to partial collapses (see (3)).

Let's go back to the partition function written in the grand canonical formalism

$$\begin{aligned} Z_\Lambda^{(N)}(\beta,\lambda) &= \sum_{n=0}^{\infty} \frac{\lambda^n}{n!} \sum_{\sigma_1\ldots\sigma_n} \int_{\Lambda^n} d\xi_1\ldots d\xi_n \, e^{-\alpha \sum_{i<j} \sigma_i\sigma_j C^{(N)}(\xi_i,\xi_j)} \\ &= \sum_{p=0}^{\infty}\sum_{q=0}^{\infty} \frac{\lambda^{p+q}}{p!\,q!} \int_{\Lambda^q} dx_1\ldots dx_q \int_{\Lambda^p} dy_1\ldots dy_p \, e^{-\beta U_{(p,q)}(x,y)} \end{aligned} \tag{18}$$

where

$$C^{(N)}(0) = \frac{1}{2\pi} \ln \gamma^N = \frac{1}{2\pi} \ln \ell_N^{-1} \tag{19}$$

and $\ell_N \equiv \gamma^{-N}$.

We can interpret the introduction of this cutoff as the assumption that the particles have linear size of order ℓ_N . Collapsing phenomena are expected in the limit : $\ell_N \longrightarrow 0$.

Let's consider now the term of the grand canonical partition function with q = p = n . (q is the number of -e particles and p that of the +e particles) and consider the contribution to $Z^{(N)}_{\Lambda,2n}$ from the configurations in which any +e particle is "near", at a distance of order ℓ_N. to a corresponding -e particle, that is dipole configurations where each dipole has momentum of order $e\ell_N$

The energy of these configurations is approximately:

$$U_{dip} \simeq -e^2 \sum_{i=1}^{n} C^{(N)}(0) = -e^2 n \frac{1}{2\pi} \ln \ell_N^{-1} = e^2 n \frac{1}{2\pi} \ln \ell_N \tag{20}$$

and the contribution to the canonical partition function is:

$$\begin{aligned} Z^{(N)}_{\Lambda(2n)dip} &\simeq \frac{\lambda^{2n}}{n!} \int_{\Lambda^n} dx_1 \dots dx_n \left(\ell_N^2\right)^n e^{-\alpha^2 \frac{n}{2\pi} \ln \ell_N} \\ &= \frac{1}{n!} \lambda^n \left(\ell_N^{(1-\alpha^2/4\pi)} \right)^{2n} |\Lambda|^n \\ &= \frac{1}{n!} \left(\lambda \gamma^{N(\frac{\alpha^2}{4\pi}-1)} \right)^{2n} |\Lambda|^n \end{aligned} \tag{21}$$

therefore this term looks like a contribution to the grand canonical partition function of a free dipole gas of activity $(\lambda \; \gamma^{N(\alpha^2/4\pi-1)})^2$ and with dipole momentum of order : $e\gamma^{-N}$. The activity

$$\lambda^{(N)}_{dip} = \left(\lambda \gamma^{N(\alpha^2/4\pi-1)}\right)^2 \tag{22}$$

goes to ∞ as $N \to \infty$ and this means that the dipole configurations are the most important ones : moreover the dipole length $\to 0$ as $N \to \infty$ and the final result is that, as the density of the dipole gas is $\approx \lambda^{(N)}_{dip}$ and the average distance between two dipoles is of order ($\lambda^{(N)}_{dip}$)$^{-1/2}$ the ratio between the average dipole length and the average dipole distance is

$$O\left(\ell_N / (\lambda^{(N)}_{dip})^{1/2} \right) \propto \gamma^{-N} \gamma^{N\frac{1}{2}\left(\frac{\alpha^2}{4\pi}-1\right)^2} = \gamma^{N\left(\frac{\alpha^2}{4\pi}-2\right)} \to 0 \qquad (23)$$

$\forall \alpha^2 < 8\pi$, which means that we can consider these dipoles configurations as describing a free dipole gas of infinite activity. This argument has an obvious generalization. We can consider a neutral multipole made by 2n particles, ℓ_N being the diameter of it (its linear size). Its energy will be approximately:

$$U_{(2n)} \simeq -e^2\left[n^2 - \frac{n(n-1)}{2} - \frac{n(n-1)}{2} \right] \frac{1}{2\pi} \ln \ell_N^{-1}$$

$$\simeq -e^2 n \frac{1}{2\pi} \ln \ell_N^{-1} = e^2 n \frac{1}{2\pi} \ln \ell_N \qquad (24)$$

The phase space of this 2n-multipole will be: $(\ell_N^2)^{2n-1}$ and its contribution to Z

$$Z_{(2n-\text{multipole})} \simeq \lambda^{2n} \int_\Lambda dx \, (\ell_N^2)^{2n-1} e^{-\frac{\alpha}{2\pi} \ln \ell_N \, n}$$

$$= \lambda^{2n} \left(\ell_N^2 \, \ell_N^{-\frac{\alpha^2}{4\pi}} \right)^{2n} \ell_N^{-2} |\Lambda|$$

$$= \lambda^{2n} \left(\gamma^{N\left(\frac{\alpha^2}{4\pi}-2\right)2n} \gamma^{2N} \right) |\Lambda| \qquad (25)$$

$$= \left(\lambda^{2n} \gamma^{N\left(\frac{\alpha^2}{4\pi} - 2 + \frac{1}{n}\right)2n} \right) |\Lambda|$$

Therefore

$$\lambda^{(N)}_{(2n-\text{clusters})} = \left(\lambda^{2n} \; \gamma^{N\left(\frac{\alpha^2}{4\pi} - 2\left(1-\frac{1}{2n}\right)\right)2n} \right) \tag{26}$$

These (2n) -neutral clusters start to give an important contribution when their activity as $N \to \infty$ diverges and they do not really contribute when as $N \to \infty$ it tends to 0 .

We can therefore conclude :

We have collapse in 2n-clusters when $\alpha^2 > \alpha^2_{2n} = 8\pi\left(1-\frac{1}{2n}\right)$ which was the result stated in the introduction. of course also the collapse in all smaller clusters takes place when $\alpha^2 > \alpha^2_{2n}$.

This is an heuristic argument. What is really reasonable is that these thresholds play really a role in the study of the superrenormalisability of the Sine-Gordon theory.

The ultraviolet stability of the massive Sine-Gordon field theory [3]

It amounts to prove the following inequality

$$e^{-cst|\Lambda|} \leq \int d\mu(\varphi^{[\leq N]}) \; e^{V_\Lambda^{(N)}} \leq e^{cst|\Lambda|} \tag{27}$$

where

$$V_\Lambda^{(N)} = 2\lambda \int_\Lambda :\cos\alpha\,\varphi^{[\leq N]}(\xi): \, d^2\xi \; + \sum_{K=1} C_{2K}^{(N)} \tag{28}$$

and $C_{2K}^{(N)}$ are constants (field-independent) diverging as $N \to \infty$.

The technique used to prove inequalities (28) is that of the (R.G.) in the form we used it to study the u.v. stability of the Φ_3^4 theory [9].

Let's sketch the general ideas:

we write

$$C^{(N)}(p) = \frac{1}{p^2+1} - \frac{1}{p^2+\gamma^{2(N+1)}} = \sum_{K=0}^{N}\left(\frac{1}{p^2+\gamma^{2K}} - \frac{1}{p^2+\gamma^{2(K+1)}}\right)$$

$$\equiv \sum_{K=0}^{N} \tilde{C}^{(K)}(p) \tag{29}$$

Therefore we can think to $\varphi^{[\leq N]}$ as a sum of N independent gaussian fields each one with covariance $\tilde{C}^{(K)}(\xi,\eta)$

$$\varphi^{[\leq N]}(\xi) = \sum_{K=0}^{N} \tilde{\varphi}^{(K)}(\xi) \tag{30}$$

Inspection of the covariances $\tilde{C}^{(K)}$ shows that the fields $\tilde{\varphi}^{(K)}$ are, apart a rescaling, equally distributed

$$\tilde{\varphi}^{(K)}(\xi) = \tilde{\varphi}^{(0)}(\gamma^K \xi) \; . \tag{31}$$

Moreover they have the following nice properties which make significant their introduction :

a) they are approximately constant on a scale γ^{-K}

b) they are approximately independent on the same scale, which means that their covariance decays exponentially on this scale.

The measure $d\mu(\varphi^{[\leq N]})$ is equivalent to the product of gaussian measures with respect to the fields $\tilde{\varphi}^{(K)}$

$$d\mu(\varphi^{[\leq N]}) = \prod_{K=0}^{N} P(d\tilde{\varphi}^{(K)}) \tag{32}$$

The basic idea of the (R.G) is, here, the following one : try to calculate the partition function iteratively performing the integrations on the different fields one after another.

This program will work if essentially we are in a situation approximately of this kind ; the following inequalities hold :

$$e^{\tilde{V}_\Lambda^{(K-1)} - R_\Lambda^{(K)}} \leq \int P(d\tilde{\varphi}^{(K)})\, e^{\tilde{V}_\Lambda^{(K)}} \leq e^{\tilde{V}_\Lambda^{(K-1)} + R_\Lambda^{(K)}} \tag{33}$$

with $\tilde{V}_\Lambda^{(K)}$: the effective potential (to a fixed order in λ) and $R_\Lambda^{(K)} = \delta^{(K)} |\Lambda|$ with the property :

$$\sum_{K=0}^{\infty} \delta^{(K)} < +\infty \tag{34}$$

and of course $\tilde{V}_\Lambda^{(N)} = V_{\Lambda,0}^{(N)}$

If (33) is true for any K , $\delta^{(K)}$ satisfying (34), we can conclude that

$$e^{-Cst|\Lambda|} \leq Z^{(N)}(\beta,\lambda) \leq e^{Cst|\Lambda|} \tag{35}$$

which is our desired result.

Let's show how to prove (35) below $4\Pi = \alpha^2{}_1$, which is easy enough and how problems arise as soon as we overcome this first threshold: Remembering the scale properties of the fields

$$\tilde{\varphi}^{(K)}(\xi) = \tilde{\varphi}^{(0)}(\gamma^K \xi) \equiv z_{\gamma^K \xi} \tag{36}$$

we can always integrate with respect to the measure P(dZ) ; in term of z, $V_{\Lambda,0}$ appears :

$$V_{\Lambda,0}^{(N)} = 2\lambda \int_\Lambda :\cos\alpha\, \varphi_\xi^{[\le N]}: d\xi$$

$$= 2\lambda\, e^{\frac{\alpha^2}{2} C^{(N)}(0)} \int_\Lambda \cos\alpha\left(\varphi_\xi^{[\le N-1]} + z_{\gamma^N\xi}\right) d\xi$$

$$= 2\, \gamma^{\frac{\alpha^2}{4\pi}} \left(\lambda\, \gamma^{(\frac{\alpha^2}{4\pi}-2)N}\right) \int_{\gamma^N\Lambda} \cos\left(\varphi_{\gamma^{-N}\xi'}^{[\le N-1]} + z_{\xi'}\right) d\xi' \tag{37}$$

remembering that

$$C^{(N)}(0) = \frac{1}{2\pi}(N+1)\ln\gamma \tag{38}$$

Therefore the effective coupling constant is in this case:

$$\lambda_N = \lambda\, \gamma^{(\frac{\alpha^2}{4\pi}-2)N} \tag{39}$$

and the interaction is over a region $\gamma^N\Lambda$.

Equ. (39) is essentially the asymptotic freedom of the theory (obvious since the theory is superrenormalizable); in fact $\lambda_N \longrightarrow 0$ as $N \to \infty$ $\forall\, \alpha^2 < 8\pi$.

Equ. (39) allows us to prove (35) easily for $\alpha^2 < 4\pi$. Let's compute with the cumulant expression :

$$\int P(d\varphi^{(N)})\, e^{V_{0,\Lambda}^{(N)}} \underset{(\ge)}{\le} e^{\mathcal{E}_{(N)}(V_{0,\Lambda}^{(N)}) + \sum_{k=2}^{M} \frac{1}{k!} \tilde{\mathcal{E}}^T_{(N)}(V_{0,\Lambda;k}^{(N)}) \underset{(-)}{+} R_\Lambda^{(N)}} \tag{40}$$

with the usual notations. Let's assume M=1, then $\tilde{\mathcal{E}}_{(N)}(V_{0,\Lambda}^{(N)}) = V_{0,\Lambda}^{(N-1)}$

therefore :

$$\int P(d\tilde{\varphi}^{(N)})\, e^{V_{0,\Lambda}^{(N)}} \underset{(\geq)}{\leq} e^{V_{0,\Lambda}^{(N-1)}} \underset{(-)}{+} R_\Lambda^{(N)} \tag{41}$$

$R_\Lambda^{(N)}$ is proportional to $\lambda_N^2 |\Lambda| \gamma^{2N}$:

$$R_\Lambda^{(N)} \simeq \lambda_N^2 \gamma^{2N} |\Lambda| = \lambda^2 \gamma^{(\frac{\alpha^2}{2\pi}-4)N} \gamma^{2N} |\Lambda|$$

$$= \lambda^2 \gamma^{(\frac{\alpha^2}{2\pi}-2)N} |\Lambda| = \delta^{(N)} |\Lambda| \tag{42}$$

and if $\alpha^2 < 4\pi$

$$\sum_{K=0}^{N} \delta^{(K)} \leq \sum_{K=0}^{\infty} \delta^{(K)} < +\infty \tag{43}$$

<u>Remark</u>

The proof that $R_\Lambda^{(N)}$ is proportional to $|\Lambda|$ is simple in this case and more technical if $\alpha^2 > 4\pi$; our proof has been called "integration grid technique" and has been carefully discussed in [9] [10]. We won't discuss it, explicitely, here.

The results (41), (42) and (43) show us that this computation does not produce the desired estimate if $\alpha^2 = 4\pi + \epsilon$.

The obvious idea will then be to compute the l. h. s. of (40) in a more precise way, that is truncating the cumulant expression at higher orders.

Let's suppose we go up to the second order in λ obtaining:

$$\int P(d\tilde{\varphi}^{(N)})\, e^{V_{\Lambda,0}^{(N)}} = e^{\tilde{\mathcal{E}}(V_{\Lambda,0}^{(N)}) + \frac{1}{2}\tilde{\mathcal{E}}_N^T(V_{\Lambda,0;2}^{(N)})} + R_\Lambda^{(N)} \tag{44}$$

with $\quad R_\Lambda^{(N)} = c\, \lambda_N^3 \gamma^{2N} |\Lambda| = c\, \lambda^3 \gamma^{3(\frac{\alpha^2}{4\pi}-2)N} \gamma^{2N} |\Lambda|$

therefore if

a) C is a constant uniform in N,

b) $\frac{3\alpha^2}{4\pi} - 4 < 0 \iff \alpha^2 < \frac{16}{3}\pi = \alpha_3^2$

we have again a summable remainder. Unfortunately statement a) is not true as we see very easily:

let's observe that C will be proportional to the maximum value of $\tilde{V}_\Lambda^{(K)}$ for a generic K as the remainder $R_\Lambda^{(K)}$ will be obtained estimating the following expression:

$$R_\Lambda^{(K)} = \frac{\partial^3}{\partial \tau^3} \ln \int P(d\tilde{\varphi}^{(K)})\, e^{\tau \tilde{V}_\Lambda^{(K)}} \Big|_{\tau=\tau^*} \tag{46}$$

and it is clear that $R_\Lambda^{(K)} \le C \sup_\varphi |\tilde{V}_\Lambda^{(K)}|$

Therefore it is needed that, provided $\tilde{\varphi}^{(K)}$ and φ be enough regular, $\tilde{V}_\Lambda^{(K)}$ can be bounded uniformly in N.

This unfortunately does not happen as a simple computation gives:

$$\begin{aligned} \tilde{V}_\Lambda^{(K)} &= \tilde{\mathcal{E}}_{K+1}(\tilde{V}_\Lambda^{(K+1)}) + \frac{1}{2!}\tilde{\mathcal{E}}^T_{K+1}(\tilde{V}_\Lambda^{(K+1)};2) \\ &= V_{\Lambda,0}^{(K)} + \mu^2 \int_{\Lambda\times\Lambda} d^2\xi\, d^2\eta \left(e^{-\alpha^2 C^{(N)}(\xi,\eta)} - e^{-\alpha^2 C^{(K)}(\xi,\eta)}\right) :\cos\alpha\left(\varphi^{[\le K]}_{(\xi)} + \varphi^{[\le K]}_{(\eta)}\right): \\ &+ \mu^2 \int_{\Lambda\times\Lambda} d^2\xi\, d^2\eta \left(e^{\alpha^2 C^{(N)}(\xi,\eta)} - e^{\alpha^2 C^{(K)}(\xi,\eta)}\right) :\cos\alpha\left(\varphi^{[\le K]}_{(\xi)} - \varphi^{[\le K]}_{(\eta)}\right): \end{aligned} \tag{47}$$

Let's consider the second term which is the dangerous one and corresponds to a neutral contribution : an easy estimate gives:

$$\begin{aligned} |W^{(K)}_{Q=0}| &\le \mu^2 |\Lambda|\, \gamma^{\frac{\alpha^2}{2\pi}K} \int_\Lambda d^2(\xi-\eta) \left(e^{\alpha^2 C^{(N)}(\xi,\eta)} - e^{\alpha^2 C^{(K)}(\xi,\eta)}\right) e^{-\alpha^2 C^{(K)}(\xi,\eta)} \\ &\le \mu^2 |\Lambda|\, \gamma^{\frac{\alpha^2}{2\pi}K} \sum_{\ell=K}^{N} \gamma^{\frac{\alpha^2}{2\pi}\ell}\, \gamma^{-2\ell}\, \gamma^{-\frac{\alpha^2}{2\pi}K} \end{aligned}$$

$$\le \mu^2 \gamma^{\frac{\alpha^2}{2\pi}K - 2K} |\Lambda| \sum_{\ell=0}^{N-K} \gamma^{\frac{\alpha^2}{2\pi}(\ell-K)} \gamma^{-2(\ell-K)}$$

$$\le \mu^2 \gamma^{2(\frac{\alpha^2}{4\pi}-2)K} \gamma^{2K} |\Lambda| \sum_{\ell=0}^{N-K} \gamma^{(\frac{\alpha^2}{2\pi}-2)(\ell-K)}$$

$$\le \mu_K^2 \gamma^{2K} |\Lambda| \sum_{\ell=0}^{N-K} \gamma^{(\frac{\alpha^2}{2\pi}-2)(\ell-K)} = C(N)\, \mu_K^2 \gamma^{2k} |\Lambda| \tag{48}$$

and it is clear that if $\alpha^2 > 4\Pi$. C is not uniformly bounded in N.

Therefore, as this C enters in the remainder, we cannot reproduce the computation for $\alpha^2 < 4\Pi$ in the $\alpha^2 > 4\Pi$ case just increasing the order of cumulant expansion.

Here it is, nevertheless, easy to see a way out. In fact looking at the expression of $W^{(K)}_{Q=0}$ (the third term of $\tilde{V}^{(K)}_\Lambda$)

$$W^{(K)}_{Q=0} = \mu^2 \int_{\Lambda\times\Lambda} d^2\xi\, d^2\eta \left(e^{\alpha^2 C^{(N)}(\xi,\eta)} - e^{\alpha^2 C^{(K)}(\xi,\eta)}\right) :\cos\alpha\left(\varphi^{[\le K]}(\xi) - \varphi^{[\le K]}(\eta)\right):$$

$$= \mu^2 \sum_{\ell=K}^{N-1} \int_{\Lambda\times\Lambda} d^2\xi\, d^2\eta \left(e^{\alpha^2 C^{(\ell+1)}(\xi,\eta)} - e^{\alpha^2 C^{(\ell)}(\xi,\eta)}\right) :\cos\alpha\left(\varphi^{[\le K]}(\xi) - \varphi^{[\le K]}(\eta)\right): \tag{49}$$

it is easy to realize that all the problems arise as

$\sum_{\ell=K}^{N-1} \int d^2(\xi-\eta) \left(e^{\alpha^2 C^{(\ell+1)}(\xi,\eta)} - e^{\alpha^2 C^{(\ell)}(\xi,\eta)}\right)$ behaves as

$$\sum_{\ell=K}^{N-1} \int d^2(\xi-\eta)\, \gamma^{\frac{\alpha^2}{2\pi}\ell}\, e^{-\gamma^\ell|\xi-\eta|} \approx \sum_{\ell=K}^{N-1} \gamma^{\frac{\alpha^2}{2\pi}\ell} \gamma^{-2\ell} \tag{50}$$

which is divergent as $N \to \infty$ if $\alpha^2 > 4\Pi$.

They will be drastically different if instead of $:\cos\alpha(\varphi^{[\le K]}(\xi) - \varphi^{[\le K]}(\eta)):$

we had $:\cos\alpha\,(\varphi^{[\le K]}_{(\xi)} - \varphi^{[\le K]}_{(\eta)}) - 1:$

In this case due to the fact that $\varphi^{[\le K]}$ is Holder-continuous of order $1-\varepsilon$ ($\varepsilon>0$), which means that, with probability 1, given $\varphi^{[\le K]}$ $\exists\, B$ such that

$$|\varphi^{[\le K]}_{(\xi)} - \varphi^{[\le K]}_{(\eta)}| \le B\,(\gamma^K|\xi-\eta|)^{1-\varepsilon} \tag{51}$$

remembering that the estimates (50) cannot be signicantly improved in the region $|\xi-\eta| < \gamma^{-K}$, if we perform the substitution

$$:\cos\alpha\,(\varphi^{[\le K]}_{(\xi)} - \varphi^{[\le K]}_{(\eta)}): \;\longrightarrow\; :\cos\alpha\,(\varphi^{[\le K]}_{(\xi)} - \varphi^{[\le K]}_{(\eta)}) - 1: \tag{52}$$

then

$$W^{(K)}_{Q=0} \longrightarrow \tilde{W}^{(K)}_{Q=0} = \mu^2 \int_{\Lambda\times\Lambda} d^2\xi\, d^2\eta\, \left(e^{\alpha^2 C^{(N)}_{(\xi,\eta)}} - e^{\alpha^2 C^{(K)}_{(\xi,\eta)}}\right) :\cos\alpha\,\Delta\varphi^{[\le K]}_{(\xi,\eta)} - 1: \tag{53}$$

which satisfies (provided the field $\varphi^{[\le K]}$ be regular) the following estimate:

$$|\tilde{W}^{(K)}_{Q=0}| \le \mu^2 |\Lambda|\, \gamma^{\frac{\alpha^2}{2\pi}K} \int_\Lambda d^2(\xi-\eta)\left(e^{\alpha^2 C^{(N)}_{(\xi,\eta)}} - e^{\alpha^2 C^{(K)}_{(\xi,\eta)}}\right) e^{-\alpha^2 C^{(K)}_{(\xi,\eta)}} B^2 (\gamma^K|\xi-\eta|)^{2(1-\varepsilon)}$$

$$\le \mu^2 \gamma^{\frac{\alpha^2}{2\pi}K} |\Lambda| \sum_{\ell=K}^{N} \gamma^{\frac{\alpha^2}{2\pi}\ell}\, \gamma^{-2\ell}\, \gamma^{-\frac{2\alpha^2}{2\pi}K}\, \gamma^{-2(\ell-K)(1-\varepsilon)}\, B^2$$

$$\le \mu^2 \gamma^{(\frac{\alpha^2}{2\pi}-4)K} (\gamma^{2K}|\Lambda|)\, B^2 \sum_{\ell=K}^{N} \gamma^{\frac{\alpha^2}{2\pi}(\ell-K)}\, \gamma^{-(\ell-K)(4-2\varepsilon)}$$

$$\le C\, B^2 \mu_K^2 (\gamma^{2K}|\Lambda|) \tag{54}$$

with C N-independent which was exactly the result we wanted.

The prescriptions (52), (53) seem an "ad hoc" one but it is not so. In fact if, instead of starting from the interaction

$$V_{\Lambda,0}^{(N)} = 2\lambda \int_{\Lambda} :\cos\alpha\,\varphi^{[\leq N]}(\xi): d^2\xi \tag{55}$$

we would have started from the following one:

$$V_{\Lambda}^{(N)} = V_{\Lambda,0}^{(N)} - \frac{1}{2!}\,\mathcal{E}^{T}\left(V_{\Lambda,0}^{(N)};2\right) \equiv V_{\Lambda,0}^{(N)} - C_2^{(N)} \tag{56}$$

where $\mathcal{E}^T$ is now the truncated expectation with respect to the global field and not to a single-frequency field. In the expression of $\tilde{V}_{\Lambda}^{(K)}$ (see eq. (47)) we would have got $\tilde{W}_{Q=0}^{(K)}$ instead of $W_{Q=0}^{(K)}$. The fact that adding the counterterm (constant) $C_2^{(N)}$ is equivalent to the prescription previsouly discussed is easily proved ; In fact:

$$\begin{aligned} C_2^{(N)} &= \mu^2 \int_{\Lambda\times\Lambda} d^2\xi\, d^2\eta \left(e^{\alpha^2 C^{(N)}(\xi,\eta)} - 1\right) \\ &= \sum_{\ell=-1}^{N-1} \mu^2 \int_{\Lambda\times\Lambda} d^2\xi\, d^2\eta \left(e^{\alpha^2 C^{(\ell+1)}(\xi,\eta)} - e^{\alpha^2 C^{(\ell)}(\xi,\eta)}\right) \\ &= \sum_{\ell=K}^{N-1} \mu^2 \int_{\Lambda\times\Lambda} d^2\xi\, d^2\eta \left(e^{\alpha^2 C^{(\ell+1)}(\xi,\eta)} - e^{\alpha^2 C^{(\ell)}(\xi,\eta)}\right) + C_2^{(K)} \end{aligned} \tag{57}$$

and the first part of the r.h.s. of (57) just matches together with $W_{Q=0}^{(K)}$ giving rise to $\tilde{W}_{Q=0}^{(K)}$.

It can therefore be proven that the substraction of this counterterm just allows us to get again inequalities (36), provided $V_{\Lambda,0}^{(N)}$ is substituted by $V_{\Lambda}^{(N)}$. In the interval of α^2 :

$$\left[4\pi, \tfrac{16}{3}\pi\right) = \left[\alpha_2^2, \alpha_3^2\right) \tag{58}$$

The interesting and non trivial thing is that this procedure can be iterated to reach larger values of α^2.

We have the following result : let's define:

$$\alpha_n^2 = 8\pi\left(1-\frac{1}{n}\right) \tag{59}$$

Each time α^2 overcomes an even n=2K threshold the corresponding counterterm:

$$C_{2K} \equiv \frac{1}{2K!}\,\mathcal{E}^T\left(V_{\Lambda,0}^{(N)}\,;2K\right) \tag{60}$$

has to be substracted.

Each time α^2 overcomes an odd n=2K+1 threshold we don't have to substract any counterterm but only to perform the cumulant expression up to the order : 2K+1.

The interpretation from the point of view of statistical mechanics is, in our opinion, the one discussed before.

. Let's make some remarks which seem to me necessary to construct completely the technical part of this result:

Remarks

a) As in Φ_3^4 -theory, we have to assume that $\varphi^{[\leq K]}$ is regular (Hölder-Cont.). Then the need of regularity plays the role of the need of boundedness in Φ_3^4 .

This has to be achieved to prove the upper bound and therefore we have to exclude the regions (small) where this does not happen.

b) An important and technically complicated part is to prove that the remainder of the cumulant expansion is really $\approx|\Lambda|$ which means that it has the right dependence on the volume.

This is not easy to obtain : it is possible, essentially, because the effective potential although non local is nearly local and is proved with a technic which has been called "integration grid technic".

c) It could be interesting to examine how the definition of statistical mechanics observables change above 4π to sustain our interpretations. (See F. Nicolo [3]).

Some technical remarks on the ultraviolet stability proof

Let's discuss separately the lower and the upper bound.

Lower bound $(\alpha^2 > 4\pi)$

Let's make the following observation : let's assume we have defined

$$\tilde{V}_\Lambda^{(N)} = V_\Lambda^{(N)} = V_{0,\Lambda}^{(N)} - C_2^{(N)} \tag{1}$$

$$\tilde{V}_\Lambda^{(K)} = \sum_{\ell=1}^{2} \frac{1}{\ell!} \tilde{\mathcal{E}}^T_{K+1} (\tilde{V}_\Lambda^{(K+1)}; \ell) \Big|_{O(\mu^2)} \tag{2}$$

The idea is that:

$$\int P(d\tilde{\varphi}^{(K+1)})\, e^{\tilde{V}_\Lambda^{(K+1)}} \geqslant e^{\tilde{V}_\Lambda^{(K)} - R_\Lambda^{(K)}} \tag{3}$$

where $$R_\Lambda^{(K)} = \frac{\partial^3}{\partial \tau^3} \ln \int P(d\tilde{\varphi}^{(K+1)})\, e^{\tau \tilde{V}_\Lambda^{(K+1)}} \Big|_{\tau=\tau^*} \tag{4}$$

therefore

$$R_\Lambda^{(K)} \approx \int P(d\tilde{\varphi}^{K+1})\, \tilde{V}_\Lambda^{(K+1)\,3}\, e^{\tau^* \tilde{V}_\Lambda^{(K+1)}}$$

has to be bounded by $\max |\tilde{V}_\Lambda^{(K+1)}|^3$: the problem is the following:

$$\tilde{V}_\Lambda^{(K+1)} = V_{0,\Lambda}^{(K+1)} - C_{2,\Lambda}^{(K+1)} + W_{Q\neq 0}^{(K+1)} + \tilde{W}_{Q=0}^{(K+1)}$$

and to bound it as we need we have that $W_{Q=0}^{(K+1)}$ has to be bounded which is true provided the field $\varphi^{[\leq K+1]}$ be Holder-continuous with coefficient B . This is true in probability with probability $1-\epsilon(B)$, ($\epsilon(B) \longrightarrow 0$ as B $\longrightarrow$) but here we want this property for any sample field $\varphi^{[\leq K+1]}$. As we are considering the lower bound we can set characteristic functions.

Therefore the lower bound has the following form:

$$\int P(d\varphi^{[\le N]})\, e^{V_\Lambda^{(N)}} \geqslant \int P(d\varphi^{[\le N]}) \prod_{K=0}^{N} \chi^{B_K}(\varphi^{[\le K]})\, e^{V_\Lambda^{(N)}}$$
$$= \int P(d\varphi^{[\le N-1]}) \prod_{K=0}^{N-1} \chi^{B_K}(\varphi^{[\le K]}) \int P(d\tilde{\varphi}^{(N)})\, \chi^{B_{N-1}}(\varphi^{[\le N-1]})\, \chi^{B_N}(\varphi^{[\le N]})\, e^{V_\Lambda^{(N)}}$$

as $\chi^{B_{N-1}}(\varphi^{[\le N-1]})\, \chi^{B_N}(\varphi^{[\le N]}) \geqslant \chi^{B_{N-1}}(\varphi^{[\le N-1]})\, \chi^{b_N}(\tilde{\varphi}^{(N)})$

We have to estimate

$$\chi^{B_{N-1}}(\varphi^{[\le N-1]}) \int P(d\tilde{\varphi}^{(N)})\, \chi^{b_N}(\tilde{\varphi}^{(N)})\, e^{\tilde{V}_\Lambda^{(N)}} \geqslant e^{\tilde{V}_\Lambda^{(N-1)} - c R_\Lambda^{(N)}}$$

and then prove that it is possible to iterate. therefore that

$$\chi^{B_{K-1}}(\varphi^{[\le K-1]}) \int P(d\tilde{\varphi}^{(K)})\, \chi^{b_K}(\tilde{\varphi}^{(K)})\, e^{\tilde{V}_\Lambda^{(K)}} \geqslant e^{\tilde{V}_\Lambda^{(K-1)} - R_\Lambda^{(K)}}$$

$\forall\, K \in \{0, 1, \dots, N\}$.

<u>For the upper bound we argue in this way</u>

Let's define $\mathcal{D}_K(\varphi^{[\le K]})$ the regions where the sample $\varphi^{[\le K]}$ is "rough".

Let's define $\hat{V}_\Lambda^{(K)} = \tilde{V}_{\Lambda\setminus\mathcal{D}_K}^{(K)}$

where $R_K(\tilde{\varphi}^{(K)})$ is the region where $\tilde{\varphi}^{(K)}$ is rough and $\hat{R}_K = R_K \cup$ (corridors around it).

Let's prove that

a) $\tilde{V}_\Lambda^{(N)^{-1}} \le \hat{V}_\Lambda^{(N)^{-1}}$

b) $\hat{V}_\Lambda^{(K)} \le H_{\Lambda\setminus\hat{R}_K}^{(K)} + c\, \lambda_{(K)}\, B_K^2\, \gamma^{2K}\, |\hat{R}_K \cap \Lambda|$

c) $\sum_{\ell=1}^{N} \frac{1}{\ell!}\, \tilde{\mathcal{E}}_K^T(H_\Lambda^{(K)}; \ell) \le \hat{V}_\Lambda^{(K-1)} + \tilde{R}_\Lambda^{(K)}$

a) It is true by "positivity" of $\tilde{W}_{Q=0}^{(N-1)}$

b) It is more complicated but it still uses the positivity of $\tilde{W}^{(\;)}$

c) It is the more hard part to prove (see [3] for a careful discussion).

Let's assume that a), b), c) have been proved, then one proceeds in that way:

Let's define Q_{N-1} : pavement of tesserae of side size $\gamma^{-(N-1)}$ and

$$\chi^{b_{N-1}}_{R_{N-1}} = \chi(\tilde{\varphi}^{(N-1)} \text{ not "rough" in } R_{N-1}), \quad \overset{\circ}{\chi} \equiv 1-\chi ,$$

$$1 = \sum_{R_{N-1} \subset Q_{N-1}} \overset{\circ}{\chi}{}^{b_{N-1}}_{R_{N-1}} \chi^{b_{N-1}}_{R^c_{N-1}}$$

where R_{N-1} are an arbitrary family of tesserae $\subset Q_{N-1}$

$$\int P(d\tilde{\varphi}^{(N-1)})\, e^{\hat{V}_\Lambda^{(N-1)}} \le \sum_{R_{N-1} \subset Q_{N-1}} \int P(d\tilde{\varphi}^{(N-1)})\, \overset{\circ}{\chi}{}^{b_{N-1}}_{R_{N-1}} \chi^{b_{N-1}}_{R^c_{N-1}}\, e^{H^{(N-1)}_{\Lambda \setminus \hat{R}_{N-1}}} \cdot e^{c\lambda_{(N-1)} B^2_{N-1} \gamma^{2(N-1)} |\hat{R}_{N-1} \cap \Lambda|}$$

$$\le \sum_{R_{N-1} \subset Q_{N-1}} e^{c\lambda_{(N-1)} B^2_{N-1} \gamma^{2(N-1)} |\hat{R}_{N-1} \cap \Lambda|} \int P(d\tilde{\varphi}^{(N-1)})\, \overset{\circ}{\chi}{}^{b_{N-1}}_{R_{N-1}} \chi^{b_{N-1}}_{R^c_{N-1}}\, e^{H^{(N-1)}_{\Lambda \setminus \hat{R}_{N-1}}}$$

as, using e),

$$\int P(d\tilde{\varphi}^{(N-1)})\, \overset{\circ}{\chi}{}^{b_{N-1}}_{R_{N-1}} \chi^{b_{N-1}}_{R^c_{N-1}}\, e^{H^{(N-1)}_{\Lambda \setminus \hat{R}_{N-1}}} \le e^{\hat{V}_\Lambda^{(N-2)} + \hat{R}_\Lambda^{(N-1)} + R_\Lambda^{(N-1)}} \left(\int dP(\tilde{\varphi}^{(N-1)})\, \chi^{b_{N-1}}_{R^c_{N-1}} \right)^{1/2}$$

we have

$$\int P(d\tilde{\varphi}^{(N-1)})\, e^{\hat{V}_\Lambda^{(N-1)}} \le \sum_{R_{N-1} \subset Q_{N-1}} e^{c\lambda_{(N-1)} B^2_{N-1} \gamma^{2(N-1)} |\hat{R}_{N-1} \cap \Lambda|} \left(\int P(d\tilde{\varphi}^{(N-1)})\, \overset{\circ}{\chi}{}^{b_{N-1}}_{R_{N-1}} \right)^{1/2} \left(e^{\hat{V}_\Lambda^{(N-2)} + c R_\Lambda^{(N-1)}} \right)$$

and this sum is controlled using that the regions where the fields are "rough" have a small probability.

4 - THE INFRARED PROBLEM

General formalism

Let's discuss shortly the general formalism, to treat a pure Coulomb gas (in the limit of ultraviolet and infrared cutoffs removed). The needed physical hypothesis to do that is neutrality. Screening must follow from the theory.

Let's define $\psi^{(-Q,N)}$ when Q refers to the infrared cutoff and N refers to the ultraviolet cutoff.

Let the covariance of $\psi^{(-Q,N)}$ be:

$$\hat{C}^{(-Q,N)} = \frac{1}{p^2+\gamma^{-2Q}} - \frac{1}{p^2+\gamma^{2N}} \tag{61}$$

Let's define the following partition function:

$$Z_I^{(-Q,N)} = \int d\mu(\psi^{(-Q,N)})\, e^{2\mu \int_I :\cos\alpha\psi^{(-Q,N)}_{(x)}:_{u.v.} d^2x} \tag{62}$$

Let's explain what $::_{u.v.}$ means:

$$\hat{C}^{(-Q,N)} = \left(\frac{1}{p^2+\gamma^{-2Q}} - \frac{1}{p^2+\gamma^{2L}}\right) + \left(\frac{1}{p^2+\gamma^{2L}} - \frac{1}{p^2+\gamma^{2N}}\right) \tag{63}$$

$$= \hat{C}^{(-Q,L)} + \hat{C}^{(L,N)} \quad , \quad L > -Q$$

therefore

$$\psi^{(-Q,N)} = \psi^{(-Q,L)} + \psi^{(L,Q)} \tag{64}$$

$$d\mu(\psi^{(-Q,N)}) = d\mu(\psi^{(-Q,L)})\, d\mu(\psi^{(L,N)}) \tag{65}$$

It is clear that $\psi^{(L,N)}$ describes the "ultraviolet part" of $\psi^{(-Q,N)}$ and $\psi^{(-Q,L)}$ describes the "infrared part".

In fact $\psi^{(L,N)}$ has a mass $\approx \gamma^{L}$. giving a decay $e^{-\gamma^{L}|\xi-\eta|}$ and an ultraviolet cutoff $\approx \gamma^{N}$; $\psi^{(-Q,L)}$ has an ultraviolet cutoff $\approx \gamma^{L}$ and a mass $\approx \gamma^{-Q}$.

$: e^{i\alpha \psi^{(-Q,N)}} : \equiv e^{i\alpha\psi^{(-Q,L)}} : e^{i\alpha\psi^{(L,N)}} :$ where : : relies on the measure $d\mu(\psi^{(L,N)})$.

$$Z_I^{(-Q,N)} = \sum_{n=0}^{\infty} \frac{\mu^n}{n!} \sum_{\varepsilon_1\dots\varepsilon_n} \int_{I^n} dx_1\dots dx_n \left\{ \int d\mu(\psi^{(-Q,N)}) \prod_{i=1}^{n} : e^{i\alpha\varepsilon_i \psi^{(-Q,N)}_{(x_i)}} :_{u.v.} \right\} \tag{66}$$

$$\{ \quad \} = \int d\mu(\psi^{(-Q,L)}) \prod_{i=1}^{n} e^{i\alpha\varepsilon_i \psi^{(-Q,L)}_{(x_i)}} \int d\mu(\psi^{(L,N)}) \prod_{i=1}^{n} : e^{i\alpha\varepsilon_i \psi^{(L,N)}_{(x_i)}} :$$

$$= \int d\mu(\psi^{(-Q,L)}) \prod_{i=1}^{n} e^{i\alpha\varepsilon_i \psi^{(-Q,L)}_{(x_i)}} \, e^{-\frac{\alpha^2}{2}\sum_{i,j}\varepsilon_i\varepsilon_j C^{(L,N)}_{(x_i,x_j)}} \, e^{\frac{\alpha^2}{2}\sum_{i=1}^{n} C^{(L,N)}_{(0)}}$$

$$= e^{\frac{\alpha^2}{2}\sum_{i=1}^{n} C^{(L,N)}_{(0)}} \, e^{-\frac{\alpha^2}{2}\sum_{i,j}\varepsilon_i\varepsilon_j C^{(-Q,N)}_{(x_i,x_j)}} \tag{67}$$

and therefore

$$Z_I^{(-Q,N)} = \sum_{n=0}^{\infty} \frac{1}{n!} \mu^n \left(e^{\frac{\alpha^2}{2} C^{(L,N)}_{(0)}}\right)^n \sum_{\varepsilon_1\dots\varepsilon_n} \int_{I^n} dx_1\dots dx_n \, e^{-\frac{\alpha^2}{2}\sum_{i,j}\varepsilon_i\varepsilon_j C^{(-Q,N)}_{(x_i,x_j)}} \tag{68}$$

Let's perform the infrared limit $Q \longrightarrow \infty$:

keeping N fixed (ultraviolet cutoff fixed) ; $C^{(-Q,N)}_{(x)}$ is not defined as $Q \to \infty$

In fact

$$\lim_{Q\to\infty} \int d^2p \, e^{ipx} \frac{(\gamma^{2N} - \gamma^{-2Q})}{(p^2+\gamma^{-2Q})(p^2+\gamma^{2N})} \text{ "=" } \int d^2p \, \frac{e^{ipx}\,\gamma^{2N}}{p^2(p^2+\gamma^{2N})} = \infty \tag{69}$$

But $\sum_{i,j} C^{(-Q,N)}_{(x_i,x_j)} \varepsilon_i \varepsilon_j$ can have a meaning in the $Q \to \infty$ limit if $\sum \varepsilon_i = 0$ (neutrality). In fact :

$$\lim_{Q\to\infty} \sum_{i,j} \varepsilon_i \varepsilon_j C^{(-Q,N)}_{(x_i,x_j)} \text{ "=" } \int d^2p \sum_{i,j} \varepsilon_i \varepsilon_j \frac{e^{ip\cdot(x_i-x_j)}}{p^2(p^2+\gamma^{2N})} \gamma^{2N} \qquad (70)$$

and if $p=0$, $\sum_{i,j} \varepsilon_i \varepsilon_j e^{ip(x_i-x_j)}$ has a zero, provided $\sum \varepsilon_i = 0$ which makes (70) finite. If $\sum \varepsilon_i \neq 0$ (not neutral) then $\sum_{i,j} \varepsilon_i \varepsilon_j C^{(-Q,N)}_{(x_i,x_j)} \xrightarrow[Q\to\infty]{} \infty$

$$\lim_{Q\to\infty} e^{-\frac{\alpha^2}{2}\sum_{i,j} \varepsilon_i \varepsilon_j C^{(-Q,N)}_{(x_i,x_j)}} = \begin{cases} 0 & \text{if } \sum_{i=1}^{m} \varepsilon_i \neq 0 \\ e^{-\frac{\alpha^2}{2}\sum_{i\neq j} \varepsilon_i \varepsilon_j \left(C^{(-\infty,N)}_{(x_i,x_j)} - C^{(-\infty,N)}_{(0,0)} \right)} & \end{cases} \qquad (71)$$

where $C^{(-\infty,N)}_{(x_i,x_j)} - C^{(-\infty,N)}_{(0)}$ is a symbolic notation for

$$\lim_{Q\to\infty} \left[C^{(-Q,N)}_{(x_i,x_j)} - C^{(-Q,N)}_{(x_i,x_i)} \right] \qquad (72)$$

(71) shows that only neutral configurations are contributing to $Z_I^{(-Q,N)}$ in the infrared limit.

Therefore $\lim_{Q\to\infty} Z_I^{(-Q,N)}$ defines the partition function of a neutral gas (of charged partcles) with a 2-body interaction

$$V^{(N)}_{(x_i,x_j)} = \lim_{Q\to\infty} \int d^2p \frac{(\gamma^{2N}-\gamma^{-2Q})}{(p^2+\gamma^{-2Q})(p^2+\gamma^{2N})} \left(e^{ip(x_i-x_j)} - 1 \right)$$

$$= \int d^2p \frac{\gamma^{2N}}{p^2(p^2+\gamma^{2N})} \left(e^{ip(x_i-x_j)} - 1 \right) \qquad (73)$$

At large distance

$$V^{(N)}_{(x_i,x_j)} \simeq -\frac{1}{2\pi} \ln |x_i - x_j| \qquad (74)$$

Reinserting in (68) $(m\, C^{(L,N)}_{(0)})$ we can rewrite:

$$Z_I^{(-Q,L,N)} = \sum_{n=0}^{\infty} \frac{1}{n!} \mu^n \sum_{\varepsilon_1 \ldots \varepsilon_n} \int_{I^n} dx_1 \ldots dx_n \left\{ \left(e^{\frac{\alpha^2}{2} C_{(0)}^{(L,N)}} \right)^n e^{-\frac{\alpha^2}{2} \sum_{ij} \varepsilon_i \varepsilon_j C^{(-Q,N)}_{(x_i,x_j)}} \right\} \qquad (75)$$

and

$$\lim_{Q\to\infty} \{ \quad \} = \begin{cases} 0 & \text{if } \sum_{i=1}^{n} \varepsilon_i \neq 0 \\ e^{-\frac{\alpha^2}{2} \sum_{i\neq j} \varepsilon_i \varepsilon_j \left(C^{(-Q,N)}_{(x_i,x_j)} - C^{(-Q,N)}_{(0)} + C^{(L,N)}_{(0)} \right)} = e^{-\frac{\alpha^2}{2} \sum_{i\neq j} V^{(L,N)}_{(x_i,x_j)}} & \text{if } \sum_{i=1}^{n} \varepsilon_i = 0 \end{cases} \qquad (76)$$

where
$$V^{(L,N)}(x_i,x_j) = \lim_{Q\to\infty} \left(C^{(-Q,N)}_{(x_i,x_j)} - C^{(-Q,L)}_{(0)} \right) \equiv \lim_{N\to\infty} V^{(-Q,L,N)}_{(x_i,x_j)} \qquad (77)$$

where

$$C^{(-Q,N)}_{(x_i,x_j)} - C^{(-Q,L)}_{(0)}$$

$$= (2\pi)^{-2} \int d^2p \left(e^{ip(x_i-x_j)} - 1 \right) \left(\frac{1}{p^2+\gamma^{-2Q}} - \frac{1}{p^2+\gamma^{2L}} \right)$$

$$+ (2\pi)^{-2} \int d^2p \; e^{ip(x_i-x_j)} \left(\frac{1}{p^2+\gamma^{2L}} - \frac{1}{p^2+\gamma^{2N}} \right) \qquad (78)$$

As $Q \longrightarrow \infty$ the first integral is finite for $|p| = 0$ and $|p| \longrightarrow \infty$
As $N \longrightarrow \infty$ the second integral is finite for $x \neq 0$.
Therefore the first integral behaves as $-1/2\pi \ln \gamma^L |x_i - x_j|$ as $|x_i - x_j| \longrightarrow \infty$, the second integral behaves as $-1/2\pi \ln \gamma^L |x_i - x_j|$ as $|x_i - x_j| \longrightarrow 0$.
Let's remember that γ^L has the dimension $[\ell^{-1}]$.
Let's define therefore

$$\gamma^L = r_0^{-1} \qquad (79)$$

Therefore it follows easily that

$$\lim_{(Q,N)\to\infty} v^{(-Q,L,N)}_{(x)} = 1/2\pi \ln\left(\frac{r_0}{|x|}\right) \tag{80}$$

To fix L means to fix an intermediate scale (a visible length scale)

$$\begin{aligned} \gamma^{-L} &= r_0 \\ \gamma^{-N} &= d \end{aligned} \tag{81}$$

d is the ultraviolet length scale: then $C^{(L,N)}_{(0)} = 1/2\pi \ln r_0/d$

Therefore we can write:

$$Z_I^{(-Q,N)} = \int d\mu(\psi^{(-Q,N)})\, e^{2\left(\mu\left(\frac{r_0}{d}\right)^{\alpha^2/4\pi}\right)\int \cos\alpha\, \psi^{(-Q,N)}_{(x)}\, d^2x} \tag{82}$$

We have discussed the connection between the massive Sine-Gordon theory and the Yukawa gas. We have also given some hints on the way to extend this connection to the true Coulomb gas.

Now we want to write the partition function for the neutral Coulomb gas system with a fixed ultaviolet cutoff.

Let's define

$$Z_I^{(R)}(\lambda,\beta) = \int d\mu(\psi^{(R)})\, e^{2\lambda\int_I \cos\alpha\, \psi^{(R)}_{(x)}\, d^2x} \tag{83}$$

where $\psi^{(R)}$ is a g.r.f. with covariance (we call here Q=R)

$$C_{\psi^{(R)}} = \frac{1}{p^2+\gamma^{-2R}} - \frac{1}{p^2+\gamma^2} \tag{84}$$

From what just discussed, we have that $\lim_{R\to\infty} Z_I^{(R)}(\lambda,\beta) \equiv Z_I^{\infty}(\lambda,\beta)$ is the partition function for the Coulomb gas in the grand canonical formalism, with an ultraviolet regulator.

The idea is now that of studying $Z_I^{(R)}$ with the R.G. and then to prove results uniform in R. Therefore :

$$\psi_\xi^{(R)} = \sum_{K=0}^{R} \tilde{\psi}_\xi^{(K)} \tag{85}$$

and $\tilde{\psi}_\xi^{(K)}$ has covariance

$$C^{(K)}_{(x-y)} = (2\pi)^{-2} \int d^2p \; e^{ip(x-y)} \left(\frac{1}{p^2+\gamma^{-2K}} - \frac{1}{p^2+\gamma^{-2(K-1)}} \right) \tag{86}$$

$$C_\psi = \sum_{K=0}^{R} C^{(K)} \tag{87}$$

The fields ψ are again identically distributed, but differently from the ultraviolet case now those with lower K have a decay on a shorter scale. In fact

$$\tilde{\psi}_\xi^{(K)} = \tilde{\psi}^{(1)}_{\gamma^{-K}\xi} \tag{88}$$

Therefore $\tilde{\psi}^{(K)}$ with K>> describe the infrared components of the field $\psi^{(R)}$ while those with finite K around 0 the ultraviolet and visible components of the field ψ.

To study the thermodynamic properties of the Coulomb gas we have to proceed in this way:

1) Consider a certain number of frequencies 1,2,....K and the corresponding fields $\tilde{\psi}^{(1)}, \tilde{\psi}^{(2)}, \ldots, \tilde{\psi}^{(K)}$

2) Compute the effective potential $\tilde{V}^{(K)}$ that we obtain by integrating over all the lowest frequencies K+1, ...R and which of course will depend on R. $\tilde{V}^{(K)}$ is defined in this way

$$e^{\tilde{V}^{(K)}(\tilde{\psi}^{(1)},\ldots,\tilde{\psi}^{(K)})} \equiv \int e^{\lambda \int_\Lambda \cos\alpha\, \psi^{(R)}_{(\xi)}\, d^2\xi} \; P(d\tilde{\psi}^{(R)}) \ldots P(d\tilde{\psi}^{(K+1)}) \tag{89}$$

3) Send $R \longrightarrow \infty$. The effective potential we have in this way tells us how the Coulomb interaction plays its role over the observables of "frequencies" associated to the fields $\tilde{\psi}^{(1)}, \dots, \tilde{\psi}^{(K)}$. Of course after that $R \longrightarrow \infty$ the highest integer K can be as large as we like.

<u>Remark</u>

We have:

$$e^{\tilde{V}_I^{(K)}(\tilde{\psi}^{(1)},\dots,\tilde{\psi}^{(K)})} = \int P(d\tilde{\psi}^{(K+1)})\dots P(d\tilde{\psi}^{(R)})\, e^{2\lambda\int_I \cos\alpha\psi^{(R)}_{\xi} d^2\xi} \tag{90}$$

$$2\lambda\int_I \cos\alpha\psi_\xi^{(R)} d^2\xi = \lambda\sum_{\sigma=\pm 1}\int e^{i\alpha\sigma\psi_\xi^{(R)}} d^2\xi \equiv V_I(R) \tag{91}$$

Let's define

$$\psi_\xi^{(K+1,R)} \equiv \sum_{J=K+1}^{R} \tilde{\psi}_\xi^{(J)} \tag{92}$$

therefore

$$\psi_\xi^{(R)} \equiv \psi_\xi^{(0,R)} = \psi_\xi^{(K)} + \psi_\xi^{(K+1,R)} \tag{93}$$

$$V_I^{(R)} = \sum_\sigma \int_I (\lambda e^{i\alpha\sigma\psi_\xi^{(K)}})\, e^{i\alpha\sigma\psi_\xi^{(K+1,R)}} d^2\xi \tag{94}$$

Let's define

$$\lambda_\sigma(\xi) \equiv \lambda e^{i\alpha\sigma\psi_\xi^{(K)}} \equiv \lambda_\sigma(\psi_\xi^{(K)}) \tag{95}$$

over $\psi_\xi^{(K)}$ I am not going to integrate and therefore λ_σ is just an activity depending on ξ. If now one tries to compute (90) one is again in situation which is formally of the same type as that discussed in the ultraviolet case.

$$e^{\tilde{V}_I^{(K)}(\psi^{(K)})} = \int P(d\tilde{\psi}^{(K+1)}) \dots P(d\tilde{\psi}^{(R)})\, e^{\sum_\sigma \int \lambda\sigma(\xi)\, e^{i\alpha\sigma\psi_\xi^{(K+1,R)}} d^2\xi}$$

is formally a partition function for the following field theory:

$$\int P(d\psi^{(K+1,R)})\, e^{\sum_\sigma \int_I \lambda_\sigma(\xi)\, e^{i\alpha\sigma\psi_\xi^{(K+1,R)}} d^2\xi} = Z_I^{(R)} \qquad (96)$$

To compute $\tilde{V}_I^{(K)}(\psi^{(K)})$ is just to compute the "pressure" for this partition function.

If we put K=−1 then this is the <u>true</u> partition function.

Two remarks now:

a) One has to compute (90) with the same technics as before, which means that one computes (90) integrating one step after another and trying to control the remainder at each step.

Let's assume for the moment we are able to do it ; we will discuss it in some detail for the pressure we get the following expression for $\tilde{V}_I^{(K)}$:

$$\tilde{V}_I^{(K)}(\psi^{(K)}) = \lim_{R\to\infty} \sum_{q=1}^{\infty} \frac{1}{q!}\, \mathcal{E}^T_{\psi^{(K+1,R)}}\left(V_I^{(R)}(\psi^{(K)}, \psi^{(K+1,R)}); q\right) \qquad (97)$$

Formal cumulant expansion. The following facts are true ([8][10]):

a) each term of this series is finite as $R\to\infty$, but the odd order terms go to 0 (neutrality)

a') each term of this series describes the contribution to the interaction typical of a multipole.

b) for $\alpha^2>8\pi$ this series is asymptotic to $\tilde{V}_I^{(K)}$ which means that with probability 1

$$\left|V_I^{(K)}(\psi^{(K)}) - \lim_{R\to\infty}\sum_{q=1}^{M}\frac{1}{q!}\, C_q(\psi^{(K)})\right| \le O(\lambda^{M+\tau}) \qquad (98)$$

with $\tau>0$

c) for $\alpha^2<8\pi$ we have the following results:

the term $C_{2q}(\psi^{(K)})$ describing the contribution of a 2q-multipole to the effective potential is in the limit $R \longrightarrow \infty$ finite as $\alpha^2>\alpha^2_{2q}$ and infinite as $\alpha^2<\alpha^2_{2q}$. Therefore the asymptotic expansion becomes

$$\left| V_I^{(K)}(\psi^{(K)}) - \lim_{R\to\infty} \sum_{q=1}^{M(\alpha)} \frac{1}{q!} C_q(\psi^{(K)}) \right| \leq O(\lambda^{M(\alpha)+\tau}) \tag{99}$$

and $M(\alpha) \leq 2(n-1)$ if $\alpha^2 \in [\alpha^2_{2(n-1)}, \alpha^2_{2n}]$

which exactly means that the description of the effective interaction as sum of all size multipoles fails below 8π but partially survives.

<u>The Mayer expression for the pressure, $\alpha^2 \geq 8\pi$</u>

now we want to estimate the pressure of the Coulomb gas which means $\tilde{V}^{(K)}(\psi^{(K)})$ for K=-1 ; therefore we want to compute:

$$\int d\mu(\psi^{(R)})\, e^{2\lambda \int_I \cos\alpha\, \psi_\xi^{(R)} d^2\xi} = Z_I^{(R)}(\lambda,\beta) \tag{100}$$

in the $R \longrightarrow \infty$ limit.

Let's define $\varphi_x^{[\leq R]} = \psi_{\gamma R_x}^{(R)}$

$$\psi_x^{(R)} \longleftrightarrow \frac{1}{p^2+\gamma^{-2R}} - \frac{1}{p^2+\gamma^2}$$

$$\varphi_x^{[\leq R]} \longleftrightarrow \frac{1}{p^2+1} - \frac{1}{p^2+\gamma^{2(R+1)}} \tag{101}$$

$$\int d\mu(\psi^{(R)}) = \int d\mu(\varphi^{[\leq R]}) \ldots$$

$$\lambda \int_I \cos\alpha\, \psi_\xi^{(R)} d^2\xi = \lambda\, \gamma^{2R} \int_{\gamma^{-R}I} \cos\alpha\, \psi_{\gamma^{2R}x}^{(R)}\, d^2x \tag{103}$$

$$\cos\alpha\, \psi^{(R)}_{\gamma^{2R}x} = \cos\alpha\, \varphi_x^{[\le R]} = :\cos\alpha\, \varphi_x^{[\le R]}: < \cos\alpha\, \varphi_x^{[\le R]} >$$

$$= :\cos\alpha\, \varphi_x^{[\le R]}: \; e^{-\frac{\alpha^2}{2} C^{(R)}_{(0)}} = \gamma^{-\frac{\alpha^2}{4\pi} R} :\cos\alpha\, \varphi_x^{[\le R]}: \tag{104}$$

therefore

$$\lambda \int_I \cos\alpha\, \psi_\xi^{(R)} d^2\xi = \left(\lambda\, \gamma^{(2-\frac{\alpha^2}{4\pi})R}\right) \int_{\gamma^{-R}I} :\cos\alpha\, \varphi_x^{[\le R]}: d^2x$$

$$= \lambda(R) \int_{\Lambda(R)} :\cos\alpha\, \varphi_x^{[\le R]}: d^2x \equiv V^{(R)}_{0,\Lambda(R)} \tag{105}$$

therefore the partition function is now exactly the same as in the ultraviolet problem for the Yakawa gas except that:

a) $\lambda \longrightarrow \lambda(R) = \lambda\, \gamma^{(2-\frac{\alpha^2}{4\pi})R}$

b) $\Lambda \longrightarrow \Lambda(R) = \gamma^{-R} I$ (106)

c) $\alpha^2 > 8\pi$

It is clear that, as we know there is no ultraviolet stability above 8π, the only possibility that the theory exists here is that $\lambda(R)$ and $\Lambda(R)$ match together in the right way to compensate every divergence.

This must be true as in the infrared case:

$$\int d\mu(\psi^{(R)})\, e^{2\lambda \int_I \cos\alpha\, \psi_\xi^{(R)} d^2\xi} \le e^{2\lambda |I|} \tag{107}$$

is trivial.

The point is that in this case one wants to prove the asymptoticity of the Mayer series for the pressure which means to control the cumulant expansion at any order and to have a non trivial estimate on its remainder.

To do that one has to proceed in this way : as in the ultraviolet case define

$$C_{2q} = \mathcal{E}^T_{\varphi^{[\le R]}}\left(V^{(R)}_{0,\Lambda(R)} ; 2q \right) \tag{108}$$

and rewrite the partition function in the following way:

$$Z_I^{(R)}(\lambda,\beta) = e^{\sum_{q=1}^{M} \frac{C_{2q}^{(R)}}{(2q)!}} \int d\mu(\varphi^{[\leq R]}) \; e^{\left[V_{0,N(R)}^{(R)} - \sum_{q=1}^{M} \frac{1}{(2q)!} C_{2q}^{(R)} \right]} \tag{109}$$

then one can prove the two following facts:

I) all the coefficients $C_{2q}^{(R)}$ are finite, in the $R \longrightarrow \infty$ limit.

II) The integration in (109) with this "renormalized" potential is such that, for a mechanism very similar to that working in the ultraviolet case, produces a remainder which is finite and has the right properties in λ and $|I|$.

The proof goes on with the same R.G. techniques used in the ultraviolet case.

(I) and (II) allow us to conclude that for any integer M, $\alpha^2>8\pi$, provided λ is small enough, (M dependent), there exists a constant Co, such that

$$\left| p(\lambda,\beta) - \sum_{q=1}^{M} \frac{1}{(2q)!} \bar{C}_{2q}(\lambda,\beta) \right| \leq C_0 \, \lambda^{M+\delta} \tag{110}$$

when

$$\bar{C}_{2q} \equiv \lim_{|I|\to\infty} \lim_{R\to\infty} \frac{C_{2q}^{(R)}}{|I|}$$

REFERENCES

[1] DYSON F., LENARD A., Stability of matter I. II J. Math. Phys., 8, 423, 1967; 9, 698, 1968.

LIEB E., The stability of matter. Rev. Mol. Phys. 48, 553.

[2] FROHLICH J., Com. Math. Phys. 47, 233, 1976.

[3] BENFATTO G., GALLAVOTTI G., NICOLO F. Com. Math. Phys. 83, 387, 1982
NICOLO F., Com. Math. Phys., 88, 581, 1983.

[4] LIEB E., LEBOWITZ J.L., Adv. Math. 9, 316, 1972.

[5] BRYDGES D. Com. Math. Phys., 58, 313, 1978

BRYDGES D., FEDERBUSH P., Com. Math. Phys. 73, 197, 1980.

IMBRIE J., Com. Math. Phys., 87, 515, 1983.

[6] FROHLICH J., SPENCER T., J. Stat. Phys. 24; 617, 1981 ; Com. Math. Phys.

[7] FROHLICH J., PARK Y.M., Com. Math. Phys., 59, 235, 1978.

[8] GALLAVOTTI G., LEBOWITZ J., NICOLO F. (to appear) "The screening phase transition in the two dimensional Coulomb gas".

[9] BENFATTO G., CASSANDRO M., GALLAVOTTI G., NICOLO F., OVLIVIERI B., PRESUTTI E., SCACCIATELLI E., Com. Math. Phys. 71, 95 (1980).

BENFATTO G., GALLAVOTTI G., NICOLO F. J. Funct. Analysis 36, 343, 1980.

[10] GALLAVOTTI G., NICOLO F., to appear.

INSTANTON CONTRIBUTION TO THE $\mathbb{C}P_1$ -MODEL WITH PERIODIC BOUNDARY CONDITIONS

J.L. RICHARD

Centre de Physique Théorique *
C.N.R.S. - Luminy - Case 907
13288 Marseille Cedex 9
FRANCE

1 - Introduction.

Semi-classical approximation of quantum systems through the saddle point method has been developped during the last years in view of a semi-classical treatment of Q.C.D. Let us recall that this method has been tested quite successfully in elementary quantum mechanical systems. For the standard double-well problem it has been shown that (complex) saddle points indeed generate the W.K.B. result [1] and that the so-called dilute gas approximation just mimics the structure of the saddle points. In the octic double-well problem the dilute gas approximation does not work [2], but the proper saddle point method is in good agreement with the W.K.B. result [3].
In order to better control this method in quantum field theory, instanton contribution for the non-linear 0(3) model has been extensively studied starting with the works of Berg and Lüscher and Fateev, Frolov and Schwartz [4]. In their approach, they imposed "spherical" boundary conditions which amounts to define the classical model on the two dimensional sphere. Their result has been analysed by Patrascioiu and Rouet. [5] and Rouet [6]. If the saddle point method is to be taken seriously one is led to think that in the infinite volume limit the behaviour of the semi-classical model is independent of the boundary conditions. This is the main reason why we apply the saddle point method starting with periodic boundary conditions for the classical model. This computation has been already done in [7]. However the analysis of the results was not achieved. In particular, the isospin invariance was not considered which play a central role if we wants to interpret the results in terms of an ensemble of charged particles.
In the present paper, we continue this analysis. To be complete we felt

---

* Laboratoire Propre du C.N.R.S.

necessary to recall some facts especially on the instanton manifold which were only suggested in [7]. This is why the classical O(3)model on the torus is presented is Section 2. Section 2.1 deals with the classical field configurations and their topological properties. In section 2.2, the instanton manifold is described in a way to be used in the saddle point method. In section 2.3, the action is introduced together with a "Riemanian metric". In section 3, we recall the computation already performed adding corrections which in fact changes our previous result. Sections 3.4 and 3.5 deal with isospin invariance which allows us to get an expression for the mean value of scalar observables to be connected with a Coulomb gas. In the last section comments are given on the infinite volume limit.

2- The classical Euclidean $\mathbb{C}P_1$ Model on the Torus

2.1 - The Field Configurations.

The classical field configurations are mappings from the two dimensional torus $\mathbb{T}_2 \equiv \mathbb{R}^2/\mathbb{Z}^2$ into the sphere S_2 (i.e. $\mathbb{C}P_1$). Let us denote by n such a mapping. As a mapping from $\mathbb{R}^2$ into the unit vectors in $\mathbb{R}^3$, it has the following properties

$$\vec{n}(x_1 + l, x_2 + m) = \vec{n}(x_1, x_2) \qquad l, m \in \mathbb{Z}$$
$$\vec{n}(x)^2 = 1 \tag{2.1}$$

We know that two maps are homotopic if and only if they have the same degree [8], (the so called topological charge). The configuration space is then decomposed into subspaces $\mathcal{N}_k$ of maps with degree k. A smooth map $\gamma : \mathbb{R} \times \mathbb{T}_2 \to S_2$ defines a path in $\mathcal{N}_k$ by $t \longmapsto n_t = \gamma(t, \cdot)$. Let $n_o \in \mathcal{N}_k$. The tangent space $T_{no}(\mathcal{N}_k)$ at no is defined as

$$T_{no}(\mathcal{N}_k) = \left\{ m : \mathbb{T}_2 \to \mathbb{R}^3 \ ; \ \vec{m}(x).\vec{n_o}(x) = 0 \quad \forall x \in \mathbb{T}_2 \right\}$$

Given any path $t \longmapsto n_t$ through n_o (at t = 0), one has

$$\frac{d}{dt} n_t \Big|_{t=0} \in T_{no} (\mathcal{N}_k)$$

Conversely, given any $m \in T_{no}(\mathcal{N}_k)$, there exists a path $t \to n_t$ such that

$$\frac{d}{dt} n_t \Big|_{t=0} = m$$

The classical action $S = \mathcal{N} \to \mathbb{R}$ is defined as

$$S(n) = \frac{1}{2f} \int_{\mathbb{T}_2} d^2x \sum_{\mu=1}^{2} \partial_\mu \vec{n}(x) \cdot \partial_\mu \vec{n}(x) \tag{2.2}$$

Due to conformal invariance of the classical model, the action is independent of the size L of the torus, this is why L = 1 in this section. Let us recall also that the degree of a map n given by

$$Q(n) = \frac{1}{8\pi}\int_{\mathbb{T}_2} d^2x \;\; \varepsilon^{\mu\nu}\; \vec{n}\cdot(\partial_\mu \vec{n} \wedge \partial_\nu \vec{n}) \qquad (2.3)$$

is the pull-back on $\mathbb{T}_2$ of the normed volume element of S_2. It is well known that

$$S(n) \geq \frac{4\pi}{f}|Q(n)| \qquad (2.4)$$

and that $S(n) = \frac{4\pi}{f}|Q(n)|$ if and only if the map n is either holomorphic or anti-holomorphic. Holomorphic and anti-holomorphic maps (named instantons and anti-instantons) are the only critical points of the action. More precisely, if $t \longmapsto n_t$ is a path in $\mathcal{N}_k$ then

$$\frac{d}{dt} S(n_t)\Big|_{t=0} = \frac{\delta S}{\delta n_o}(\dot{n}_o) \;,\; \dot{n}_o = \frac{d}{dt} n_t\Big|_{t=0}$$

vanishes if and only if no is such a map. Finally at every critical point one defines a bilinear map (the Hessian of S) on $T_{n_o}(\mathcal{N}_k)$ defined by

$$\frac{d^2}{dt} S(n_t)\Big|_{t=0} \equiv \frac{\delta^2 S}{\delta n_o{}^2}(\dot{n}_o,\dot{n}_o) \qquad (2.5)$$

2.2 - The Manifold of Instantons

Using holomorphic coordinates for the sphere S_2 (App.A) that is, defining :

$$u(x) = (\varphi_+^{-1} \circ n)(x) \text{ for the x's such that } n(x) \in S_2-\{e_3\}$$
$$v(x) = (\varphi_-^{-1} \circ n)(x) \text{ for the x's such that } n(x) \in S_2-\{-e_3\}$$

and using $z = x_1 + i x_2$ the instanton solutions are such that

$$\partial_{\bar{z}} u = 0 \quad , \quad \partial_{\bar{z}} v = 0 \qquad (2.6)$$

They are elliptic functions and can be expressed [9] in terms of the σ -functions with half periods 1/2 and i/2 having the covariance property

$$\sigma(z+w) = (-1)^{m+n+mn}\; e^{\pi \bar{w}(z+\frac{w}{2})}\sigma(z) \;,\; w=n+im \in \mathbb{Z}+i\mathbb{Z} \qquad (2.7)$$

More precisely, an instanton solution of degree k is given by

$$u = c \prod_{i=1}^{k} \frac{\sigma(z-a_i)}{\sigma(z-b_i)} \qquad (2.8)$$

where C is a non vanishing complex parameter while the ai's and bi's are such that

$$\sum_{i=1}^{k} a_i = \sum_{i=1}^{k} b_i \;,\quad a_i \neq b_j + w \quad \forall i,j=1,2,\dots,k \quad w \in \mathbb{Z}+i\mathbb{Z} \qquad (2.9)$$

For later convenience we define

$$r = \frac{1}{k}\sum_{i=1}^{k} a_i = \frac{1}{k}\sum_{i=1}^{k} b_i$$

Two sets of parameters (c,a,b) and (c',a',b') define the same solution if and only if there exist

a) $\mu_i\ ,\ \lambda_i \in \mathbb{Z} + i\,\mathbb{Z} \qquad i = 1, 2, \ldots k$ such that $\sum_{i=1}^{k} \mu_i = \sum_{i=1}^{k} \lambda_i$

b) ε and δ permutations of $\{1,2,\ldots,k\}$

such that

$$a_i' = a_{\varepsilon i} + \mu_{\varepsilon i}$$

$$b_i' = b_{\delta i} + \lambda_{\delta i}$$

$$c' = c\,(-1)^{\sum_{i=1}^{k} (R\lambda_i I\lambda_i - R\mu_i I\mu_i)}\; e^{-\pi \sum_{i=1}^{k} [\bar{\mu}_i a_i - \bar{\lambda}_i b_i + \frac{|\mu_i|^2 - |\lambda_i|^2}{2}}$$

where $R\lambda$ and $I\lambda$ denote the real and imaginary part of λ respectively. Hence one deduces that an instanton solution of degree k is entirely determined by

i) a set $[a_i, b_i] \in T_2^{k-1} \times T_2^{k-1} \qquad a_i \neq b_j$

ii) a number $r \in T_2$

iii) a complex number $c \neq 0$

the solution being given by Equ. (2.8) where

$$a_k = kr - \sum_{i=1}^{k-1} a_i \qquad\qquad b_k = kr - \sum_{i=1}^{k-1} b_i$$

up to permutations.

In order to treat correctly the parameter c one needs to look at the action of the isospin group SU_2 on the instanton manifold. The action of SU_2 reads

$$u \longrightarrow \frac{\alpha u + \beta}{-\bar{\beta} u + \bar{\alpha}} \qquad |\alpha|^2 + |\beta|^2 = 1 \qquad (2.10)$$

The function $\alpha u + \beta$ is an elliptic function with k poles $\{b_i\}$ and therefore has k zeros $\{a'_i\}$ such that $\sum b_i = \sum a_i$. So we put

$$\alpha c \prod_{i=1}^{k} \sigma(z - a_i) + \beta \prod_{i=1}^{k} \sigma(z - b_i) = c'_1 \prod_{i=1}^{k} \sigma(z - a'_i)$$

$$-\bar{\beta} c \prod_{i=1}^{k} \sigma(z - a_i) + \bar{\alpha} \prod_{i=1}^{k} \sigma(z - b_i) = c'_2 \prod_{i=1}^{k} \sigma(z - b'_i) \qquad (2.11)$$

where c'_1 and c'_2 are non vanishing complex numbers and

$$\sum a'_i = \sum b'_i = \sum a_i = \sum b_i = kr \qquad (2.12)$$

We see that r is invariant under the action of SU_2. The above equations are valid for any z. For a given z_0, we therefore obtain a solution $u'(z)$ such that

$$u'(z_0) \equiv \frac{\alpha u(z) + \beta}{-\bar{\beta} u(z) + \bar{\alpha}} = 1 \qquad (2.13)$$

from a solution u(z). Conversely, from the set of solutions given by

$$u(z) = \prod_{i=1}^{k} \frac{\sigma(z_0-b_i)}{\sigma(z_0-a_i)} \frac{\sigma(z-a_i)}{\sigma(z-b_i)} \tag{2.14}$$

one generates all the instanton solutions by the action of the SU_2 group. If one imposes only that

$$\sum_{i=1}^{k} (a_i-b_i) \in \mathbb{Z}+i\,\mathbb{Z} \tag{2.15}$$

then these solutions reads

$$u(z) = e^{\pi\left[\sum_{i=1}^{k}(\bar{a}_i-\bar{b}_i)\right](z-z_0)} \prod_{i=1}^{k} \frac{\sigma(z_0-b_i)\,\sigma(z-a_i)}{\sigma(z_0-a_i)\,\sigma(z-b_i)} \tag{2.16}$$

We will use this parametrization in our final result.

2.3 - The Hessian of the Action ; Riemannian Metric.

Let us denote by $\dot{u}$ and $\dot{v}$ the components of a tangent vector at an instanton solution n with components u and v. One gets

$$\dot{u} = -u^2\,\dot{v} \tag{2.17}$$

when $z \neq a_i$ and $z \neq b_i$. The Hessian $\frac{\delta^2 S}{\delta n^2}(\dot{n},\dot{n})$ reads

$$\frac{\delta^2 S}{\delta n^2}(\dot{n},\dot{n}) = \frac{1}{f}\int_{\mathbb{T}_2} d^2x \,\frac{|\partial_{\bar{z}}\dot{u}|^2}{(1+|u|^2)^2} \tag{2.18}$$

or as well

$$\frac{\delta^2 S}{\delta n^2}(\dot{n},\dot{n}) = \frac{1}{f}\int_{\mathbb{T}_2} d^2x \,\frac{|\partial_{\bar{z}}\dot{v}|^2}{(1+|v|^2)^2} \tag{2.19}$$

Hence, defining

$$\begin{aligned} x &= q^{-1} \prod_{i=1}^{k} \sigma(z-b_i)^2\,\dot{u}_0 \qquad \text{when } z \neq b_i \\ \xi &= q^{-1} \prod_{i=1}^{k} \sigma(z-a_i)^2\,\dot{v}_0 \qquad \text{when } z \neq a_i \end{aligned} \tag{2.20}$$

where

$$q = |c|^2 \prod_{i=1}^{k} |\sigma(z-a_i)|^2 + \prod_{i=1}^{k} |\sigma(z-b_i)|^2|^2 \tag{2.21}$$

one checks that

$$\xi = \chi \qquad \text{when } z \neq b_i \text{ and } z \neq a_i$$

so that in termsof these new functions the Hessian can be written up independently of the choice of coordinates,

$$\frac{\delta^2 S}{\delta n^2}(\dot{n},\dot{n}) = \frac{1}{f}\int_{\mathbb{T}_2} d^2x \,|(\partial_{\bar{z}} + \partial_{\bar{z}}\log q)\,\chi|^2 \tag{2.22}$$

This is the change of variables already introduced in Ref. [7].

From the periodicity property of the σ-function Equ. (2.7), we obtain

$$\chi(z+w,\bar{z}+\bar{w}) = e^{\pi k[\bar{w}(z-\gamma)-w(\bar{z}-\bar{\gamma})]}\,\chi(z,\bar{z}),\, w \in \mathbb{Z} + i\,\mathbb{Z} \qquad (2.23)$$

As a sesqulinear form, $\frac{\delta^2 S}{\delta \eta^2}$ will be defined on the space of square integrable functions with periodicity property (2.23). One then defines on this space the following operators

$$T_q = \partial_{\bar{z}} + \partial_{\bar{z}} \log q \qquad (2.24)$$

$$T_q^{+} = -\partial_z + \partial_z \log q \qquad (2.25)$$

Introducing

$$q_o = e^{\pi k|z-\gamma|^2} \qquad (2.26)$$

we write

$$T_q = T_{q_o} + \partial_{\bar{z}} \log q/q_o \qquad (2.27)$$

$$T_q^{+} = T_{q_o} + \partial_z \log q/q_o \qquad (2.28)$$

where now the "perturbation" $\partial \log q/q_o$ is seen to be periodic. Moreover let us list the following identities

$$\begin{aligned} T_q &= q_o\, q^{-1}\, T_{q_o}\, q\, q_o^{-1} \\ P_{q_o}\, q\, q_o^{-1}\, P_q &= q\, q_o^{-1}\, P_q \end{aligned} \qquad (2.29)$$

where P_q and P_{q_o} are the orthogonal projection operators on the zero mode subspaces of $\Delta_q = T_q^{+} T_q$ and $\Delta_{q_o} = T_{q_o}^{+} T_{q_o}$ respectively. Finally, a Riemannian metric is introduced in the "neighborhood" of the instanton solution u induced by the metric on the sphere S_2. It reads

$$ds^2 = \int_{\mathbb{T}_2} d^2x \, \frac{|\dot{u}|^2}{(1+|u|^2)^2} \qquad (2.30)$$

Decomposing $\dot{u}$ into components along the zero modes $\frac{\partial u}{\partial \lambda_i}$ and orthogonal modes τ one gets

$$ds^2 = \sum_{i,j=1}^{k} \mathcal{N}_{ij}\, d\bar{\lambda}_i\, d\lambda_j + \int_{\mathbb{T}_2} d^2x \, \frac{|\tau|^2}{(1+|u|^2)^2} \qquad (2.31)$$

so that using the change of variable Equ. (2.20) for τ

$$\chi = q^{-1} \prod_{i=1}^{k} \sigma\,(z-b_i)^2\, \tau \qquad (2.32)$$

one arrives at

$$ds^2 = \mathcal{N}_{ij}\, d\bar{\lambda}_i\, d\lambda_j + \int_{\mathbb{T}_2} d^2x \, \bar{\chi}\chi \qquad (2.33)$$

where

$$\mathcal{N}_{ij} = \int_{\mathbb{T}_2} d^2x \, [1+|u|^2]^{-2} \, \frac{\partial \bar{u}}{\partial \lambda_i} \frac{\partial u}{\partial \lambda_j} \qquad (2.34)$$

the λ_i's denoting a set of parameters defining the solution u.

3- Functional Integral ; Instanton contribution.

3.1 - Functional Integral.

Writting formally the expectation value of an observable Θ as

$$\langle\Theta\rangle = Z^{-1}\int_{\mathcal{N}} \mathcal{D}(n)\Theta(n)\; e^{-S(x)} \tag{3.1}$$

where $\mathcal{D}(n)$ denotes the formal measure

$$\mathcal{D}(n) = \prod_{x \in T_2} dn(x) \tag{3.2}$$

$dn(x)$ being the volume element of S_2.
The integral over $\mathcal{N}$ is decomposed into a sum over all the topological sectors and an integral over each sector in which the saddle point method is to be used [10].
To be short, let us just say that $\langle\Theta\rangle$ is written as

$$\langle\Theta\rangle = Z^{-1}\sum_k \frac{1}{(k!)^2}\, e^{-\frac{4\pi}{f}k} \int \prod_{i=1}^{2k} d^2\lambda_i \; \det\mathcal{N}\; \Theta(n_{inst}) \times \int \mathcal{D}\chi\; e^{-\frac{8}{f}(\chi,\,[T_q^+ T_q + P_q]\chi)} \tag{3.3}$$

according to section 2.3.
In order to regularise the Gaussian integral we introduce Pauli-Villars regulators so that

$$\langle\Theta\rangle = Z^{-1}\sum_k \frac{e^{-\frac{4\pi}{f}k}}{(k!)^2}\left(\frac{8}{\pi f}\right)^{2k} \int \prod_{i=1}^{k} d^2\lambda_i\; \det\mathcal{N}\; \Theta(n_{inst})\, e^{-\Gamma_{reg}} \tag{3.4}$$

where

$$\Gamma_{reg}(q) = Tr\left\{\log[\Delta_q + P_q] + \sum_{i=1}^{2} e_i \log(\Delta_q + \Gamma_i^2)L^2\right\} \tag{3.5}$$

Note that the regularisation procedure breaks the conformal invariance and therefore makes the size L of the torus a parameter in the theory.
We chose

$$e_1 = 1;\; e_2 = -2\;;\; M_1^2 = 2M^2\;;\; M_2^2 = M^2 \tag{3.6}$$

3.2 - The computation of Γ_{reg}

This computation has been already performed in Ref. [7]. To be complete we just recall the steps of the calculus.
a) Assuming q arbitrary, for instance defining $q_t = (1-t)q_0+tq_1$ where q_1 corresponds to Equ.(2.21), one computes first $\delta\,\Gamma_{reg}(q)$.
By using algebraic properties only, one gets

$$\delta\Gamma_{reg} = \int_{\mathbb{T}_2} d^2x \quad \delta\,\partial_z \log q \,[\, G_q^+ + \sum_{i=1}^{2} e_i\, G_+^{M_i} \,](x,x) + \text{c.c.} \tag{3.7}$$

where $G^+q(x,x')$ is the kernel of the operator G^+q satisfying

$$Tq^+\, G^+q = \mathbb{1} - pq \quad , \quad G^+q\, Pq = 0 \tag{3.8}$$

and $G_+^M(x,x')$ is the kernel of the operator $Tq\,[\Delta q + M^2]^{-1}$.
Let us emphasize that the function $[G^+q + \sum_{i=1}^{2} e_i\, G_+^{M_i}](x,x)$ has to be periodic. One can easily see that from Equ. (B.7) of Ref. [7] our result has not this property. This is due to an incorrect expression of G^+q at coinciding arguments for which the main singularity has been only retained. A complete computation leads to the following result (see Append. B).

$$\delta\Gamma_{reg} = -\frac{2}{\pi}\int_{\mathbb{T}_2} d^2x \,\{\delta\,\partial_z \log q \;\; \partial_{\bar z} \log q/q_o + \partial_z \log q/q_o\, \delta\,\partial_{\bar z} \log q\}$$
$$+ 4k \int d^2x \;\; \delta\log q \tag{3.9}$$
$$+ \delta\,\mathrm{Tr} \log N(q)$$

where $N(q)$ is the matrix given by

$$N(q)_{ij} = \int d^2x\; \bar{X}_{(i)}\, X_{(j)} \quad i,j = 1,2,..2k \tag{3.10}$$

$X_{(i)}$ denoting the zero eigen modes of Δq (Equ.(3.15) of Ref.[7])
Hence, having computed $\Gamma_{reg}(q_o)$, one gets

$$\Gamma_{reg}(q) = -\frac{2}{\pi}\int d^2x\, |\,\partial_z \log q/q_o|^2 + 4k\int d^2x \,\log\frac{q}{q_o} + \mathrm{Tr}\, \log N(q)$$
$$- k \log M^2 L^2 - k \log k/4 \tag{3.11}$$

to be compared with Equ.(3.27) of Ref.[7].
b) The computation of Γ_{reg} corresponding to the instanton solutions follows quite easily referring to Equ. (3.36) of Ref.[7]. As one can see the function $I(c,a,b)$ does not enter any more which changes rather drastically the final result. We get

$$\Gamma_{reg}(\text{inst}) = \sum_{i,j=1}^{k} \log|\sigma(a_i-b_j)|^2 - \sum_{i<j}^{k} \log|\sigma(a_i-a_j)|^2 - \sum_{i<j} \log|\sigma(b_i-b_j)|^2$$
$$+ \log|c|^2 + \log\det\mathcal{N} - k\log M^2L^2$$
$$+ \mu(k) \tag{3.12}$$

where

$$\mu(k) = 2k - k\log k/4 + 2k^2\int d^2x\, \log|\sigma(z)|^2\, e^{-\pi|z|^2} - \log|\lambda(k)|^2 \tag{3.13}$$

It has been shown elsewhere (see Ref.[11]) that

$$\mu(k) = -2\alpha k + c^{\,\underline{st}} \tag{3.14}$$

α being a numerical constant.

3.3 - The expression of $\langle\Theta\rangle$

Using (c,γ,a,b) to parametrize the instanton, the expectation value (3.3) reads

$$\langle\Theta\rangle = Z^{-1} \sum_k \frac{L^{2k}\,\xi^{2k}}{(k!)^2} \int \frac{d^2c}{|c|^2}\, k^2 d^2\gamma \prod_{i=1}^{k-1} d^2a_i\, d^2b_i \;\frac{\prod_{i<j} |\sigma(a_i-b_j)|^2\, |\sigma(b_i-b_j)|^2}{\prod_{i,j} |\sigma(a_i-b_j)|^2}\, \Theta(u) \tag{3.15}$$

where

$$\xi = \frac{8M}{\pi f}\, e^{-2\pi/f}\, e^{\alpha}$$

$$a_k = -\sum_{i=1}^{k-1} a_i + k\gamma \tag{3.16}$$

$$b_k = -\sum_{i=1}^{k-1} b_i + k\gamma$$

The integral over the a'_is and b_i's being taken on the torus. The integration over γ is also taken on the torus. The c-integration is to be looked at in connection with the isospin invariance of the problem as it was suggested in Section 2.2.

3.4 - Isospin Invariance ; the k = 2 case

The k = 2 case is quite instructive in dealing with isospin invariance. Due to the constraint $\Sigma a_i = \Sigma b_i$, we use the new variables a and b defined through

$$a_1 = a + \gamma \qquad a_2 = -a + \gamma \qquad a \neq \pm b \tag{3.17}$$

$$b_2 = b + \gamma \qquad b_2 = -b + \gamma$$

Using the equations (2.11) the action of SU_2 on the instanton manifold reads

$$\gamma' = (\alpha\gamma + \beta)/(-\bar{\beta}\gamma + \bar{\alpha})$$

$$\mathcal{P}(a') = [\alpha\gamma\,\mathcal{P}(a) + \beta\,\mathcal{P}(b)]\,/\,(\alpha\gamma + \beta) \tag{3.18}$$

$$\mathcal{P}(b') = [-\bar{\beta}\gamma\,\mathcal{P}(a) + \bar{\alpha}\,\mathcal{P}(b)]\,/\,(-\bar{\beta}\gamma + \bar{\alpha})$$

where $\gamma = c\sigma(a)^2/\sigma(b)^2$, $\mathcal{P}$ denoting the Weisstrass function with half periods $1/2$ and $i/2$. From these equations, one constructs 3 independent SU_2 invariants namely,

$$I_0 = \frac{|\gamma|^2\, |\mathcal{P}(a)|^2 + |\mathcal{P}(b)|^2}{(|\gamma|^2 + 1)} \tag{3.19}$$

$$I_1 + iI_2 = \frac{|\gamma|^2\, \mathcal{P}(a) + \mathcal{P}(b)}{(|\gamma|^2 + 1)} \tag{3.20}$$

Hence, the invariant

$$J = I_0 - |I_1 + i\, I_2|^2 \tag{3.21}$$

is positive unless c is zero or infinity or $\mathcal{P}(a) = \mathcal{P}(b)$ that is $a = \pm b$. Therefore the range of integration which is SU_2 invariant is defined to be

$$\mathcal{D}_\varepsilon = \{(c,a,b) \mid J \geqslant \varepsilon\} \tag{3.22}$$

To deal with the c-integration, we shall perform a change of variables using the SU_2 invariance. According to Eqs (3.18), we put

$$c\,\frac{\sigma(a)^2}{\sigma(b)^2} = \frac{\bar{\alpha} - \beta}{\alpha + \bar{\beta}} \qquad (\text{i.e } \gamma' = 1)$$

$$\mathcal{P}(a') = \alpha(\bar{\alpha} - \beta)\mathcal{P}(a) + \beta(\alpha + \bar{\beta})\mathcal{P}(b) \tag{3.23}$$

$$\mathcal{P}(b') = \bar{\beta}(\beta - \alpha)\mathcal{P}(a) + \bar{\alpha}(\alpha + \bar{\beta})\mathcal{P}(b)$$

where

$$\alpha = \cos\theta\, e^{i\varphi/2}, \quad \beta = \sin\theta\, e^{-i\varphi/2}, \quad -\frac{\pi}{4} < \theta < \frac{\pi}{4}, \quad 0 \leqslant \varphi \leqslant 2\pi \tag{3.24}$$

Then going from the variables (c,a,b) to the variables (θ,φ,a',b') one gets

$$\frac{d^2c}{|c|^2}\, d^2a_1\, d^2b_1 \frac{\prod_{i<j} |\sigma(a_i - a_j)|^2\, |\sigma(b_i - b_j)|^2}{\prod_{i,j} |\sigma(a_i - b_j)|^2} =$$

$$= \cos 2\theta\, d\theta\, d\varphi\, d a_1'\, d b_1' \quad \frac{\prod |\sigma(a_i' - a_j')|^2\, |\sigma(b_i' - b_j')|^2}{\prod |\sigma(a_i' - b_j')|^2}.$$

The invariant domain of integration now reads

$$\mathcal{D}_\varepsilon' = \{(a', b') \mid |\mathcal{P}(a') - \mathcal{P}(b')|^2 \geqslant \varepsilon\} \tag{3.26}$$

which is to be interpreted as a hard core condition.

For k arbitrary, one checks that the measure occuring into the expression (3.15) of $\langle 0 \rangle$ is invariant under isospin transformations. At every step of the calculation, this can be proved by noticing the following main property : let q be given by equ.(2.21) and q' be the corresponding quantity after an isospin transformation ; then, according to Equs(2.11) one deduces that q and q' differs only by a multiplicative constant (which is $|c'_2|^2$ in our notation). In particular, the invariance of $\Gamma_{reg}(q)$ given in (3.5) becomes quite obvious.

As suggested at the end of Section 2.2, the integration over the parameter c can be performed by using the isospin invariance. It was shown that defining

$$\gamma = c \prod_{i=1}^{k} \frac{\sigma(z_0 - a_i)}{\sigma(z_0 - b_i)}$$

there always exists an SU_2 transformation sending γ into $\gamma'=1$. Hence performing a change of variable induced by the well known Fadeev-Popov trick, the integration over c or equivalently over γ is replaced by an integration over the SU_2 group. As a consequence, the instanton solutions to be used are then those which have a fixed given direction at $z = z_o$ that is, such that $u(z_o) = 1$.

3.5- Final Expression

So after integration over the SU_2 group, it is easy to see that $\langle\theta\rangle$ reads

$$\langle\theta\rangle = Z^{-1} \sum_k \frac{(L^2 \xi^2)^{k-1}}{[(k-1)!]^2} \int \prod_{i=1}^{k-1} d^2a_i\, d^2b_i\; d^2r\; e^{-4\pi W_k(a,b)}\, \theta(u) \tag{3.27}$$

where now the instanton solution u is given by Equ.(2.14).
Introducing the Coulomb potential on the torus (up to a constant), namely

$$v(x) = -\frac{1}{4\pi} \log \left\{ |\sigma(z)|^2 e^{-\pi|z|^2} \right\} \tag{3.28}$$

the potential $W_k(a,b)$ reads

$$W_k(a,b) = \sum_{i,j} v(a_i-b_j) - \sum_{i<j} v(a_i-a_j) - \sum_{i<j} v(b_i-b_j) \tag{3.29}$$

using the constraint (2.9). This potential is naturally interpreted as the Coulomb energy of a neutral gas of 2k charged particles with zero electric dipole moment.
If the constraint (2.9) is introduced explicitly in (3.27) by using an obvious change of variables, we arrive then at the final result

$$\langle\theta\rangle = Z^{-1} \sum_k \frac{(L^2 \xi^2)^k}{(k!)^2} \int_{T_2{}^{2k}} \prod_{i=1}^{k} d^2a_i\; d^2 b_i \\ .\ \delta\Big[\sum_{j=1}^{k} (a_j-b_j)\Big] \quad e^{-4\pi W_k(a,b)}\ \theta(u) \tag{3.30}$$

where the instanton solutions are given by Equ.(2.16). Let us add that integration of the a_i's and b_i's on the torus is understood with a hard core condition as exhibited in Equ.(3.26) for the k=2 case.

4 - Some comments on the final result.

In combining isospin invariance and some parametrization of the instanton manifold, we have shown that the instanton contribution to the mean value of scalar observables can be interpreted in terms of the grand canonical ensemble of a neutral Coulomb gas on a two dimensional torus with zero electric dipole moment at $\beta = 4\pi$.

An other interpretation can be given using Equ.(3.27) for <0>. In this parametrisation, it is easy to see that the potential $W_k(a,b)$ splits into two pieces. The first one corresponds to the Coulomb energy W_{k-1} of $2(k-1)$ charged particles. The second one can be interpreted as a multi body interaction. Hence, heuristic arguments might be developped along the line of Ref.[6] in which the spherical case is analysed in the infinite volume limit leading to expect that in this limit one recovers the quantum Sine-Gordon model.

However, the study of ensemble of particles interacting by long range forces has been proved no so easy even in the standard problems such as the Coulomb gas. We therefore leave the problem of the infinite volume limit quite open. Let us mention only that the hard core condition such as Equ.(3.26) plays the role of an ultra violet cut-off. The stability properties of such a system have been extensively studied by F. Nicolo [12]. We refer the reader to this work and references contained in it.

Appendix A - Stereographic coordinates.

In connection with Section 2.2, let us recall the holomorphic chart we have choosen for the sphere S_2. Let us denote by (e_1, e_2, e_3) an orthonormal basis in R^3. Define

$$\varphi_\pm \quad \mathbb{C} \longrightarrow S_2 - \{\pm e_3\}$$

by

$$\varphi_+(u) = \left(\frac{2u}{1+|u|^2} \, , \, \frac{|u|^2-1}{|u|^2+1} \right) \equiv (n^1 + i n^2 \, , \, n^3)$$

and

$$\varphi_-(v) = \left(\frac{2\bar{v}}{1+|w|^2} \, , \, \frac{1-|v|^2}{|w|^2+1} \right) \equiv (n^1 + i n^2 \, , \, n^3)$$

Then for $u \neq 0$

$$(\varphi_-^{-1} \cdot \varphi_+)(u) = \frac{1}{u} = v$$

Appendix B - On the Computation of Green's function

The equation

$$T_q^+ G_q^+ = \mathbb{1} - P_q \text{ with } G_q^{+} P_q = 0$$

is solved by putting

$$G^+_q = q\, q_o^{-1} \; G^+_{q_o}\, q_o\, q^{-1} \,(\mathbb{1} - P_q)$$

where $G^+_{q_o}$ is the solution of

$$T^+_{q_o} G^+_{q_o} = \mathbb{1} - P_{q_o}$$

such that its adjoint $G^-_{q_o} = (G^+_{q_o})^+$ satisfies

$$T_q G^-_{q_o} = \mathbb{1}$$

In terms of differential equations one has to solve

$$(-\partial_z + \partial_z \log q_o)\, G^+_{q_o}(x,x') = \delta(x-x') - P_{q_o}(x,x')$$

$$(\partial_{\bar z} + \partial_{\bar z} \log q_o)\, \overline{G^+_{q_o}(x',x)} = \delta(x-x')$$

Since we are insterested in the local behaviour $G^+_{q_o}(x,x')$ we introduce the following Ansatz

$$G^+_{q_o}(x,x') = \left[-\frac{1}{\pi}\, \partial_z \log|z-z'| + H(x,x') \right] e^{-\pi k[(\bar z - \bar z')z - (z-z')\bar z']}$$

the exponential factor taking care of the periodicity property of $G^+_{q_o}$ when x and x' are close together. The function H(x,x') is regular and periodic. Then we get for H the following equations

$$2k + \partial_{\bar z} H(x,x') = P_{q_o}(x,x') + O(|x-x'|)$$
$$\partial_{\bar z'} H(x,x') = 0 + O(|x-x'|)$$

These equations need not be solved but will be used in our next computation. Note that the H-term in the expression of G^+_q was absent in our previous computation Ref. [7].
The expression of G^{+M}_q is borrowed from Ref.[7] Putting things together, one gets

$$\left[G^+_q + \sum e_i\, G^{+Mi}_q\right](x,x') = -\frac{2}{\pi}\, \partial_{\bar z} \log \frac{q}{q_o} + H(z,\bar z)$$
$$- q\, q_o^{-1}\, G^+_{q_o}\, q_o q^{-1}\, P_q$$

Hence

$$\mathrm{Tr}\left\{ \delta \partial_z \log q \left[G^+_q + \sum e_i\, G^{+Mi}_q\right] \right\}$$
$$= \int d^2x\, \delta\partial_z \log q \left[-\frac{2}{\pi}\, \partial_{\bar z} \log \frac{q}{q_o} + H(z,\bar z)\right] - \mathrm{Tr}\left\{\delta T^+_q\, q\, q_o^{-1} G^+_{q_o}\, q_o q^{-1} P_q\right\}$$

Since

$$\mathrm{Tr}\left\{\delta T_q^+ q q_o^{-1} G^+_{q_o}\, q_o q^{-1} P_q\right\} = \mathrm{Tr}\left\{\delta \log q\, T_q^+ q q_o^{-1} G^+_{q_o} q_o q^{-1} P_q\right\}$$

One gets

$$\mathrm{Tr}\left\{\delta T_q^+ q\, q_o^{-1} G^+_{q_o}\, q_o q^{-1} P_q\right\} = \mathrm{Tr}\left\{\delta \log q\, P_q\right\} - \mathrm{Tr}\left\{\delta \log q\, P_{q_o}\right\}$$

Now by integrating by part the H term and using the H-equation, one arrives at

$$\mathrm{Tr}\left\{\delta\partial_z \log q\left[G^+_q + \sum e_i\, G^{+Mi}_q\right]\right\}$$
$$= -\frac{2}{\pi} \int d^2x\, \delta\, \partial_z \log q\, \partial_{\bar z} \log \frac{q}{} - \mathrm{Tr}\left\{\delta \log q\, P_q\right\} + 2k \int d^2x\, \delta \log q$$

Moreover, it has already been shown that

$$\mathrm{Tr}\left\{\delta \log q\, P_q\right\} = -\frac{1}{2}\,\delta\, \mathrm{Tr} \log N(q)$$

Acknowledgements

The author thanks P. Iglesias for very fruitfull discussions. Thanks are due to A. Rouet for a stimulating controversy and enlightening discussions.

References.

| 1| J.L. RICHARD and A. ROUET Nucl.Phys. B185, 47 (1981).
A. LAPEDES and E. MOTTOLA Nucl.Phys. B203, 58 (1982).

| 2| A. PATRASCIOIU Phys.Rev. D17, 2764 (1978).

| 3| E. MOTTOLA and A. ROUET Phys. Lett. 119B, 162 (1982).

| 4| B. BERG and M. LUSCHER Com.Math.Phys. 69, 57 (1979).
V.A. FATEEV, I.V. FROLOV and A.S. SCHWARTZ Nucl.Phys.B154,57(1979).

| 5| A. PATRASCIOIU and A. ROUET Lett.Nuovo Cim. 35, 107 (1982).

| 6| A. ROUET Phys. Let. 124B, 379 (1983).

| 7| J.L. RICHARD and A. ROUET Nucl.Phys. B211, 447 (1983).

| 8| W. GREUB, S. HALPERN, R. VANSTONE Connections, Curvature and Cohomology Vol.I (Academic Press, 1972).

| 9| ERDELY, MAGNUS, OBERHETTINGER, TRICOMI,Higher Transcendantal Functions, Vol.2 (Mc Graw-Hill, 1953).

|10| A.S. SCHWARTZ Com.Math.Phys. 64, 233 (1979).

|11| P. CHIAPETTA and J.L. RICHARD Nucl.Phys. B211, 465 (1983).

|12| F. NICOLO This book.

STOCHASTIC QUANTIZATION AND GRAVITY

Helmut Rumpf
Institut für Theoretische Physik
Universität Wien
A-1090 Vienna, Austria

Introduction

Stochastic quantization is an unpretentious theory as far as its formal apparatus is concerned: It establishes a direct link between the classical equations of motion of a dynamical system and its quantum description. The method by which this is accomplished is based on probabilistic rather than on algebraic concepts. These properties give stochastic quantization a flavor of universality and yet distinction from standard (i.e. canonical or path integral) quantization methods. It appears powerful enough to make a breach into unexplored territory. Quantum gravity certainly has to be sought in this territory. Its investigation by stochastic methods has been initiated only very recently. In the following therefore we can give only a very preliminary account of the application of stochastic quantization to the gravitational field. Not surprisingly, gravity presents problems to the stochastic quantization program that ordinary gauge theories do not. To put the peculiarities of the gravitational field into proper perspective it will be useful to recall the most important results of the stochastic quantization of gauge fields. They have therefore also been incorporated into these lecture notes (see especially Sec. III.3). Moreover we have tried to make the presentation as pedagogical as possible. Thus we start in Section I from Nelson's formulation of quantum mechanics as Newtonian stochastic mechanics and only then introduce the Parisi-Wu stochastic quantization scheme on which all the later discussion will be based. Owing to the fact that the Euclidean gravitational action is not bounded from below the Parisi-Wu scheme is not applicable to the gravitational field. In Section II we present a generalization of the scheme that is applicable to fields in physical (i.e. Lorentzian) space-time and treat the free linearized gravitational field in this manner. The most remarkable result of this is the noncausal propagation of conformal gravitons. Moreover the concept of stochastic gauge-fixing is introduced and a complete discussion of all the covariant gauges is given. A special symmetry relating two classes of covariant gauges is exhibited. Finally Section III contains some preliminary remarks on full nonlinear gravity. In particular we argue that in contrast to gauge fields the stochastic gravitational field cannot be transformed to a Gaussian process.

As a consequence the stochastic analogue of the background field method becomes very complicated. The nontrivial modification occurs already in first order perturbation theory and casts doubt on the validity of some results obtained recently for the perturbations of flat space.

I. Preliminaries

1.1 Stochastic Mechanics

In 1966 Nelson [1] proposed a completely classical interpretation of the Schrödinger equation, which was based on stochastic Newtonian mechanics. He considered the random motion of a point particle whose position x(t) obeys the stochastic differential equation

$$\dot{x}(t) = v_+(x(t),t) + \left(\frac{\hbar}{2m}\right)^{1/2} \eta(t) . \tag{1.1}$$

This so-called Langevin equation defines the stochastic process (or random function) x(t) in terms of another, simpler stochastic process $\eta(t)$, which is called Gaussian white noise. The latter may be characterized by the expectation values

$$\langle\eta(t)\rangle_\eta = 0 , \qquad \langle\eta_i(t_1)\eta_j(t_2)\rangle = 2\delta_{ij}\delta(t_1-t_2) \tag{1.2}$$

$$\langle\eta(t_1) \ldots \eta(t_{2n+1})\rangle_\eta = 0 \tag{1.3}$$

$$\langle\eta(t_1) \ldots \eta(t_{2n})\rangle_\eta = \sum_{\substack{\text{possible}\\ \text{comb. of}\\ \text{pairs}}} \ \prod_{\text{pairs}} \langle\eta(t_i)\eta(t_j)\rangle_\eta . \tag{1.4}$$

Equations (1.3) and (1.4) imply that all correlations of order higher than two vanish (this is meant by the "Gaussian" property of the process). Formally the equations (1.2) - (1.4) are implied by the following path integral definition of the expectation value of a functional $F[\eta]$:

$$<F[\eta]>_\eta = \int d[\eta]F[\eta]\, e^{-\frac{1}{4}\int dt\eta^2} / \int d[\eta]\, e^{-\frac{1}{4}\int dt\eta^2} \quad . \tag{1.5}$$

Strictly speaking $\eta(t)$ is a generalized stochastic process, its sample space consisting of distributions rather than functions (this may be concluded from eq. (1.2)). Therefore mathematicians prefer to write (1.1) in the form

$$dx(t) = v_+dt + \sqrt{\hbar/2m}\, dw(t) \tag{1.1'}$$

where w is the Wiener process which is formally related to η by

$$w(t) = \int^t dt'\ \eta(t') \quad . \tag{1.6}$$

The sample space of w consists of continuous functions and admits the definition of a well-defined probability distribution called Wiener measure. Since the subspace of differentiable functions has measure 0, the process $\dot{w}(t)$ can be given meaning only in the distributional sense. This has to be kept in mind when operating with η instead of w, as will be done throughout these lecture notes. A similar remark applies to the purely formal path integral appearing in (1.5). Note, however, that the integration of (1.1) is unique because η appears in it multiplied only by a constant factor. Thus the Langevin equation is free from the so-called Ito-Stratonovich ambiguity [2].

An alternative characterization of the stochastic process x(t) defined by (1.1) can be given in terms of a probability density $\rho(x,t)$ via

$$<f(x(t))>_\eta = \int \rho(x,t) f(x) d^3x \quad . \tag{1.7}$$

It follows from (1.1) that $\rho(x,t)$ has to obey the Fokker-Planck-equation

$$\frac{\partial\rho}{\partial t} = -\, div(\rho\ v_+) + \frac{\hbar}{2m}\Delta\rho \quad . \tag{1.8}$$

Physically speaking equation (1.1) defines the kinematics of the process considered by Nelson. Note that this kinematics is essentially different from that of the historical Langevin equation describing

Brownian motion with friction in the presence of an external force F:

$$m\ddot{x} = -\alpha\dot{x} + F(x) + \left(\frac{3\alpha kT}{m}\right)^{1/2}\eta . \qquad (1.9)$$

It is rather of the type that describes the limiting case of large friction coefficient α of the process defined by (1.9) (the so-called Ornstein-Uhlenbeck process). This limiting case corresponds to the approximate theory of Brownian motion due to Einstein and Smoluchowski, where

$$\dot{x} = v_+(x,t) + D^{1/2}\eta \qquad (1.10)$$

(D is the diffusion coefficient $kT/m\alpha$). The dynamics of the latter theory is defined by setting v_+ equal to $F/\alpha m$, i.e. the velocity imparted to the particle by the external force. It is only at this point (apart from the choice of the diffusion constant in (1.1)) that Nelson made a different ansatz. He required that v_+ be determined by the Newtonian equation of motion, i.e. the friction is set equal to zero.

As $x(t)$ is not differentiable, the implementation of Newton's law requires a substitute for the notion of time derivative. There are in fact two such substitutes: The <u>mean forward derivative</u> $D_+x(t)$ is defined by the conditional expectation value

$$D_+x(t) = \lim_{\Delta t\to 0+} \left\langle \frac{x(t+\Delta t) - x(t)}{\Delta t} \right\rangle_{x(t)=x} \qquad (1.11)$$

and the <u>mean backward derivative</u> $D_-x(t)$ is defined as the mean forward derivative of the time reversed process $\hat{x}(t) = x(-t)$. $D_+x = v_+$ (cf. (1.1)) and $D_-x \equiv v_-$ are again stochastic processes. The appropriate generalization of the Newtonian equation of motion is

$$\frac{m}{2}(D_+D_- + D_-D_+)x = F . \qquad (1.12)$$

Equation (1.12) implies a nonlinear equation for the unknown functions v_+, v_-. However, if

$$F = -\nabla V(x) \qquad (1.13)$$

and

$$\frac{1}{2}(v_+ + v_-) = \frac{1}{m}\nabla S(x,t) \qquad (1.14)$$

then this nonlinear equation may be transformed into a linear one by introducing

$$\psi(x,t) = \sqrt{\rho(x,t)}\, e^{\frac{i}{\hbar} S(x,t)} \quad . \tag{1.15}$$

The linear equation obeyed by ψ,

$$i\hbar \frac{\partial \psi}{\partial t} = - \frac{\hbar^2}{2m} \Delta\psi + V\psi \equiv H\psi \tag{1.16}$$

is just the Schrödinger equation (with $\hbar = h/2\pi$, where h is Planck's constant)! More precisely, with every quantum state ψ there is associated a stochastic process x(t) which at any time yields as averages of arbitrary functions f(x) the corresponding quantum mechanical expectation values. Note the restriction, in this statement, to functions of x (there is no phase space for the classical stochastic process) and to equal-time correlations. Thus the equivalence of stochastic Newtonian mechanics and quantum mechanics is certainly limited in a formal sense (this is why von Neumann's theorem on the impossibility of hidden variables is circumvented), but it may still hold for all measurements that can actually be performed. We refer the reader to Ref. [1] for a discussion of this interesting question.

Now the above restriction to equal-time correlations looks rather forbidding to a generalization of Nelson's scheme to field theory, where, of course, correlations of fields with different parameter values (i.e. at different space-time points) play an essential role. This, however, is not quite so. Even in stochastic mechanics, if H possesses a ground state ψ_0, the following general property of the ground state process can be shown [3]:

$$\langle x(0)\, x(t) \rangle_\eta = (\psi_0, x\, e^{-H|t|/\hbar}\, x\, \psi_0) \quad . \tag{1.17}$$

Here (,) denotes the scalar product in the quantum-mechanical Hilbert space. Equation (1.17) is remarkable because it relates the classical stochastic process to quantum mechanics with imaginary time parameter. In 1973 it was shown by Guerra and Ruggiero [4] that a similar relationship exists also for the free scalar field in Minkowski space: Euclidean (scalar) quantum field theory is the ground state process of stochastic (scalar) field theory in Minkowski space. To show this a Hamiltonian formulation of the classical field theory has to be adopted. We will not go into the details of this because this is not the approach that we shall eventually take in the stochastic quantization of the gravitational field. Let us only mention that the above relationship is not strictly valid for the electromagnetic [5] and linearized gravitational field [6]. The reason is that only the magnetic field components (or the "magnetic" components of the Riemann tensor, respectively) become unique stochastic processes yielding the correct correlation functions, whereas the electric field components split into couples E_+, E_- involving the "time derivatives" D_+, D_-. Moreover the processes are not Markovian [6,7] (i.e. they are not determined by an initial condition). They satisfy Osterwalder-Schrader positivity, however. In the case of finite temperature the equivalence with Euclidean quantum field theory is lost even for the scalar field [8]. Finally we remark that a Nelson-type stochastic theory of spin-$\frac{1}{2}$ particles (coupled to an external electromagnetic field) employing an invariant evolution parameter was proposed recently [9].

1.2 Stochastic Quantization

A possible way out of the Euclidean non-covariance of the stochastic field theory described in the preceding section would be the introduction of an invariant evolution parameter on which the field depends additionally. It appears that conventional quantum field theory is imbedded in the quantized version of the five-parameter field theory [10]. It can be expected that the application of Nelson's scheme to the five-parameter field will associate a stochastic field in Minkowski space (with invariant evolution parameter) with every wave functional of the Schrödinger representation of the quantized field.

Nowadays physicists prefer, for well-known reasons, to define quantum field theory in terms of path integrals rather than using the Schrödinger formulation. Usually path integrals can be given a rigorous mathematical interpretation after the quantum field has been Wick-rotated into the Euclidean sector of complexified Minkowski space. There the path integral measure is a positive probability distribution and hence the Euclidean quantum field is an ordinary stochastic process whose correlation functions are the Euclidean Green functions. Formally we have

$$\langle F[\Phi]\rangle = \int d[\Phi]\ F[\Phi]\ P[\Phi] \tag{1.18}$$

$$P[\Phi] = e^{-S[\Phi]}/\int d[\Phi]\ e^{-S[\Phi]} \tag{1.19}$$

where $S[\Phi]$ is the Euclidean action. (We have set $\hbar = 1$ in (1.19) and will continue to do so.) If Φ is a free Bose field, S is a quadratic form. Hence in this case $P[\Phi]$ may be considered as an infinite-dimensional generalization of the Boltzmann distribution, Φ playing the role of a "velocity". Now in nature thermal equilibrium is always the result of a stochastic relaxation process, and one may imagine that the "equilibrium distribution" (1.19) arises from such a process, taking place in a fictitious time t, in the limit $t \to \infty$. The simplest equation describing the approach to this equilibrium is the Langevin equation

$$\frac{\partial}{\partial t} \Phi(x,t) = - \frac{\delta S[\Phi]}{\delta \Phi(x,t)} + \eta(x,t) \tag{1.20}$$

where η is a Gaussian white noise in 5 dimensions:

$$<\eta(x,t)>_\eta = 0 \tag{1.21}$$

$$<\eta(x,t)\eta(x',t')> = 2\delta^{(4)}(x-x')\delta(t-t') \tag{1.22}$$

etc. (cf. (1.2) - (1.4)).

Equation (1.20) is the basis of the stochastic quantization scheme of Parisi and Wu [11]. Its intuition is closer to the original Langevin approach to Brownian motion (eq. (1.9) than to the theory of Einstein and Smoluchowski. But of course the analogy with Brownian motion is of a more formal nature here than in Nelson's scheme, where it may be taken literally. As to the interpretation of the fictitious time t, it has been pointed out already in [11] that it corresponds roughly to the computer time in Monte Carlo simulations.

As in (1.7) one may introduce a probability distribution $P[\Phi,t]$ for the process $\Phi(x,t)$ defined by

$$<F[\Phi(x,t)]>_\eta = \int d[\Phi]\, F[\Phi(x)]\, P[\Phi,t] \tag{1.23}$$

where $F[\Phi]$ is an arbitrary functional of Φ. The Fokker-Planck equation (cf. (1.8)) for $P[\Phi,t]$ is

$$\frac{dP}{dt} = \int d^4x \left[\frac{\delta^2}{\delta\Phi(x)^2} + \frac{\delta}{\delta\Phi(x)} \frac{\delta S}{\delta\Phi(x)}\right] P \tag{1.24}$$

$$= - 2\, e^{-\frac{1}{2} S}\, \hat{H}_{FP}(P\, e^{\frac{1}{2} S}) \tag{1.25}$$

where $\hat{H}_{FP}$ is the Fokker-Planck Hamiltonian

$$\hat{H}_{FP} = \frac{1}{2} \int d^4x\, \hat{R}^{\dagger}\, \hat{R} \geq 0 \tag{1.26}$$

$$\hat{R} = \frac{\delta}{\delta\Phi(x)} + \frac{1}{2}\frac{\delta S}{\delta\Phi(x)} \; . \tag{1.27}$$

Note that $\psi_o = \exp(-\frac{1}{2}S)$ is the ground state of $\hat{H}_{FP}$. This fact allows to prove formally the existence of the correct equilibrium limit in stochastic quantization: Assuming that the spectrum of $\hat{H}_{FP}$ is discrete and has a "mass gap" ($\lambda_1 > 0$), we have

$$P[\Phi,0]\, e^{\frac{1}{2}S[\Phi]} = \sum_{o}^{\infty} c_n\, \psi_n[\Phi] \tag{1.28}$$

with

$$\hat{H}_{FP}\,\psi_n = \lambda_n\,\psi_n \, , \qquad 0 = \lambda_o < \lambda_1 \; . \tag{1.29}$$

From (1.25) we then conclude

$$P[\Phi,t] = \sum_{o}^{\infty} c_n\, e^{-2\lambda_n t}\, \psi_n[\Phi]\psi_o[\Phi] \xrightarrow{t\to\infty} c_o\, \psi_o^2[\Phi] = c_o\, e^{-S[\Phi]} \; . \tag{1.30}$$

Thus the formal limit is proportional to (1.19) as desired.

Notwithstanding the fact that the Nelson and Parisi-Wu methods aim at different aspects of (what is hoped to be) the same theory, one may formally employ the Nelson scheme also to give directly the relaxation processes of Parisi and Wu. One simply has to start from the classical action

$$S_{FP}[\Phi(x,t)] = \int dt d^4x \left[\frac{1}{2}\dot{\Phi}^2 - \frac{1}{8}\left(\frac{\delta S}{\delta\Phi(x,t)}\right)^2 + \frac{1}{4}\frac{\delta^2 S}{\delta\Phi(x,t)^2}\right] \tag{1.31}$$

whose quantum Hamiltonian is $\hat{H}_{FP}$. The Wick-rotated (with respect to t) version of S_{FP} appears in the generating functional for the correlations of the process $\Phi(x,t)$ defined by (1.20) [12]. Exponentiating a functional determinant that appears in the functional integral defining the generating functional by introducing Grassmann variables and Berezin integration, the total action in the generating functional becomes supersymmetric [12,13]. The origin of this "hidden" supersymmetry can be read off directly from (1.20): This equation may be interpreted as defining a transformation from the field $\Phi(x,t)$ to the Gaussian field $\eta(x,t)$, i.e. a so-called Nicolai mapping [14]. The existence of such a mapping is characteristic for supersymmetric theories. Note that the supersymmetry encountered here is not of the relativistic type, but is generated by "square roots" of the generator of translations in the fictitious time t.

Why should one replace Euclidean field theory by the seemingly more complicated framework embodied in (1.20)? At least three reasons can be given:

(i) Stochastic quantization does not assume a Hamiltonian or even Lagrangian formulation of the classical theory, but starts directly from the classical equations of motion (this is why we called it "unpretentious" in the introduction).

(ii) A new invariant and non-perturbative regularization scheme [15] is provided by replacing the 2-point correlation function by

$$\langle \eta_A(t)\ \eta_B(t')\rangle_\eta = 2\delta_{AB}\ \alpha_\Lambda(t-t') \qquad (1.32)$$

where α_Λ is a smooth function that tends to $\delta(t-t')$ for $\Lambda \to \infty$. (Unfortunately, although stochastic regularization respects all the global and local symmetries of the theory, the conserved currents and Ward identities differ from the standard ones by non-local terms.)

(iii) No gauge-fixing and no associated Faddeev-Popov ghosts are needed in the stochastic quantization of gauge fields. This was the main motivation of the initial papers on the Parisi-Wu method.

For the sake of fairness it has to be stated that all these advantages are of a conceptual nature and they do not necessarily imply that practical calculations become easier in stochastic quantization. We conclude this general discussion by mentioning that there is a direct relationship between stochastic quantization and two other interesting new approaches to field theory. One is the functional formulation using the Gibbs average of De Alfaro, Fubini and Furlan [16], where also a fifth parameter is introduced. It was shown to be a consequence of stochastic quantization by Gozzi [17]. On the other hand stochastic quantization may be considered as arising from the microcanonical ensemble in field theory [18].

Finally we treat as a concrete example the free Maxwell field A_a. In this case the Langevin equation reads

$$\frac{\partial A_a(k,t)}{\partial t} = -\,k^2 T_{ab}\ A_b + \eta_a(k,t) \qquad (1.33)$$

$$T_{ab} \equiv \delta_{ab} - \frac{k_a k_b}{k^2}\ . \qquad (1.34)$$

The only non-vanishing correlation of the white noise η is

$$\langle \eta_a(k,t)\eta_b(k',t')\rangle_\eta = 2(2\pi)^4\delta_{ab}\delta^{(4)}(k+k')\delta(t-t')\ . \qquad (1.35)$$

The solution of (2.34) subject to the initial condition $A_a(t=0) = 0$ is

$$A_a(k,t) = \int_0^t H_{ab}(k,t-t')\eta_b(k,t')dt' \qquad (1.36)$$

where the heat kernel H_{ab} is given by

$$H(k,t) = e^{-tk^2T} = T\, e^{-k^2t} + L \qquad (1.37)$$

$$L_{ab} := \frac{k_a k_b}{k^2} \, . \qquad (1.38)$$

The 2-point correlation function, or stochastic propagator, may be calculated from (1.37) using the semi-group property

$$H(k,t)H(k,t') = H(k,t+t') \qquad (1.39)$$

and (1.36):

$$D_{ab}(k,t;k',t') := \langle A_a(k,t)A_b(k',t')\rangle_\eta = 2(2\pi)^4\delta^{(4)}(k+k')\int_0^{\min(t,t')} d\tau H_{ab}(k,t+t'-2\tau) \, . \qquad (1.40)$$

From this we obtain the propagator

$$\langle A_a(k)A_b(k')\rangle = \lim_{t\to\infty} \langle A_a(k,t)A_b(k',t)\rangle_\eta =$$

$$= (2\pi)^4\delta^{(4)}(k+k')\left(\frac{1}{k^2} T_{ab} + \infty^2 L_{ab}\right) \, . \qquad (1.41)$$

The divergence linear in t (which has dimension $(\text{length})^2$) is due to the random walk of the longitudinal part of A implied by (1.34). Note that it drops out of gauge-invariant expectation values, e.g. $\langle F_{ab}(x)F_{cd}(x')\rangle$, which are obviously reproduced by stochastic quantization in the correct way. In the non-abelian case the random walk of LA effectively restores unitarity, at least this is what is indicated by the results for gauge-invariant quantities in perturbation theory [19]. Note that although we have not fixed the gauge in (1.34), the invariant finite part of the propagator (1.42) appears in the Landau gauge.

II. Linearized Gravity

2.1 Classical Euclidean Theory

We start from the Einstein-Hilbert action in Euclidean (more precisely: Riemannian) space-time,

$$S_{EH}[g_{ab}] = -\frac{1}{2\kappa}\int d^4x\,\sqrt{g}\,R[g_{ab}] \tag{2.1}$$

$$\kappa = \frac{8\pi G}{c^3} \tag{2.2}$$

where R is the curvature scalar of the positive-definite metric $g_{ab}(x)$. Expanding the metric around flat space-time,

$$g_{ab} = \delta_{ab} + 2\sqrt{\kappa}\,\psi_{ab} \tag{2.3}$$

one obtains in lowest order the quadratic action

$$S^{(o)}_{EH}[\psi_{ab}] = \frac{1}{2}\int d^4x\,\psi_{ab}\,V_{abcd}\,\psi_{cd} \tag{2.4}$$

describing a helicity-2 field with gauge invariance

$$\psi_{ab} \to \psi_{ab} + \partial_a\Lambda_b + \partial_b\Lambda_a \tag{2.5}$$

where Λ_a corresponds to an infinitesimal coordinate transformation. The operator

$$\begin{aligned} V_{abcd} &= k^2(\mathbb{1}_{abcd} - \delta_{ab}\delta_{cd}) + (k_a k_b \delta_{cd} + \delta_{ab}k_c k_d) - \\ &- \frac{1}{2}(k_a k_c \delta_{bd} + k_a k_d \delta_{bc} + k_b k_c \delta_{ad} + k_b k_a \delta_{cd}) \end{aligned} \tag{2.6}$$

with

$$\mathbb{1}_{abcd} = \frac{1}{2}(\delta_{ac}\delta_{bd} + \delta_{ad}\delta_{bc}) \tag{2.7}$$

is not positive. It is not even bounded from below, as may be seen upon the introduction of a complete set of spin projection operators [20] P^2, P^1, P^o, $P^{o'}$ obeying

$$P^A\,P^B = \delta^{AB}\,P^B\,, \qquad \sum P^A = \mathbb{1} \tag{2.8}$$

($\mathbb{1}$ denoting the unit operator on the space of symmetric tensor fields with components (2.7)). In terms of these,

$$V = k^2(P^2 - 2P^{o'}) \ . \tag{2.9}$$

Instead of defining the P^A explicitly we write down the generic elements $\psi^{(A)}$ of the subspaces onto which they project: $\psi^{(2)}$ has (massive) spin 2, $\psi^{(1)}$ and $\psi^{(o)}$ are pure gauge modes,

$$\psi^{(1)}_{ab} = k_a \Lambda^T_b + k_b \Lambda^T_a \tag{2.10}$$

$$\psi^{(o)}_{ab} = k_a k_b \Lambda \tag{2.11}$$

(Λ^T_a denoting a transversal vector field), and

$$\psi^{(o')}_{ab} = (\delta_{ab} - \frac{k_a k_b}{k^2})\Lambda \tag{2.12}$$

comprises the conformal degrees of freedom. The latter make the action "bottomless", just as in the full theory. This fact is well-known from the difficulties it presents to the path integral approach to Euclidean quantum gravity [21,22].

2.2 Standard Quantization

When formulating linearized quantum gravity in Euclidean space-time one has to bear in mind that due to the difficulties mentioned at the end of the last paragraph it is at least not more rigorously defined there than in Minkowski space. One should therefore consider the results stated below as a mere transcription of those obtained in Minkowski space, where the quantization of the ψ field presents no difficulties at the formal level. Thus e.g. reflection positivity of gauge-invariant observables is automatically ensured.

The standard procedure to eliminate the zero modes of V is, as in any gauge theory, to impose a gauge condition. We shall consider here the most general linear, covariant and local condition

$$C^{(\lambda)}_a \equiv \partial_c \psi_{ac} - \lambda \partial_a \psi_{cc} = 0 \ , \qquad \lambda \neq 1 \ , \tag{2.13}$$

which in the full theory corresponds to

$$(g^\lambda g^{ab})_{,b} = 0 \tag{2.14}$$

and yields in the case $\lambda = 1/2$ the harmonic coordinate condition

$$\Box_g \, x^a = 0 \tag{2.15}$$

($\Box_g$ denoting the Laplace-Beltrami operator of the metric g_{ab}). The usual way to implement the gauge condition is to add a gauge fixing term

$$L_{gf}^{(\lambda,\alpha)} = \alpha^{-1} \, C_a^{(\lambda)} \, C_a^{(\lambda)} \tag{2.16}$$

to the classical Lagrangian, thus defining a "quantum" action

$$S_\Omega^{(o)} = S_{EH}^{(o)} + S_{gf}^{(o)} . \tag{2.17}$$

We have left out the ghost part of $S_\Omega^{(o)}$ on the right hand side of (2.17), since in the non-interacting theory considered here the ghost fields decouple from ψ. The general form of $S_\Omega^{(o)}$ is

$$S_\Omega^{(o)} = \frac{1}{2} \int d^4x \; \psi \; V^{(\lambda,\alpha)} \; \psi \tag{2.18}$$

with $V^{(\lambda,\alpha)}$ invertible. Thus there exists a propagator $K^{(\lambda,\alpha)} = [V^{(\lambda,\alpha)}]^{-1}$ which in momentum space is given by

$$\begin{aligned} K_{abcd}^{(\lambda,\alpha)} &= \frac{1}{2k^2}(\delta_{ac}\delta_{bd} + \delta_{ad}\delta_{bc} - \delta_{ab}\delta_{cd}) + \frac{2\lambda-1}{2\lambda-2} \, \frac{k_a k_b \delta_{cd} + \delta_{ab} k_c k_d}{k^4} + \\ &+ (\alpha-1) \, \frac{1}{2k^4} \, (k_a k_c \delta_{bd} + k_a k_d \delta_{bc} + k_b k_c \delta_{ad} + k_b k_d \delta_{ac}) + \\ &+ \frac{1-3\alpha+8\lambda\alpha-4\lambda^2-4\lambda^2\alpha-1}{(\lambda-1)^2} \, \frac{k_a k_b k_c k_d}{k^4} . \end{aligned} \tag{2.19}$$

The gauge-invariant part of (2.19),

$$K^{(1/2,1)} = \frac{1}{2k^2}(\delta_{ac}\delta_{bd} + \delta_{ad}\delta_{bc} - \delta_{ab}\delta_{cd}) \tag{2.20}$$

will be referred to as the propagator in the "Feynman gauge", whereas the parameter choice $\lambda = 0$, $\alpha = 0$ with

$$K^{(o,o)} = \frac{1}{k^2}(P^2 - \frac{1}{2} \, P^{o'}) \tag{2.21}$$

will be called "Landau gauge". The primary objects of physical interest, the "quantum averages" (corresponding to the expectation value of the chronological product in Minkowski space) of gauge-invariant observables, may be calculated from any of the propagators. In the linear field theory considered here the only non-trivial gauge-invariant quantities are $\langle R_{abcd}(x) \, R_{ijk\ell}(x') \rangle$ with

$$R_{abcd} = 4k_{[a}\,\psi_{b][c}\,k_{d]} \tag{2.22}$$

the linearized Riemann tensor.

2.3 Parisi-Wu Quantization

The tensorial version of the Langevin equation (1.20) reads in momentum space

$$\frac{\partial}{\partial t}\,\psi_{ab}(k,t) = -\,V_{abcd}(k,t) + \eta_{ab}(k,t) \tag{2.23}$$

$$\langle\eta_{ab}(k,t)\,\eta_{cd}(k',t')\rangle_\eta = 2(2\pi)^4\,\mathbb{1}_{abcd}\,\delta^{(4)}(k+k')\,\delta(t-t')\ . \tag{2.24}$$

Equation (2.23) may be solved by an integral equation (cf. (1.36)) involving the heat kernel

$$H(k,t) = e^{-k^2t}\,P^2 + e^{2k^2t}\,P^{0'} + P^1 + P^0\ . \tag{2.25}$$

Because of the exponentially growing factor of $P^{0'}$ in this expression the equal time correlation function $\langle\psi(k,t)\psi(k',t)\rangle_\eta$ diverges exponentially in the limit $t \to \infty$, which was supposed to give the graviton propagator. This divergence affects the physical degrees of freedom $\psi^{(0')}$ and is hence more serious than the linear divergence of the photon propagator (1.41). The latter divergence was a pure gauge term and linear in t, and has its counterpart in the linear divergences proportional to P^1 and P^0 that arise from the last two terms in (2.25). The reason for the new type of divergence encountered here is, of course, the fact that the "equilibrium distribution" $\exp(-S^{(0)}_{EH})$ is non-normalizable (it still is a formal solution of the Fokker-Planck equation associated with (2.23)). Thus the Parisi-Wu formalism breaks down in the case of the linearized Euclidean gravitational field exactly for the same reason as the (unmodified) path integral formalism, namely because of the indefiniteness of the Euclidean action.

2.4 Stochastic Quantization in Physical Space-Time [23,24]

It is possible to modify the Parisi-Wu prescription in such a way that it is directly applicable to fields in Minkowski space. As a matter of fact, from the physical point of view this version of stochastic quantization is even more appealing than the Euclidean one, though the mathematics becomes more formal here.

For a given field ψ in Minkowski space with action S, consider the stochastic process defined by the Langevin equation

$$\frac{\partial\psi_A(x,t)}{\partial t} = i\,\frac{\delta S[\psi]}{\delta\psi^A(x,t)} + \eta_A(x,t) \tag{2.26}$$

$$<\eta_A(x,t)\eta_B(x',t')> = C_{AB}\delta^{(4)}(x-x')\delta(t-t') \;. \tag{2.27}$$

The correlations of η_A read formally as in the Euclidean case, but with the symbol δ_{ab} now representing the Minkowski metric. As a consequence the covariance matrix C_{AB} is no longer positive definite (except for the scalar field). Therefore η_A cannot in general be a real random field. It can be defined in the most straightforward manner by multiplying some components of the corresponding Euclidean process by the factor i. Alternatively one can define η_A as $\eta_A = i^{1/2}\zeta_A$ where ζ_A has a complex probability distribution given by the formal path integral measure $\prod_A D[\zeta_A]\,\exp(\frac{i}{2}\int dt d^4x \zeta_A C^{-1}_{AB}\zeta_B)$. The latter definition, though more formal, works also in curved space-time.

The most conspicuous property of eq. (2.26) is that it defines a complex process $\psi_A(x,t)$ even if the underlying classical field $\psi_A(x)$ is real. As a complex term $i\delta S/\delta\psi$ no longer represents a "friction force", an equilibrium limit does not exist in the ordinary sense. However, this limit does exist in the sense that the equal-time correlation functions converge for $t \to \infty$ if interpreted as tempered distributions. In the case of the linearized gravitational field we have

$$D(k,t;k',t) \equiv <\psi(k,t)\psi(k',t)>_\eta = i(2\pi)^4\delta^{(4)}(k+k')[\frac{1}{k^2}(P^2 - \frac{1}{2}P^{o'}) -$$
$$- 2i(P^1 + P^o)t - \frac{1}{k^2}(e^{2ik^2t}P^2 - \frac{1}{2}e^{-4ik^2t}P^{o'})] \;. \tag{2.28}$$

We now recall the weak limit relations of distribution theory,

$$\lim_{t\to\infty} e^{ixt} = 0 \tag{2.29}$$

$$\lim_{t\to\infty} P\,\frac{1}{x}\,e^{ixt} = \pi\delta(x) \tag{2.30}$$

((2.29) is essentially the Riemann-Lebesgue lemma, "P" in (2.30) denotes the principal value). Making use of them we obtain

$$\lim_{t\to\infty} D(k,t;k',t) = i(2\pi)^4\delta^{(4)}(k+k')\left[\frac{P^2}{k^2+i0} - \frac{1}{2}\frac{P^{o'}}{k^2-i0} - i\infty^2(P^1 + P^o)\right] . \tag{2.31}$$

We thus obtain from stochastic quantization in Minkowski space a propagator whose finite part coincides with a certain Minkowskian version of the propagator $K^{(o,o)}$ given by (2.21), the divergent part being a pure gauge. This is very similar to what we found in the example of the Maxwell field discussed at the end of Sec. 1.2. The reason for the appearance of the Landau gauge in both cases is the following. $K^{(o,o)}$ can be obtained as the invariant finite part of $\lim_{m^2\to 0} K(m^2)$ where $K(m^2)$ is the propagator of the massive extension of the theory defined by adding $-\frac{m^2}{2}\psi^2$ to the Lagrangian, i.e.

$$K(m^2) = (V - m^2\mathbb{1})^{-1} = -i\int_0^\infty dt\ e^{-im^2t}\ e^{iVt} . \tag{2.32}$$

Since the Schrödinger kernel exp(iVt) is the Minkowskian analog of the heat kernel H(t), the generalized versions of eqs. (1.40), (1.41) imply that taking the limit $t \to \infty$ in (2.31) is identical with taking the limit $m^2 \to 0$ in (2.32) (note that m^2 is the variable conjugate to t via the Fourier transform).

At this point we would like to make the following side remarks:

(i) The "naive" massive extension $V \to V - m^2\mathbb{1}$ considered above introduces a spin-0 tachyon (associated with the projector $P^{o'}$). The physical massive extension of the theory, describing pure massive spin 2, is defined by $V_{abcd} - m^2(\mathbb{1}_{abcd} - \delta_{ab}\delta_{cd})$, involving the so-called Fierz-Pauli mass term. However the latter theory does not yield gravity in the limit $m^2 \to 0$, a fact known as the van Dam-Veltman [25] mass discontinuity.

(ii) For any linear field theory defined by a self-adjoint operator V stochastic quantization in Minkowski space yields $(A = (a;x))$

$$\lim_{t\to\infty} \langle\Phi_A(t)\Phi_B(t)\rangle_\eta \propto -i\int_0^\infty d\tau\,(e^{i\tau V})_{AB} = (V + i0)^{-1}_{AB} \equiv K_{AB} . \tag{2.33}$$

As a consequence $K(m^2)$ is analytic in the lower half complex m^2-plane. Thus, in particular, stochastic quantization implies a unique Feynman propagator in curved space-time. The associated definition of vacuum has been studied in a variety of examples [26].

(iii) Gauges different from the Landau gauge can be obtained by assuming an initial (complex) probability distribution for $\psi(t=0)$

instead of setting $\psi(t=0) = 0$. In the gravitational case only the $(0,\alpha)$-gauges can be obtained from covariant Gaussian initial distributions.

There is one unorthodox feature in the right hand side of (2.31), namely the unusual sign of the imaginary displacement of the pole in $k^2 = 0$ with residue $P^{0'}$. This implies "noncausal" propagation, i.e. positive frequency conformal modes (2.12) are propagated by $K^{(0,0)}$ into the past rather than into the future. As the conformal modes are not pure gauge (though upon using the field equations they can be transformed into non-conformal type, thus enabling one to choose the well-known transverse-traceless gauge making the helicity-2 content of the theory manifest), the noncausal propagation affects also gauge-invariant expectation values. This is evident from the gauge-invariant part of the propagator,

$$K^{(1/2,1)}_{abcd}(k) = \frac{1}{k^2+i0}\left(\frac{1}{2}\,\delta_{ac}\delta_{bd} + \frac{1}{2}\,\delta_{ad}\delta_{bc} - \frac{1}{3}\,\delta_{ab}\delta_{cd}\right) - \frac{1}{6}\,\frac{1}{k^2-i0}\,\delta_{ab}\delta_{cd}\ . \tag{2.34}$$

Noncausal propagation implies that the propagator is not the expectation value of the chronological product of fields in a pure quantum state. This is not uncommon in (Euclidean) quantum gravity, where finite temperature states occur [27]. The present context is quite different, however.

We close this paragraph by mentioning that the equivalence of the modified stochastic quantization presented in this Section with Minkowskian quantum field theory has been proven perturbatively for non-gauge theories [23]. A formal non-perturbative proof should be based on the Fokker-Planck equation (cf. eq. (1.30)). However, because of the complex nature of the process considered here there is not even a candidate for an equilibrium distribution. In this respect the following conjecture of Parisi [28] could be helpful: For every complex process $\Phi_C(x,t)$ defined by a Langevin equation (2.26) (with real probability) there exists a real process $\Phi_R(x,t)$ such that

$$\langle F[\Phi_C,t]\rangle_\eta = \int d[\Phi_R]\; P[\Phi_R,t]\; F[\Phi_R] \tag{2.35}$$

with $P[\Phi_R,t]$ complex and obeying

$$\dot{P} = \int d^4x\left[\frac{\delta}{\delta\Phi_R}\left(\frac{\delta}{\delta\Phi_R} - i\,\frac{\delta S}{\delta\Phi_R}\right)\right] P[\Phi_R,t]\ . \tag{2.36}$$

The equivalence between Φ_C and Φ_R is interesting because the complex

Fokker-Planck equation (2.36) has the equilibrium solution $P[\Phi_R] \propto \exp(iS[\Phi_R])$. Non-perturbative arguments for the equivalence between stochastic quantization and quantum field theory in Minkowski space not using the Fokker-Planck equation have been given in [23] and [29].

2.5 Stochastic Gauge-Fixing [24]

In 1981 Zwanziger [30] proposed to add an extra drift term to the right hand side of the Langevin equation that should act as a restoring force to the pure gauge degrees of freedom, resulting in the damping of the random walk of the latter and thus in a finite propagator. In order to preserve gauge invariance, the restoring force has to be tangential to the gauge orbits.

Let us illustrate this with the example of the Euclidean Maxwell field. Here the restoring force has to be of the form $k_a\Lambda(k,t)$. One is thus led to define a stochastic field $B_a(k,t)$ by the modified Langevin equation

$$\dot{B}_a = -k^2 T_{ab} B_b + k_a \Lambda(k,t) + \eta_a . \qquad (2.37)$$

Note that B_a is related to the process A_a defined by (1.33) by the generalized (i.e. t-dependent) gauge transformation

$$B_a = A_a + k_a \chi(k,t) \qquad (2.38)$$

$$\dot{\chi} = \Lambda \qquad (2.39)$$

If we substitute the most general linear, covariant and local expression for Λ,

$$\Lambda^{(\alpha)} = -\alpha^{-1} k_b B_b \qquad (\alpha > 0) \qquad (2.40)$$

in (2.37), the modified Langevin equation becomes

$$\dot{B}_{ab} = -W^{(\alpha)}_{ab} B_b + \eta_a \qquad (2.41)$$

where $W^{(\alpha)}$ is now a symmetric <u>invertible</u> operator. The limit for $t \to \infty$ of the equal-time stochastic propagator of B yields just the inverse of this operator,

$$K^{(\alpha)}_{ab} = \frac{\delta_{ab} + (\alpha-1)k_a k_b k^{-2}}{k^2} , \qquad (2.42)$$

i.e. the propagator in the covariant α-gauge.

Generalizing now to linearized gravity in Minkowski space, the stochastic gauge-fixing force has the general form $i(k_a\Lambda_b + k_b\Lambda_a)$. The modified Langevin equation

$$\dot{\Phi}_{ab} = iV_{abcd}\Phi_{cd} + i(k_a\Lambda_b + k_b\Lambda_a) + \eta_{ab} \tag{2.43}$$

defines a process Φ_{ab} related to ψ_{ab} by

$$\Phi_{ab} = \psi_{ab} + i[k_a\chi_b(k,t) + k_b\chi_a(k,t)] \tag{2.44}$$

$$\dot{\chi}_a = \Lambda_a \; . \tag{2.45}$$

The most general covariant linear stochastic gauge-fixing force is implied by

$$\Lambda_a^{(\alpha,\beta,\gamma)}(\Phi) = \alpha^{-1}[k_b\Phi_{ab} + \tfrac{\beta}{2}\, k_a\Phi_{cc} + \tfrac{\gamma}{2}\, k_a k_b k_c k^{-2}\Phi_{bc}] \tag{2.46}$$

(as we work in Minkowski space, there are no sign restrictions on α, β, γ). We have included the third, non-local, term on the r.h.s. of (2.46) because only so we will recover the ordinary covariant gauges from stochastic gauge-fixing. Eqs. (2.46) and (2.43) imply

$$\dot{\Phi}_{ab} = i\, W_{abcd}^{(\alpha,\beta,\gamma)}\, \Phi_{cd} + \eta_{ab} \; . \tag{2.47}$$

Observe that W is self-adjoint only if $\beta = 0$. Thus the gauge-fixing force is in general not a variational derivative, but constitutes a nonholonomic constraint. If W is not self-adjoint, the stochastic propagator involves $H(k,t)H^T(k',t')$. Hence one cannot appeal to (1.39), as was done in the derivation of (1.40), and the calculation of the stochastic propagator for $\beta \neq 0$ is more complicated than indicated by (1.40). Therefore $K^{(\alpha,\beta,\gamma)}$ is equal to $(W^{(\alpha,\beta,\gamma)} + i0)^{-1}$ only if $\beta = 0$.

The stochastic (α,β,γ) gauges yield more general propagators than the ordinary (λ,α) gauges defined by (2.17), (2.16). We can always identify a given (λ,α) gauge with a certain stochastic gauge, but not vice versa. Specifically we have, if $\alpha \neq 0$,

$$\beta = \frac{2}{\lambda-1}(4\lambda^2 - 2\lambda + 1 - \alpha) \tag{2.48}$$

$$\gamma = \frac{2\lambda(4\lambda - \alpha - 1)}{1 - \lambda} \; . \tag{2.49}$$

Consider now the case $\alpha = 1$, $\beta = \gamma = 0$. Then eqs. (2.48), (2.49) have two solutions for λ: $\lambda = 0$ and $\lambda = 1/2$. This shows that $K^{(1,0,0)}$ is not well-defined. More precisely: If one takes the limit $\beta = -\frac{\gamma}{2} \to 0$ of $K^{(1,\beta,\gamma)}$, then one obtains $K^{(1/2,1)}$, all other approaches to $\beta = \gamma = 0$ yield $K^{(0,1)}$. Still, loosely speaking, the ordinary gauges (0,1) and (1/2,1) correspond to the same stochastic gauge.

The reason is that the field transformation

$$A: \psi_{ab} \to (A\psi)_{ab} = \psi_{ab} - \frac{1}{2}\delta_{ab}\psi_{cc} \tag{2.50}$$

which transforms $C_a^{(0)}$ of (2.13) into $C_a^{(1/2)}$, is an involution and an isometry with respect to the metric $\mathbb{1}_{abcd}$ in the space of symmetric tensor fields:

$$A = A^{-1} = A^T . \tag{2.51}$$

Since the covariance matrix of η_{ab} is proportional to $\mathbb{1}_{abcd}$, $A\eta$ has the same correlations as η. Therefore one obtains the same stochastic quantization prescription if one starts from the Langevin equation

$$\dot{\hat{\psi}} = i\,\frac{\delta S[\psi(\hat{\psi})]}{\delta\hat{\psi}} + \eta \tag{2.52}$$

if $\hat{\psi} = A\psi$. Note that this is not the case if $\hat{\psi}_{ab} = \psi_{ab} - \lambda\delta_{ab}\psi_{cc}$ with $\lambda \neq 1/2$ (if $\lambda \neq 1/4$, these transformations are still isometries, but no longer involutory). We have thus touched the problem of the definition of the "natural" field variable, which will be pursued further in the next Section.

III. Nonlinear Stochastic Gravity

3.1 Langevin Equation

In this section we consider the full gravitational field. We want to promote the classical metric $g_{ab}(x)$ to a stochastic process $g_{ab}(x,t)$. There appears to be a unique Langevin equation generalizing the Parisi-Wu ansatz to this case. Our guiding principle will be manifest covariance with respect to field redefinitions (which may be considered as coordinate transformations in field configuration space). For conciseness we shall adopt the notation of De Witt [31] and represent a general stochastic field by $\Phi^A(t)$, where the index $A = (a,b,\ldots;x)$ comprises component indices as well as the space-time coordinate on which the field depends. By covariance, then, the general form of the

Langevin equation for Φ^A must be

$$\dot{\Phi}^A = i\, G^{AB} \frac{\delta S[\Phi]}{\delta \Phi^B} + \eta^A(t) \quad . \tag{3.1}$$

Here G^{AB} is the inverse of a field metric tensor G_{AB} that is required by covariance. (Note that the Φ^A have to be considered as coordinates in field configuration space; therefore $\dot{\Phi}^A$ is a vector, and so must be η^A, while S is a scalar.) The process η is defined by the formal path integral

$$\langle F[\eta]\rangle_\eta = \int d[\eta] F[\eta] \exp\left(-\frac{1}{4}\int dt G_{AB}\eta^A\eta^B\right) / \int d[\eta] \exp\left(-\frac{1}{4}\int dt G_{AB}\eta^A\eta^B\right). \tag{3.2}$$

As the metric G_{AB} will in general be indefinite, the integration contour for every η^A in $\int d[\eta]$ should be rotated by 45° into the complex plane (i.e. $\eta = \pm i^{1/2}\zeta$ and the integration is over real ζ; cf. the discussion in 2.4 following eq. (2.27)). Note that (3.2) implies $\langle \eta^A(t)\eta^B(t')\rangle_\eta = 2G^{AB}\delta(t-t')$ only if G_{AB} is independent of Φ, and hence of η. For non-gravitational fields there exists a field coordinate system (defining the "natural field variables") for which G_{AB} is indeed independent of Φ. Note that the problem of finding these variables is essentially the same as the well-known problem of defining the correct path integral measure in standard quantum field theory.

In the case of the gravitational field $\Phi^A = g_{ab}(x)$ the most general local field metric is known [31] to be

$$G^{AB} = \frac{C}{2\sqrt{-g}}[g_{ac}(x)g_{bd}(x')+g_{bc}(x)g_{ad}(x')+\lambda g_{ab}(x)g_{cd}(x')]\delta^{(4)}(x-x') \tag{3.3}$$

where $\lambda \neq -1/2$ and C is an arbitrary constant. Substituting this into (3.1) and recalling that

$$\frac{\delta S_{EH}}{\delta g_{ab}} = -\sqrt{-g}\,(R^{ab} - \frac{1}{2} g^{ab} R) \tag{3.4}$$

we obtain

$$\dot{g}_{ab} = -iC(R_{ab} - \frac{1+\lambda}{2} g_{ab} R) + \eta_{ab} \tag{3.5}$$

$$= -iC(-g)^{(\lambda-1)/2} \frac{\delta S}{\delta[(-g)^{\lambda/2} g^{ab}]} + \eta_{ab} \; . \tag{3.6}$$

The choice of constants that yields the standard quantization of the linearized field is

$$C = 1, \qquad \lambda = 0 . \tag{3.7}$$

Adhering to this choice we obtain the following definition of the stochastic gravitational field:

$$\dot{g}_{ab} = -i(R_{ab} - \frac{1}{2} g_{ab} R) + \eta_{ab} \tag{3.8}$$

$$\langle F[g]\rangle_\eta = N\int d[\eta]F[g[\eta]]\exp(-\frac{1}{4}\int d^4xdt\sqrt{-g}g^{ac}[\eta]g^{bd}[\eta]\eta_{ab}\eta_{cd}) . \tag{3.9}$$

The stochastic source η_{ab} is genuinely non-Gaussian in the case of the gravitational field. This was first noted in [32], where auxiliary fields were introduced in terms of which the process becomes Gaussian. But in general, i.e. if also non-gravitational fields are present, an infinite hierarchy of such auxiliary fields will be necessary. This is true even in the simple case of a scalar particle moving in an external gravitational field. Let us consider this prototype of non-Gaussian stochastic dynamics in some detail. We choose the action

$$S[x(s)] = \frac{1}{2} \int ds\, g_{ab}(x) \frac{dx^a}{ds} \frac{dx^b}{ds} \tag{3.10}$$

which gives s the meaning of an affine parameter (proportional to proper time for timelike world lines) in the resulting geodesic equation of motion. The action (3.10) may also be considered as defining a non-linear σ-model in $0+1$ dimensions. The covariant Langevin equation for the stochastic variable $x(s,t)$ is

$$\frac{\partial}{\partial t} x^a(s,t) = ig^{ab}(x) \frac{\delta S}{\delta x^b(s,t)} + \eta^a(s,t) = i \frac{D^2x^a}{ds^2} + \eta^a . \tag{3.11}$$

D/ds denoting the absolute derivative. The path integral measure is $d[\eta]\exp -\frac{1}{4}\{\int dsdtg_{ab}(x[\eta])\eta^a\eta^b\}$ (requiring again a rotation of the η integration contours into the complex plane). The only way of getting rid of the non-Gaussian character of η is by introducing redundant variables. The most attractive way of doing this appears to be the following: Imbed the curved space-time M_4 in an N-dimensional pseudo-Euclidean manifold M_N such that the pseudo-Euclidean metric δ_{AB} $(A,B = 1,\dots,N)$ induces the metric g_{ab} in M_4. Consider a particle in M_N whose coordinate $X^A(s)$ is confined to the 4-dimensional submanifold $M_4 = \{X \in M_N | F^i(X) = 0,\ i = 1,\dots,N-4\}$. Its Lagrangian is

$$L = \frac{1}{2} \delta_{AB} \frac{dX^A}{ds} \frac{dX^B}{ds} + \lambda_i F^i(X) \tag{3.12}$$

where the λ_i are Lagrange multipliers. The Langevin equation associated with (3.12) is

$$\frac{dX^A(s,t)}{dt} = i\left(\frac{d^2X^A}{dt^2} + \lambda_i \delta^{AB} \frac{\partial F^i}{\partial X^B}\right) + \eta^A \tag{3.13}$$

$$F^i(X(s,t)) = 0 \tag{3.14}$$

$$\langle \eta^A(s,t)\eta^B(s',t')\rangle_\eta = 2\delta^{AB}\delta(s-s')\delta(t-t') \ . \tag{3.15}$$

We are thus led to the stochastic quantization of constrained systems which was introduced recently [33]. It may prove useful also in the case of the gravitational field.

3.2 Background Field Formalism

Owing to the non-Gaussian probability measure appearing in the stochastic quantization rule (3.9) for the gravitational field, a consistent perturbation theory is complicated by the fact that the correlations themselves depend on the interaction. Splitting the stochastic metric $g_{ab}(x,t)$ into a deterministic part $g^{(0)}_{ab}$ and a fluctuating part,

$$g_{ab} = g^{(0)}_{ab}(x,t) + 2\sqrt{\kappa}\ \psi_{ab}(x,t) \tag{3.16}$$

the modified correlation will contribute already in first order (in $\kappa^{1/2}$) perturbation theory. Therefore a perturbation theory based on the Langevin equation

$$\dot{\psi}_{ab} = i g^{(0)}_{ac} g^{(0)}_{bd} (-g^{(0)})^{-1/2} \frac{\delta S_{EH}[g^{(0)}+2\kappa^{1/2}\psi]}{\delta\psi_{cd}} + \eta_{ab} \tag{3.17}$$

with

$$\langle \eta_{ab}(x,t)\eta_{cd}(x',t')\rangle_\eta = (-g^{(0)})^{-1/2}(g^{(0)}_{ac}g^{(0)}_{bd} + g^{(0)}_{bc}g^{(0)}_{ad})\delta^{(4)}(x-x')\delta(t-t') \tag{3.18}$$

$$g^{(0)}_{ab}(x,t) \xrightarrow{t\to\infty} g^{cl.}_{ab}(x) \tag{3.19}$$

cannot be expected to be equivalent to the standard quantum perturbation theory around a classical metric $g^{cl.}_{ab}$. Nevertheless this, with $g^{(0)}_{ab}(x,t) = \delta_{ab}$, has so far been the only approximation accessible to explicit calculation, and we are going to discuss it in some detail.

Splitting the gravitational action into a free and an interacting part,

$$S_{EH}[\delta + 2\kappa^{1/2}\psi] = S^{(o)}[\psi] + S^{(int)}[\psi] \tag{3.20}$$

we may write the Langevin equation (without gauge-fixing) in the form

$$\dot{\psi} - iV\psi = i\,\frac{\delta S^{(int)}}{\delta\psi} + \eta \tag{3.21}$$

with V being given by (2.6). Its solution

$$\psi(t) = \int_0^\infty H(t-\tau)\left[i\,\frac{\delta S^{(int)}}{\delta\psi(\tau)} + \eta(\tau)\right]d\tau \tag{3.22}$$

may be expanded perturbatively into a series of tree diagrams:

$\psi =$ 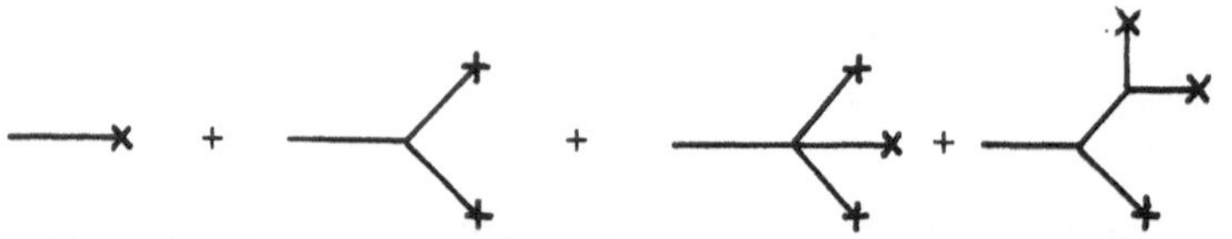 $+ \ldots$ (3.23)

Here the cross represents η and the line the Schrödinger kernel H. Note that every vertex carries a time τ which is integrated over. Eq. (3.23) implies the perturbative expansion of the stochastic average of products of ψ fields. This expansion involves the free two-point correlation function $D(t_1,t_2)$ (see (2.28)), which will be represented graphically by a crossed line,  t_1 t_2. For instance, the three-point function $\langle\psi(x_1,t_1)\psi(x_2,t_2)\psi(x_3,t_3)\rangle_\eta$ is given in lowest order by the following sum of "stochastic diagrams":

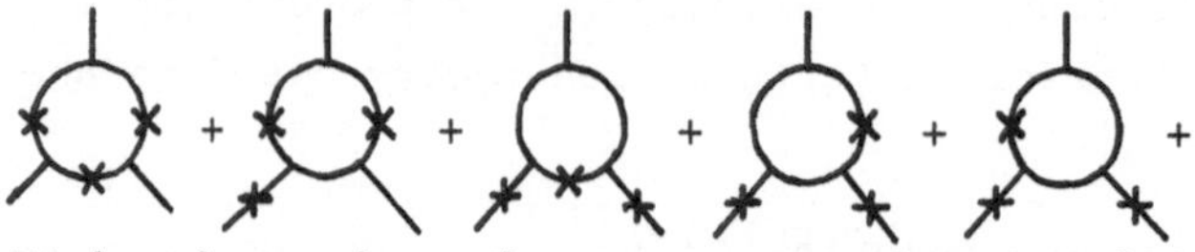 + rotations . (3.24)

It has been shown for non-gauge theories [34,23] that the sum of all stochastic diagrams with the same topology as a given Feynman diagram yields just this Feynman diagram in the limit $t \to \infty$. As we have observed already in (1.41), in the present case of a gauge theory all diagrams will actually diverge as $t \to \infty$. These divergencies will cancel, however, in gauge-invariant quantities.

For practical calculations the method of stochastic gauge-fixing is more convenient, as it yields a finite propagator. To generalize the method from the abelian case treated in 2.5 to the interacting theory, one has to take into account the non-abelian gauge transformations of ψ induced by general coordinate transformations of the full metric g:

$$\delta\psi_{ab} = \partial_a\chi_b + \partial_b\chi_a + 2\sqrt{\kappa}(\psi_{ma}\partial_b\chi_m + \psi_{mb}\partial_a\chi_m + \chi_c\partial_c\psi_{ab}) \equiv \tag{3.25}$$

$$\equiv [\int d^4y\chi_c(y)Q_c(y)]\psi_{ab}(x) \ . \tag{3.26}$$

In (3.26) we have introduced the generators of gauge transformations Q_c. According to Zwanziger's scheme [30] (see 2.5) we may add the "drift force" $i\{\int d^4y\Lambda_c^{(\alpha,\beta,\gamma)}[\psi(y,t)]Q_c(y)\}\psi_{ab}(x,t)$ to the right hand side of the Langevin equation (3.21) (in the non-abelian case this drift force is never derivable from a gauge-fixing action). The resulting modified Langevin equation has the general structure

$$\dot{\psi}_{ab} = iW_{abcd}^{(\alpha,\beta,\gamma)}\psi_{cd} + iZ_{ab}^{(\alpha,\beta,\gamma)} + i\frac{\delta S^{(int)}}{\delta\psi_{ab}} + \eta_{ab} \tag{3.27}$$

where Z_{ab} is of the form

$$Z_{ab}(k,t) \sim \int d^4k' \ Z_{abcdij}(k,k')\psi_{cd}(k',t)\psi_{ij}(k-k',t) \ . \tag{3.28}$$

Therefore the gauge-fixing force introduces an effective interaction that has the same structure as $\delta S^{(1)}/\delta\psi_{ab}$, where $S^{(1)}$ is the first order part of $S^{(int)}$. The effective interaction results in a new 3-vertex, which has to replace the old one in all stochastic diagrams. Graphically,

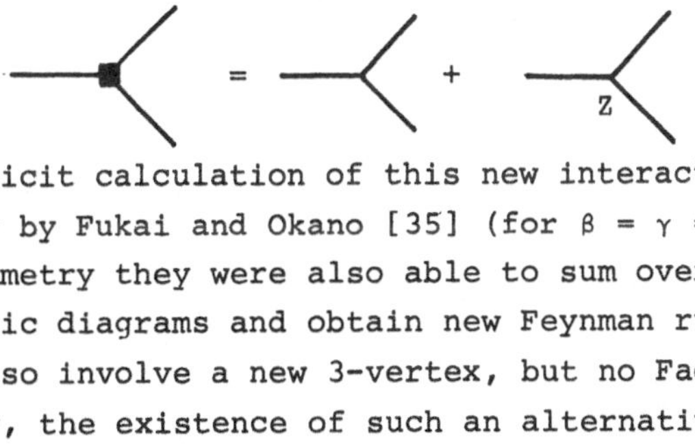

(3.29)

The explicit calculation of this new interaction has been carried out recently by Fukai and Okano [35] (for $\beta = \gamma = 0$). Exploiting the hidden supersymmetry they were also able to sum over topologically equivalent stochastic diagrams and obtain new Feynman rules for <u>ordinary</u> diagrams: These also involve a new 3-vertex, but no Faddeev-Popov ghosts. (Interestingly, the existence of such an alternative set of Feynman rules for the Yang-Mills field has been argued recently by considerations independent from stochastic quantization [36].) In view of the general considerations made at the beginning of this section, the physical significance of this result is still unclear, however.

In any discussion of perturbative quantum gravity there arises the question of the relevance of the non-renormalizability of this theory. Recently the view [37] was expressed that stochastic quantization is not applicable to non-renormalizable field theories if one wishes to include the radiative corrections and renormalize before taking the limit of infinite fictitious time. On the other hand it was observed [38] that the consequence of stochastic quantization in non-renormalizable theories is that an infinite number of stochastic sources has to be introduced. This may be considered just as a reflection of the fact that an infinite number of parameters is needed to describe a non-renormalizable theory at the quantum level.

3.3 Non-Perturbative Aspects

The outstanding question that remains regarding the general topic of this Section is: What are the physical implications of the gravitational Langevin equation (3.8), (3.9) and what is its relation to other approaches to quantum gravity? Rather than attempting to answer this question we shall give in the following an outline of the present status of knowledge concerning the corresponding questions in the case of non-abelian gauge fields. This might give an idea of what can be hoped for in the case of stochastic gravity.

We begin with a brief recapitualtion of the standard formalism used in the quantization of gauge theories, again adopting De Witt's conventions (cf. 3.1). The basic object of interest is the Schwinger average of an observable $\Theta[\Phi]$ formed from the gauge fields Φ^A,

$$\langle out|T(\Theta[\Phi])|in\rangle \propto \int d[\Phi]d[c]d[\bar{c}]\; e^{iS_{tot}[\Phi,c,\bar{c}]}\; \Theta[\Phi] \tag{3.30}$$

where c, $\bar{c}$ are the Faddeev-Popov ghost and anti-ghost fields and the total action consists of three parts:

$$S_{tot} = S_{cl}[\Phi] + S_{gf}[\Phi] + S_{ghost}[\Phi,c,\bar{c}] \tag{3.31}$$

$$S_{gf} = \tfrac{1}{2}(\alpha^{-1})_{\rho\sigma}\; C^{\rho}\; C^{\sigma} \tag{3.32}$$

$$S_{ghost} = \bar{c}_{\rho}\; F^{\rho}{}_{\sigma}[\Phi]\; c^{\sigma}\;. \tag{3.33}$$

Here ρ, σ are Lie algebra indices that include also the space-time argument x, $\alpha_{\rho\sigma}$ form a symmetric "matrix" (generalizing the gauge parameter), $C^{\rho}[\Phi]$ are the gauge conditions and $F^{\rho}{}_{\sigma}[\Phi]$ is the linear operator

$$F^{\rho}{}_{\sigma}[\Phi] = \left(\frac{\delta}{\delta\Phi^A}\, C^{\rho}[\Phi]\right)\Omega^A{}_{\sigma}[\Phi] \tag{3.34}$$

involving the generators $\Omega^A{}_\sigma$ of gauge transformations (cf. (3.26)) defined by

$$\delta\Phi^A = Q^A{}_\sigma[\Phi]\ \chi^\sigma \ . \tag{3.35}$$

The Faddeev-Popov ghosts were introduced by "exponentiating" the Faddeev-Popov determinant det $F[\Phi]$:

$$\int d[c]d[\bar{c}]\ e^{iS_{ghost}} \propto \det F[\Phi] \ . \tag{3.36}$$

For a discussion of the stochastic quantization of gauge fields we shall return to Euclidean space-time and make the following specializations:

$$\Phi^A = A^\rho{}_a(x) \tag{3.37}$$

$$\delta\Phi^A = D_a\chi^\rho \equiv \partial_a\chi^\rho + gf^\rho{}_{\sigma\gamma}A^\sigma\chi^\gamma \tag{3.38}$$

$$(\alpha^{-1})_{\rho\sigma} = \alpha^{-1}\ \delta_{\rho\sigma} \tag{3.39}$$

$$C^\rho = \partial_a\ A^\rho{}_a \ . \tag{3.40}$$

Equation (3.40) implies via (3.34)

$$S_{ghost} = \int d^4x\ \partial_a\ \bar{c}_\rho\ D_a\ c^\rho \ . \tag{3.41}$$

Consider now the unconstrained Langevin equation

$$\frac{\partial A^\rho_a}{\partial t} = -\frac{\delta S_{cl}}{\delta A_{\rho a}} + \eta^\rho{}_a \ . \tag{3.42}$$

The associated Fokker-Planck equation (1.24) has the equilibrium solution $P[A] = e^{-S_{cl}[A]}$, which is not normalizable due to the infinite volume of every gauge orbit which is integrated over in $\int d[A]P[A]$. Therefore, at the level of the probability distributions, equivalence of stochastic quantization with standard quantization can hold only with stochastic gauge-fixing. According to Zwanziger's argument (see Secs. 2.5, 3.2) the gauge-fixing force has to be of the form $(D_a\Lambda(x,A))^\rho$, hence the modified Langevin equation and the associated Fokker-Planck equation read

$$\frac{\partial A^{\rho}{}_{a}}{\partial t} = -\frac{\delta S_{cl}}{\delta A_{\rho a}} + (D_a \Lambda(x,A))^{\rho} + \eta^{\rho}{}_{a} \tag{3.43}$$

$$\dot{P}[A,t] = \int d^4x \frac{\delta}{\delta A^{\rho}{}_{a}(x)} \left[\frac{\delta}{\delta A_{\rho a}(x)} + \frac{\delta S}{\delta A_{\rho a}(x)} - (D_a \Lambda)^{\rho}\right] P[A,t] \equiv - LP \,. \tag{3.44}$$

The Fokker-Planck equation has an equilibrium solution P_o defined by

$$P_o : LP_o = 0 \tag{3.45}$$

which can indeed be made to coincide with the "Faddeev-Popov" distribution"

$$P_{FP} = N \int d[c]\, d[\bar{c}]\, e^{-S_{tot}} \tag{3.46}$$

if Λ^{ρ} is chosen to be

$$\Lambda^{\rho}(x,A) = P_{FP}^{-1} N \int d[c]d[\bar{c}]c^{\rho}(x) \int d^4y \partial_a \bar{c}^{\sigma}(y) \frac{\delta S_{tot}}{\delta A^{\sigma}{}_{a}(y)} e^{-S_{tot}} \tag{3.47}$$

[39]. But so far only formal arguments [40] have been given that the equilibrium is indeed reached, i.e. that

$$\lim_{t\to\infty} P[A,t] = P_o[A] \,. \tag{3.48}$$

Of course $P_o = P_{FP}$ can hold only if P_{FP} is positive. But it is known that this is not true because of the notorious Gribov ambiguity [41]. Let us define the "Gribov region" Ω in the gauge field configuration space by the property that $P_{FP} > 0$ there (its boundary is called the Gribov horizon). Then numerical calculations in a finite-dimensional lattice approximation indicate the following [3]: If the stochastic process A(t) with gauge-fixing force starts in Ω, then it remains there for all t. Maybe this is an instance where something new can be learned from stochastic quantization, namely that the domain of integration in the path integral should be confined to the Gribov region.

The two main obstacles to generalizing these results to the gravitational field are, of course, the indefiniteness of the action and the non-Gaussian character of the stochastic source. At present there seem to exist three possible assessments of this situation: (i) The problems of stochastic gravity are purely mathematical and will finally be overcome. (ii) Quantum gravity (whatever it means) cannot be constructed in the stochastic manner. (iii) Einstein's theory is not the correct starting point for quantization. Only the future will tell which variant is the correct one.

References

[1] E. Nelson, Phys. Rev. 150 (1966) 1079.

[2] See e.g. N.G. van Kampen, Stochastic Processes in Physics and Chemistry, North Holland 1981.

[3] E. Seiler, Stochastic Quantization and Gauge Fixing in Gauge Theories, Lectures given at 23. Internationale Universitätswochen für Kernphysik, Schladming, Austria, 1984.

[4] F. Guerra and P. Ruggiero, Phys. Rev. Lett. 31 (1973) 1022.

[5] F. Guerra and M.I. Loffredo, Lett. Nuovo Cim. 27 (1980) 41.

[6] S.C. Lim, Lett. Math. Phys. 7 (1983) 469.

[7] S.C. Lim, Phys. Lett. 135B (1984) 417.

[8] F. Guerra and M.I. Loffredo, Lett. Nuovo Cim. 30 (1981) 81.

[9] N. Cufaro Petroni, Ph. Gueret, J.-P. Vigier, Nuovo Cim. 81B (1984) 243.

[10] R. Kubo, Five-Dimensional Formulation of Quantum Field Theory with an Invariant Parameter, Hiroshima University preprint RRK 84-11.

[11] G. Parisi and Wu Yong-Shi, Sci. Sinica 24 (1981) 483.

[12] E. Gozzi, Phys. Rev. D28 (1983) 1922.

[13] R. Kirschner, Stochastic Quantization and Supersymmetry, Seminar given at 23. Internationale Universitätswochen für Kernphysik, Schladming, Austria, 1984.

[14] H. Nicolai, Phys. Lett. 89B (1980) 341; Nucl. Phys. B176 (1980) 419.

[15] J.D. Breit, S. Gupta, A. Zaks, Nucl. Phys. B233 (1984) 61.

[16] V. De Alfaro, S. Fubini, G. Furlan, Phys. Lett. 105B (1981) 462.

[17] E. Gozzi, The new functional approach to field theory by De Alfaro-Fubini and its connection to Parisi-Wu stochastic quantization, New York City College preprint HEP-83/7.

[18] D. Callaway, Phys. Lett. 145B (1984) 363.

[19] M. Namiki, I. Ohba, K. Okano and Y. Yamanaka, Progr. Theor. Phys. 69 (1983) 1580.

[20] P. van Nieuwenhuizen, Nucl. Phys. B60 (1973) 478.

[21] G.W. Gibbons, S.W. Hawking, M.J. Perry, Nucl. Phys. B138 (1978) 141.

[22] G.T. Horowitz, Quantum cosmology with a positive definite action, UCSB preprint TH-3 1984.

[23] H. Hüffel and H. Rumpf, Stochastic quantization in Minkowski space, Phys. Lett. B (1884), to appear.

[24] H. Hüffel and H. Rumpf, Stochastic quantization and gauge-fixing of the linearized gravitational field, University of Vienna preprint UWThPh-1984-30.
[25] H. van Dam, M. Veltman, Nucl. Phys. B22 (1970) 397.
[26] H. Rumpf, Phys. Rev. D28 (1983) 2946, and references cited therein.
[27] S.W. Hawking, Acausal propagation in quantum gravity, in Quantum Gravity 2, (eds. C.J. Isham, R. Penrose and D.W. Sciama), Clarendon Press, Oxford (1981).
[28] G. Parisi, Phys. Lett. 131B (1983) 393.
[29] E. Gozzi, Langevin simulation in Minkowski space, Max-Planck-Institute Munich preprint MPI-PAE/PTh 73/84.
[30] D. Zwanziger, Nucl. Phys. B192 (1981) 259.
[31] B.S. De Witt, Quantum gravity: the new synthesis, in General Relativity - an Einstein Centenary survey (eds. S.W. Hawking and W. Israel), Cambridge University Press 1979.
[32] J. Sakamoto, Progr. Theor. Phys. 70 (1983) 1424.
[33] M. Namiki, I. Ohba and K. Okano, Stochastic Quantization of Constrained Systems - General Theory and Nonlinear Sigma Model, Waseda University (Tokyo) preprint WU-HEP-84-3.
[34] W. Grimus and H. Hüffel, Z. Phys. C18 (1983) 129.
[35] T. Fukai and K. Okano, Stochastic quantization of linearized Euclidean gravity and no-ghost Feynman rules, Waseda University (Tokyo) preprint WU-HEP-8.
[36] A. Burnel, Phys. Rev. D29 (1984) 2344.
[37] J. Alfaro, Non-renormalizable theories do not have a stochastic interpretation, Laboratoire de Physique Théorique de l'Ecole Normale Supérieure preprint, 1984.
[38] R. Floreanini and O. Foda, Stochastic quantization of non-renormalizable theories, ISAS (Trieste) preprint 56/84/E.P.
[39] L. Baulieu and D. Zwanziger, Nucl. Phys. B193 (1981) 163.
[40] M. Horibe, A. Hosoya and J. Sakamoto, Progr. Theor. Phys. 70 (1983) 1636.
[41] V.N. Gribov, Nucl. Phys. B139 (1978) 1.

FEYNMAN'S CHECKERBOARD AND OTHER GAMES

T. JACOBSON

Department of Physics
University of California
Santa Barbara, CA 93106/USA

Abstract

Feynman's checkerboard path integral for the retarded Dirac propagator in 1+1 dimensions is derived and extensions to 3+1 dimensions are developed. Methods for obtaining the Feynman propagator (rather than the retarded propagator) are briefly discussed.

1. Feynman's Checkerboard

Feynman [1] gave a simple path integral representation for the retarded Dirac propagator in 1+1 dimensions. Here is one way to derive it: the Dirac equation $i\partial_t\psi = (\alpha\cdot p+\beta m)\psi$ in the Chiral representation $\alpha = \begin{pmatrix}1 & 0\\ 0 & -1\end{pmatrix}$, $\beta = \begin{pmatrix}0 & -1\\ -1 & 0\end{pmatrix}$, $\psi = \begin{pmatrix}R\\ L\end{pmatrix}$ reads

$$\begin{aligned}(\partial_t+\partial_x)R &= imL\\ (\partial_t-\partial_x)L &= imR\ .\end{aligned} \tag{1}$$

Feynman's path integral results from a particular finite differencing of (1) on a square spacetime lattice with mesh size ϵ:

$$\begin{aligned}R(n,m) &= R(n-1,m-1) + i\epsilon m\, L(n-1,m+1)\\ L(n,m) &= L(n-1,m+1) + i\epsilon m\, L(n-1,m-1)\end{aligned} \tag{2}$$

where the integers n,m label the time and space coordinates of the lattice sites. These equations (2) may be interpreted as saying that the amplitude for a particle to be at (n,m) moving toward the right is equal to the amplitude that it was at (n-1,m-1) moving toward the right plus $i\epsilon m$ times the amplitude that it was at (m-1,m+1) moving toward the left.

That one can in this manner interpret the components R,L of the wave function ψ as amplitudes for particular states of motion is related to the fact that the conserved probability current $(\psi^+\psi,\ \psi^+\alpha\psi)$ in the chiral representation reads $|R|^2(1,1)+|L|^2(1,-1)$. To put it another way, the state vectors $\begin{pmatrix}R\\ 0\end{pmatrix}$ and $\begin{pmatrix}0\\ L\end{pmatrix}$ are eigenvectors of the velocity operator α with eigenvalues +1 and -1, corresponding to motion at the speed of light to the right and left respectively.

Iterating the finite difference equations (2) $R(n,m)$ and $L(n,m)$ may be expressed as a sum over paths leading to (n,m), where a path with B bends is given an amplitude $(i\epsilon m)^B$. The retarded propagator $K(n,m)$ is a 2x2 matrix giving the amplitude $\psi_f^+ K(n,m)\psi_i$ that a state ψ_i initially localized at (0,0) would evolve to a state ψ_f at (n,m). Thus $K(n,m)$ can be expressed as

$$K_{\chi\chi'}(n,m) = \sum_B \Phi_{\chi\chi'}(n,m;B)(i\epsilon m)^B \tag{3}$$

where $\Phi_{\chi\chi'}(n,m;B)$ is the number of paths with B bends that leaves (0,0) in the direction χ' (right or left) and arrives at (n,m) in the direction χ.

The convergence of (3) to the exact continuum propagator in the limit $\epsilon \to 0$ is not a priori guaranteed; however, in [2] (3) is evaluated (in the limit $\epsilon \to 0$) and this convergence is demonstrated.

In this article we shall address the questions

1) can a path integral generalizing (3) be found for the Dirac propagator in 3+1 dimensions? and
2) is there some modification of (3) that will yield the Feynman propagator in place of the retarded (or advanced) propagator?

2. 3+1 dimensions

In 3+1 dimensions the Dirac equation in the chiral representation $\underset{\sim}{\alpha} = \begin{pmatrix} \underset{\sim}{\sigma} & 0 \\ 0 & -\underset{\sim}{\sigma} \end{pmatrix}$, $\beta = \begin{pmatrix} 0 & -1 \\ -1 & 0 \end{pmatrix}$, $\psi = \begin{pmatrix} R \\ L \end{pmatrix}$ reads

$$\begin{aligned} (\partial_t + \underset{\sim}{\sigma}\cdot\underset{\sim}{\nabla})R &= imL \\ (\partial_t - \underset{\sim}{\sigma}\cdot\underset{\sim}{\nabla})L &= imR \end{aligned} \tag{4}$$

where now R,L are two-component spinors corresponding to right and left chiralities. The conserved current is given by

$$j^\mu = j^\mu_R + j^\mu_L = (R^+R,\ R^+\underset{\sim}{\sigma}R) + (L^+L, -L^+\underset{\sim}{\sigma}L) \quad , \tag{5}$$

i.e., the current is parallel (anti-parallel) to the spin polarization vector for a right (left) chirality spinor.

Let us begin by treating the massless case, so that R and L are decoupled. The effect of the mass term is to introduce chirality switches, as in the 1+1 dimensional case, although now chirality no longer determines velocity. Indeed, the velocity can take on any direction according to the spin vector. We shall now show how spin transition amplitudes can be used to determine the amplitude for spacetime translations in a path integral for the propagator of (4).

Rather than finite differencing (4) on a hypercubical spacetime lattice, we prefer to maintain spherical symmetry. To this end we first rewrite (4) (with m=0) in an explicitly spherically symmetric manner:

$$\{\int d\Omega/4\pi)(1+\zeta\hat{\underset{\sim}{n}}\cdot\underset{\sim}{\sigma})(\partial_t+3\zeta^{-1}\hat{\underset{\sim}{n}}\cdot\underset{\sim}{\nabla})\}\ R(t,\underset{\sim}{x}) = 0 \tag{6}$$

where ζ is arbitrary and the integral is over the vectors $\hat{\underset{\sim}{n}}$ on the unit 2-sphere. (6) is easily verified using $\int d\Omega/4\pi = 1$, $\int d\Omega\hat{\underset{\sim}{n}} = 0$, $\int(d\Omega(4\pi)\hat{n}^i\hat{n}^j = (1/3)\,\delta^{ij}$. (That the correct factor is 1/3 is seen by tracing both sides.) Now we finite difference the directional derivatives

$$\begin{aligned}\epsilon(\partial_t &+ 3\zeta^{-1}\hat{\underset{\sim}{n}}\cdot\underset{\sim}{\nabla})R(t,\underset{\sim}{x})\\ &= R(t,\underset{\sim}{x}) - R(t-\epsilon,\ \underset{\sim}{x} - 3\zeta^{-1}\epsilon\hat{\underset{\sim}{n}}) + O(\epsilon^2)\end{aligned} \tag{7}$$

and insert (7) in (6) to obtain

$$R(t,\underset{\sim}{x}) = \int(d\Omega/4\pi)(1+\zeta\hat{\underset{\sim}{n}}\cdot\underset{\sim}{\sigma})R(t-\epsilon,\underset{\sim}{x}-3\zeta^{-1}\epsilon\hat{\underset{\sim}{n}}) + O(\epsilon^2) \tag{8}$$

As in the 1+1 dimensional case, (8) can be iterated to obtain a path integral approximation for the propagator - but the amplitude for a path is now the ordered product of "propagation matrices" $(1+\zeta\hat{\underset{\sim}{n}}\cdot\underset{\sim}{\sigma})$, one for each step. We have left ζ arbitrary until now to illustrate the following important point: the convergence of this iteration to the continuum propagator will obtain only for some values of ζ. For instance, if $\zeta > 3$ the exact propagator is clearly not obtained since its domain of dependence includes the past light cone whereas according to (8) $R(t,\underset{\sim}{x})$ receives contributions at each step only from the points $(t-\epsilon,\ \underset{\sim}{x}-3\zeta^{-1}\epsilon\hat{\underset{\sim}{n}})$ which fall <u>inside</u> the light cone. In fact [3], as $\epsilon \to 0$ the iteration diverges for $\zeta > 3$. For $\zeta = 3$ the contributions to $R(t,\underset{\sim}{x})$ come from points <u>on</u> the past light cone, hence it might be thought that this case would converge to the continuum propagator, but this is not so. It turns out [3] that convergence requires $\zeta \le \sqrt{3}$, i.e., the domain of dependence of the finite difference equations must be <u>larger</u> than the continuum domain. (The same phenomenon arises when hyperbolic equations are finite differenced on a <u>lattice</u> [4], where its origin is geometrically evident.)

The choice ζ=1 is convergent and leads to a particularly elegant path integral which exhibits most clearly the dual role of spin in determining transition amplitudes and propagation directions. With

$\zeta=1$, the propagation matrix associated with a step in direction $\hat{n}$ is $1+\hat{n}\cdot\sigma$, which is proportional to the projection operator $(1/2)(1+\hat{n}\cdot\sigma)=\lambda\lambda^+$ with λ a normalized two-component spinor ($\lambda^+\lambda = 1$) determined up to a phase by the condition $\hat{n} = \lambda^+\sigma\lambda$. With this notation, (8) may be written

$$R(t,x) = \int(d\Omega/2\pi)\lambda\lambda^+R(t-\epsilon,x-3\epsilon\hat{n}) + O(\epsilon^2) \qquad (9)$$

Neglecting the $O(\epsilon^2)$ contribution, (9) states that the contribution of the spinor amplitude $R(t-\epsilon,x-3\epsilon\hat{n})$ to the amplitude at (t,x) is obtained by projecting out that part of the former that is parallel to λ in spin space. In effect, <u>the amplitude for the spacetime translation is determined by the amplitude for the spin transition.</u>

Iterating (9) we obtain an approximation to the retarded propagator in the form

$$K^\epsilon(t,x) = \int\prod_{i=1}^{N}(d\Omega_i/2\pi)\lambda_N\lambda_N^+ \cdots \lambda_2\lambda_2^+\lambda_1\lambda_1^+ \;\delta(x - 3\epsilon\sum_{i=1}^{N}\lambda_i^+\sigma\lambda_i) \qquad (10)$$

where $N=t/\epsilon$. In [3] it is shown by explicit evaluation of (10) that in the limit $N \to \infty$, $\epsilon \to 0$ K^ϵ converges to the continuum propagator. Note that the ordered product of propagation matrices $\lambda_i\lambda_i^+$ can now be viewed as a product of scalar products $\lambda_{i+1}^+\lambda_i$, each giving the amplitude for a transition from spin λ_i to λ_{i+1} and, concomitantly, for the step sequence $\lambda_i^+\sigma\lambda_i \to \lambda_{i+1}^+\sigma\lambda_{i+1}$.

The propagator for left handed spinors is obtained by reversing the sign of $\lambda_i^+\sigma\lambda_i$ in the δ-function of (10). When the mass is non-zero, there is an additional possibility of switching chirality at each step, and the path integral for this case can be written [3] as

$$K_{\chi\chi'} = \int\prod_{i=1}^{N}(d\Omega_i/2\pi) \sum_{\chi_i=\pm1} \lambda_N\lambda_N^+ \cdots \lambda_1\lambda_1^+(i\epsilon m)^B\delta(x-3\epsilon\sum_{i=1}^{N}\chi_i\lambda_i^+\sigma\lambda_i) \qquad (11)$$

where $\chi_i = \pm1$, the sequence of chiralities begins with χ' and ends with χ, and B is the number of chirality switches, i.e.,

$$B = \sum_{i=1}^{N} |\chi_{i+1}-\chi_i|/2 \quad .$$

3. <u>Null steps?</u>

The steps of the path integrals (10) and (11) are <u>spacelike</u> 4-vectors of the form $\epsilon(1,3\hat{n})$, which followed from the representation (8) for the Dirac equation with $\zeta=1$. Is it possible to modify the construction so that the steps would be <u>null vectors</u>? It was already remarked that although choosing $\zeta = 3$ in (8) corresponds to null steps, the iteration yielding a path integral does not converge in this case [3]. Another idea would be to choose $\zeta = 1$, but to divide (6) by 3 and rewrite it as

$$\{ \int (d\Omega/4\pi)\ (1/3 + \hat{\mathbf{n}}\cdot\boldsymbol{\sigma})(\partial_t + n^i\partial_i)\}\ R(t,\mathbf{x}) = 0$$

which after finite differencing yields

$$R(t,\mathbf{x}) = \int (d\Omega/4\pi)(1/3+\hat{\mathbf{n}}\cdot\boldsymbol{\sigma})R(t-\epsilon,x-\epsilon\hat{\mathbf{n}}) + O(\epsilon^2) \quad . \tag{12}$$

The iteration of (12) converges in the limit $N \to \infty$, but because of the 1/3 it converges to <u>zero</u> which, while it is a solution to the Weyl equation, is clearly not the propagator!

I have found only one way to get by with null steps, and that is to <u>include steps backward in time</u>. An account of that construction (for the massless case) will now be given. The result will be a path integral built with null steps of the form $\epsilon(\pm 1,\hat{n})$, in which the propagation matrix for a step is (as before) $\lambda\lambda^{+}$, now <u>with a weight 2/3 for steps forward and 1/3 for steps backward in time.</u> One sums over the two time directions at each step as well as integrating over $\hat{\mathbf{n}}$; in addition, one sums over the total number of steps (which is no longer fixed by the time t).

The propagator we seek is the integral kernel of $(\partial_t + \boldsymbol{\sigma}\cdot\boldsymbol{\nabla} + \eta)^{-1}$, with η an infinitesimal positive number. We shall make use of the identity

$$(\partial_t+\boldsymbol{\sigma}\cdot\boldsymbol{\nabla}+\eta)^{-1} = \lim_{\epsilon\to 0}\ \epsilon \sum_{N=0}^{\infty} \exp[-N\epsilon(\partial_t+\boldsymbol{\sigma}\cdot\boldsymbol{\nabla}+\eta)] \tag{13}$$

which is verified by summing the geometric series and then expanding the exponential. Now the exponential in (13) is the evolution operator of the Weyl equation with a "fifth parameter":

$$\partial_s\psi = -(\partial_t + \sigma.\nabla+\eta)\psi \ . \tag{14}$$

We shall derive a path integral representation for this evolution operator and then sum over N as in (13) to arrive at the propagator.

In analogy with the method of Section 2 we first rewrite (14) as

$$\{ \sum_{\tau=\pm 1} \int (d\Omega/2\pi)(1+\tau/3)(1+\hat{\mathbf{n}}\cdot\boldsymbol{\sigma})(1/3\partial_s+\tau\partial_t+\mathbf{n}\cdot\boldsymbol{\nabla})\}\ \psi(s,t,\mathbf{x}) = 0 \ , \tag{15}$$

with η suppressed for notational simplicity. $\tau(=\pm 1)$ has been introduced so that we can divide through by 3 while preserving the "normalization"

$$\sum_{\tau=\pm 1}(1/2)\int(d\Omega/4\pi)(1+\tau/3)(1+\hat{\underset{\sim}{n}}\cdot\underset{\sim}{\sigma}) = 1$$

(compare with (12) and the comment following (12)). There are several ways to write (15) (e.g., replace $(1+\tau/3)(1+\hat{\underset{\sim}{n}}\cdot\underset{\sim}{\sigma})$ by $(1+\tau/3+\hat{\underset{\sim}{n}}\cdot\underset{\sim}{\sigma})$) but (15) has the virtue that the propagation matrix is still proportional to a projection $\lambda\lambda^+$.

Introducing the finite difference approximation

$$\epsilon(1/3\partial_s + \tau\partial_t + \hat{\underset{\sim}{n}}\cdot\underset{\sim}{\nabla})\psi(s,t,\underset{\sim}{x})$$
$$= \psi(s,t,\underset{\sim}{x}) - \psi(s-\epsilon/3,t-\tau\epsilon,\underset{\sim}{x}-\underset{\sim}{n}\epsilon) + O(\epsilon^2)$$

into (15) yields

$$\psi(s,t,\underset{\sim}{x}) = \sum_{\tau=\pm 1}\int(d\Omega/2\pi)G(\tau,\hat{n})\psi(s-\epsilon/3,t-\tau\epsilon,\underset{\sim}{x}-\hat{\underset{\sim}{n}}\epsilon) + O(\epsilon^2) \qquad (16)$$

with

$$G(\tau,\hat{n}) := (1+\tau/3)(1+\hat{\underset{\sim}{n}}\cdot\underset{\sim}{\sigma})/4$$
$$= \begin{cases} 2/3\lambda\lambda^+ , & \tau = +1 \\ 1/3\lambda\lambda^+ , & \tau = -1 \end{cases} \qquad (17)$$

Iterating (16) we approximate the kernel $K_N(t,\underset{\sim}{x})$ of $\exp[-(N\epsilon/3)(\partial_t+\underset{\sim}{\sigma}\cdot\underset{\sim}{\nabla})]$ as a path integral

$$K_N^\epsilon(t,\underset{\sim}{x}) = \sum_{\tau_i=\pm 1}\int\prod_{i=1}^{N}(d\Omega_i/2\pi)G(\tau_N,\hat{\underset{\sim}{n}}_N)\cdots G(\tau_1,\hat{\underset{\sim}{n}}_1)\delta((t,x)-\epsilon\sum_i(\tau_i,\hat{\underset{\sim}{n}}_i)). \qquad (18)$$

Finally, imitating (13) we conjecture that the exact retarded propagator $K(t,\underset{\sim}{x})$ is given by

$$K(t,\underset{\sim}{x}) \overset{?}{=} \lim_{\epsilon\to 0}(\epsilon/3)\sum_{N=0}^{\infty}K_N^\epsilon(t,\underset{\sim}{x}) , \qquad (19)$$

which is a path integral representation as described in the second paragraph of this section. The proof of this conjecture is given in the appendix.

4. Feynman propagator

The retarded propagator is just exp(-iHt) with H the one-particle Dirac Hamiltonian, and it propagates both positive and negative energy states forward in time. In the correct, many-particle theory positive energy states are propagated forward in time while negative energy states are propagated backward. That is, the appropriate propagator is

$$K_F = K_{ret}^{(+)} - K_{adv}^{(-)}$$
$$= (2\pi)^{-4} \int d^4p e^{ipx} (\not{p}-m+i\eta)^{-1} \tag{20}$$

where $K_{ret}^{(+)}$ ($K_{adv}^{(-)}$) is the positive (negative) frequency part of the retarded (advanced) propagator. Can the checkerboard path integral and its generalizations be modified so as to yield the Feynman propagator (20)?

One's first thought is to try again the fifth parameter representation,

$$(\not{p}-m+i\eta)^{-1} = -i\int_0^\infty ds \, \exp[i(\not{p}-m+i\eta)s] \tag{21}$$

in analogy with eq'n (13). This is unsuitable however because in Minkowski space the spatial γ-matrices are anti-Hermitian, hence $\not{p} = \gamma^\mu p_\mu$ is not Hermitian and the integral in (21) does not converge. A possible solution is to work in Euclidean space where all the γ's are Hermitian and then to analytically continue back to Minkowski space. This is the approach taken in reference [5].

Working exclusively in Minkowski space, the only way I have managed to arrive at the Feynman propagator with the sort of path integral under consideration is simply to project out the positive and negative frequency parts of K_{ret} and K_{adv} respectively via the formulas

$$K_{ret}^{(t)}(t) = (i/2\pi)\int_0^\infty dt'(t-t'+i\epsilon)^{-1}K_{ret}(t')$$
$$K_{adv}^{(-)}(t) = (-i/2\pi)\int_{-\infty}^0 dt'(t-t'-i\epsilon)^{-1}K_{adv}(t').$$

In terms of the path integral for, say, $K_{ret}^{(+)}(t,\underset{\sim}{x})$, this amounts to "weighting" paths with time lapse t' by $(i/2\pi)$ $(t-t'+i\epsilon)^{-1}$ instead of $\delta(t-t')$. In this way one can piece together K_F from the path integrals given in the previous sections. This solution is not very appealing however since the paths no longer connect two fixed points in spacetime.

Appendix: Evaluation of the path integral in (19).

The δ-function in (18) is interpreted to be a highly peaked Gaussian of width $\sim\xi^{1/2}$, given as a Fourier integral by

$\delta_\xi((t,\underset{\sim}{x})-\epsilon\Sigma(\tau_i,\hat{n}_i)) =$

$$\int \frac{d\omega d^3k}{(2\pi)^4}\, e^{i(\omega t+\underset{\sim}{k}\cdot\underset{\sim}{x})-\xi(\omega^2+k^2)}\, e^{-i\omega\epsilon\Sigma\tau_i}\, e^{-i\underset{\sim}{k}\epsilon\cdot\Sigma\hat{\underset{\sim}{n}}_i} \quad . \tag{1A}$$

At the end of all other computations we take the limit $\xi \to 0$.

After substituting (1A) in (18), the integral over $d\omega d^3k$ can be interchanged with the sum/integral over $\tau_i, \hat{\underset{\sim}{n}}_i$ to obtain

$$K_N^{\epsilon}(t,\underset{\sim}{x}) = \int \frac{d\omega d^3k}{(2\pi)^4}\, e^{i(\omega t+\underset{\sim}{k}\cdot\underset{\sim}{x})-\xi(\omega^2+k^2)}\, A_{\epsilon}^{N}(\omega,\underset{\sim}{k}) \tag{2A}$$

with

$$\begin{aligned} A_{\epsilon}(\omega,\underset{\sim}{k}) := &\sum_{\tau=\pm 1} \int (d\Omega/2\pi)\, G(\tau,\hat{\underset{\sim}{n}})\, e^{-i(\omega\tau+\underset{\sim}{k}\cdot\hat{\underset{\sim}{n}})\epsilon} \\ = &[\cos\omega\epsilon+(i/3)\sin\omega\epsilon][j_0(k\epsilon)-ij_1(k\epsilon)\hat{\underset{\sim}{k}}\cdot\underset{\sim}{\sigma}] \\ = &1-(i\epsilon/3)(\omega+\underset{\sim}{k}\cdot\underset{\sim}{\sigma}) + O(\omega^2\epsilon^2, k^2\epsilon^2) \end{aligned} \tag{3A}$$

where $k := (\underset{\sim}{k}\cdot\underset{\sim}{k})^{1/2}$, $\hat{\underset{\sim}{k}} := \underset{\sim}{k}/k$, and j_0, j_1 are spherical Bessel functions.

Now we may interchange the integral in (2A) with the sum in (19) provided $\Sigma_N A_{\epsilon}^{N}(\omega,\underset{\sim}{k})$ converges uniformly in $\omega,\underset{\sim}{k}$.† $A_{\epsilon}(\omega,\underset{\sim}{k})$ is Hermitian, and the squared moduli of its two eigenvalues are given by

$$|\lambda_{\pm}|^2 = [\cos^2\omega\epsilon + (1/9)\sin^2\omega\epsilon][j_0^{\,2}(k\epsilon)+j_1^{\,2}(k\epsilon)]$$

which are less than unity (cf. [3], Section 2.3) except at $\underset{\sim}{k}=0$, $\omega = n\pi/\epsilon, n = \text{integer}$, where $|\lambda_{\pm}| = 1$. $\Sigma_N A_{\epsilon}^{N}(\omega,\underset{\sim}{k})$ thus converges except at $\underset{\sim}{k}=0$, $\omega=n\pi/\epsilon$, but it does not converge uniformly in $\omega,\underset{\sim}{k}$. The situation is changed however when we recall the (suppressed) infinitesimal positive η that was used to define the inverse $(\partial_t+\underset{\sim}{\sigma}\cdot\underset{\sim}{\nabla}+\eta)^{-1}$ in (13). If η is maintained throughout the calculation we obtain $(1-\eta\epsilon/3)G(\tau,\hat{\underset{\sim}{n}})$ in place of $G(\tau,\hat{\underset{\sim}{n}})$ in (17) and hence $(1-\eta\epsilon/3)A_{\epsilon}(\omega,\underset{\sim}{k})$ in place of $A_{\epsilon}(\omega,\underset{\sim}{k})$ in (3A). Now the eigenvalues of $(1-\eta\epsilon/3)A_{\epsilon}(\omega,\underset{\sim}{k})$ have moduli <u>everywhere</u> less than unity, and furthermore $\Sigma_N(1-\eta\epsilon/3)\tilde{A}_{\epsilon}^{N}(\omega,\underset{\sim}{k})$ converges uniformly in $\omega,\underset{\sim}{k}$, so we may interchange the sum and integral to obtain

---

†The definitions of uniform convergence and relevant theorems are given in [6]. The theorems concerning interchange of series or limits with integrals apply to integrals with finite range of integration, and must be supplemented with arguments invoking the large ω,k behavior of $\exp[-\xi(\omega^2+k^2)]$.

$$\lim_{\epsilon\to 0}(\epsilon/3)\sum_{N=0}^{\infty} K_N^{\epsilon}(t,\underset{\sim}{x})$$

$$= \lim_{\epsilon\to 0}\int \frac{d\omega d^3k}{(2\pi)^4}\, e^{i(\omega t+\underset{\sim}{k}\cdot\underset{\sim}{x})-\xi(\omega^2+k^2)}(\epsilon/3)\left[1-(1-\eta\epsilon/3)A_{\epsilon}(\omega,\underset{\sim}{k})\right]^{-1} . \tag{4A}$$

The limit can be interchanged with the integral in (4A) provided the integrand converges to its limit uniformly in $\omega,\underset{\sim}{k}$,[†] but this is not the case since ϵ appears in $A_{\epsilon}(\omega,\underset{\sim}{k})$ always in the combinations $\omega\epsilon, k\epsilon$. On any compact range of $\omega,\underset{\sim}{k}$ however, the integrand converges uniformly, so let us break the integral into two terms, one over a very large but compact range and one over the remainder. Due to the exponential damping factor $\exp[-\xi(\omega^2+k^2)]$, the remainder can be made as small as we wish by taking the compact range large enough. Interchanging the limit with the integral is thus justified.

Now $\lim_{\epsilon\to 0}(\epsilon/3)[1-(1-\eta\epsilon/3)A_{\epsilon}(\omega,\underset{\sim}{k})]^{-1} = [i(\omega+\underset{\sim}{k}\cdot\underset{\sim}{\sigma})+\eta]^{-1}$,

so we have

$$\lim_{\xi\to 0}\lim_{\epsilon\to 0}(\epsilon/3)\sum_{N=0}^{\infty} K_N^{\epsilon}(t,\underset{\sim}{x})$$

$$= \int \frac{d\omega d^3k}{(2\pi)^4}\, e^{i(\omega t+\underset{\sim}{k}\cdot\underset{\sim}{x})}[i(\omega+\underset{\sim}{k}\cdot\underset{\sim}{\sigma})+\eta]^{-1} ,$$

which is indeed the kernel of $(\partial_t+\underset{\sim}{\sigma}\cdot\underset{\sim}{\nabla}+\eta)^{-1}$, i.e., the exact retarded propagator.

Acknowledgments

I would like to thank for their hospitality both the Groupe d'Astrophysique Relativiste at l'Observatoire de Meudon and the Laboratoire de Physique Théorique at l'Institut Henri Poincaré, Paris, where part of this research was carried out while I was a visiting chercheur associé au CNRS. This research was also supported in part at the University of Texas under NSF grant PHY 84-04931. I acknowledge the NSF grant that supports my work here at U.CSB.

References

1. Feynman, R.P. and Hibbs, A.R.: Quantum Mechanics and Path Integrals. New York: McGraw-Hill 1965: pp. 34-6.
2. Jacobson, T. and Schulman, L.S.: J. Phys. A17, 375-84 (1984).
3. Jacobson, T.: J. Phys. A17, 2433-51 (1984).
4. Bers, L., John, F. and Schechter, M.: Partial Differential Equations; Lectures in Applied Mathematics, v.3A. J. Wiley 1964, American Mathematical Society 1974.

5. Jacobson, T.: Submitted to Comm. Math. Phys.

6. Apostol, T.M.: Calculus, v.1. Lexington, MA: Xerox 1967.

LIST OF CONTRIBUTORS

E. ABDALLA
The Niels Bohr Institute, University of Copenhagen
DK - 2100 Copenhagen ϕ
Denmark

M. BANDER
Physics Dept.
University of California
IRVINE, CA 92717
U.S.A.

B. CARTER
Groupe d'Astrophysique Relativiste
D.A.F., Observatoire de Paris-Meudon
92195 Meudon Principal Cedex
France

A. CHAKRABARTI
Centre de Physique Théorique de l'Ecole Polytechnique
Plateau de Palaiseau
91128 Palaiseau Cedex
France

A. DEGASPERIS
Dipartimento di Fisica, Università di Roma, Italy
Istituto Nazionale di Fisica Nucleare, Sezione di Roma.

B. DERRIDA
Service de Physique Théorique
CEN - Saclay,
91191 Gif sur Yvette Cedex
France

H.J. DE VEGA
Laboratoire de Physique Théorique et Hautes Energies
Université Pierre et Marie Curie
Tour 16 - 1er étage, 4, place Jussieu
75230 Paris Cedex 05
France

H. EICHENHERR
ETH Hönggerberg, Theoretical Physics
CH-8093 Zürich,
Switzerland

P. FORGACS
Central Research Institute for Physics
P.O. Box 49
1525 Budapest 114
Hungary

J. GASQUI
Institut Fourier
Laboratoire de Mathématiques
B.P. 74
38402 St Martin D'Hyeres
France

B. KENT HARRISON
Department of Physics and Astronomy
Brigham Young University
Provo, Utah 84602
U.S.A.

J. HIETARINTA
Wihuri Physical Laboratory and Department of Physical Sciences
University of Turku
20500 Turku 50
Finland

C. ITZYKSON
Service de Physique Théorique
CEN - Saclay, 91191 Gif sur Yvette Cedex
France

T. JACOBSON
Department of Physics
University of California
Santa Barbara, CA 93106
U.S.A.

H.P. JAKOBSEN
Mathematics Institute
Universitetsparken 5
DK - 2100 Copenhagen
Denmark

V.G. KAC
Department of Mathematics
M.I.T.
Cambridge, Mass 02139
U.S.A.

J.M. MAILLET
Laboratoire de Physique Théorique et Hautes Energies
Université Pierre et Marie Curie
Tour 16 - 1er étage, 4 place Jussieu
75230 Paris Cedex 05
France

D. MAISON
Max-Planck-Institute für Physik und Astrophysik
Werner Heisenberg Institut für Physik
P.O. Box 40 12 12, Munich
Fed. Rep. Germany

W. NAHM
Physikalisches Institut der Universität Bonn
Nussallee 12
5300 Bonn 1
W. Germany

F. NICOLO
Dipartimento di Fisica, Università degli studi
di Roma "La Sapienza", Piazzale Aldo Moro 2,
00185 Roma
Italy

K. POHLMEYER
Department of Physics
University of Freiburg
D-7800 Freiburg,
W. Germany

J.L. RICHARD
Centre de Physique Théorique
C.N.R.S. -Luminy - Case 907
13288 Marseille Cedex 9
France

P. RUJAN
Institute für Festkörperforschung der KFA, Jülich and
Institute for Theoretical Physics, Eötvös University
Budapest
Hungary

H. RUMPF
Institut für Theoretische Physik
Universität Wien
A-1090 Vienna,
Austria

E.K. SKLYANIN
V.A. Steklov of Mathematicas Institute
Fontanka 25, Leningrad 191011
U.S.S.R.

T.T. TRUONG
Institut für Theoretische Physik
Freie universität Berlin
Arnimallee 14, D-1000 Berlin 33
W. Germany

Lecture Notes in Physics

Vol. 195: Trends and Applications of Pure Mathematics to Mechanics. Proceedings, 1983. Edited by P.G. Ciarlet and M. Roseau. V, 422 pages. 1984.

Vol. 196: WOPPLOT 83. Parallel Processing: Logic, Organization and Technology. Proceedings, 1983. Edited by J. Becker and I. Eisele. V, 189 pages. 1984.

Vol. 197: Quarks and Nuclear Structure. Proceedings, 1983. Edited by K. Bleuler. VIII, 414 pages. 1984.

Vol. 198: Recent Progress in Many-Body Theories. Proceedings, 1983. Edited by H. Kümmel and M. L. Ristig. IX, 422 pages. 1984.

Vol. 199: Recent Developments in Nonequilibrium Thermodynamics. Proceedings, 1983. Edited by J. Casas-Vázquez, D. Jou and G. Lebon. XIII, 485 pages. 1984.

Vol. 200: H.D. Zeh, Die Physik der Zeitrichtung. V, 86 Seiten. 1984.

Vol. 201: Group Theoretical Methods in Physics. Proceedings, 1983. Edited by G. Denardo, G. Ghirardi and T. Weber. XXXVII, 518 pages. 1984.

Vol. 202: Asymptotic Behavior of Mass and Spacetime Geometry. Proceedings, 1983. Edited by F. J. Flaherty. VI, 213 pages. 1984.

Vol. 203: C. Marchioro, M. Pulvirenti, Vortex Methods in Two-Dimensional Fluid Dynamics. III, 137 pages. 1984.

Vol. 204: Y. Waseda, Novel Application of Anomalous (Resonance) X-Ray Scattering for Structural Characterization of Disordered Materials. VI, 183 pages. 1984.

Vol. 205: Solutions of Einstein's Equations: Techniques and Results. Proceedings, 1983. Edited by C. Hoenselaers and W. Dietz. VI, 439 pages. 1984.

Vol. 206: Static Critical Phenomena in Inhomogeneous Systems. Edited by A. Pękalski and J. Sznajd. Proceedings, 1984. VIII, 358 pages. 1984.

Vol. 207: S. W. Koch, Dynamics of First-Order Phase Transitions in Equilibrium and Nonequilibrium Systems. III, 148 pages. 1984.

Vol. 208: Supersymmetry and Supergravity/Nonperturbative QCD. Proceedings, 1984. Edited by P. Roy and V. Singh. V, 389 pages. 1984.

Vol. 209: Mathematical and Computational Methods in Nuclear Physics. Proceedings, 1983. Edited by J. S. Dehesa, J. M. G. Gomez and A. Polls. V, 276 pages. 1984.

Vol. 210: Cellular Structures in Instabilities. Proceedings, 1983. Edited by J. E. Wesfreid and S. Zaleski. VI, 389 pages. 1984.

Vol. 211: Resonances – Models and Phenomena. Proceedings, 1984. Edited by S. Albeverio, L. S. Ferreira and L. Streit. VI, 369 pages. 1984.

Vol. 212: Gravitation, Geometry and Relativistic Physics. Proceedings, 1984. Edited by Laboratoire "Gravitation et Cosmologie Relativistes", Université Pierre et Marie Curie et C.N.R.S., Institut Henri Poincaré, Paris. VI, 336 pages. 1984.

Vol. 213: Forward Electron Ejection in Ion Collisions. Proceedings, 1984. Edited by K.O. Groeneveld, W. Meckbach and I.A. Sellin. VII, 165 pages. 1984.

Vol. 214: H. Moraal, Classical, Discrete Spin Models. VII, 251 pages. 1984.

Vol. 215: Computing in Accelerator Design and Operation. Proceedings, 1983. Edited by W. Busse and R. Zelazny. XII, 574 pages. 1984.

Vol. 216: Applications of Field Theory to Statistical Mechanics. Proceedings, 1984. Edited by L. Garrido. VIII, 352 pages. 1985.

Vol. 217: Charge Density Waves in Solids. Proceedings, 1984. Edited by Gy. Hutiray and J. Sólyom. XIV, 541 pages. 1985.

Vol. 218: Ninth International Conference on Numerical Methods in Fluid Dynamics. Edited by Soubbaramayer and J.P. Boujot. X, 612 pages. 1985.

Vol. 219: Fusion Reactions Below the Coulomb Barrier. Proceedings, 1984. Edited by S.G. Steadman. VII, 351 pages. 1985.

Vol. 220: W. Dittrich, M. Reuter, Effective Lagrangians in Quantum Electrodynamics. V, 244 pages. 1985.

Vol. 221: Quark Matter '84. Proceedings, 1984. Edited by K. Kajantie. VI, 305 pages. 1985.

Vol. 222: A. García, P. Kielanowski, The Beta Decay of Hyperons. Edited by A. Bohm. VIII, 173 pages. 1985.

Vol. 223: H. Saller, Vereinheitlichte Feldtheorien der Elementarteilchen. IX, 157 Seiten. 1985.

Vol. 224: Supernovae as Distance Indicators. Proceedings, 1984. Edited by N. Bartel. VI, 226 pages. 1985.

Vol. 225: B. Müller, The Physics of the Quark-Gluon Plasma. VII, 142 pages. 1985.

Vol. 226: Non-Linear Equations in Classical and Quantum Field Theory. Proceedings, 1983/84. Edited by N. Sanchez. VII, 400 pages. 1985.

Selected Issues from

Lecture Notes in Mathematics

Vol. 932: Analytic Theory of Continued Fractions. Proceedings, 1981. Edited by W.B. Jones, W.J. Thron, and H. Waadeland. VI, 240 pages. 1982.

Vol. 934: M. Sakai, Quadrature Domains. IV, 133 pages. 1982.

Vol. 935: R. Sot, Simple Morphisms in Algebraic Geometry. IV, 146 pages. 1982.

Vol. 936: S.M. Khaleelulla, Counterexamples in Topological Vector Spaces. XXI, 179 pages. 1982.

Vol. 937: E. Combet, Intégrales Exponentielles. VIII, 114 pages. 1982.

Vol. 938: Number Theory. Proceedings, 1981. Edited by K. Alladi. IX, 177 pages. 1982.

Vol. 942: Theory and Applications of Singular Perturbations. Proceedings, 1981. Edited by W. Eckhaus and E.M. de Jager. V, 363 pages. 1982.

Vol. 953: Iterative Solution of Nonlinear Systems of Equations. Proceedings, 1982. Edited by R. Ansorge, Th. Meis, and W. Törnig. VII, 202 pages. 1982.

Vol. 956: Group Actions and Vector Fields. Proceedings, 1981. Edited by J.B. Carrell. V, 144 pages. 1982.

Vol. 957: Differential Equations. Proceedings, 1981. Edited by D.G. de Figueiredo. VIII, 301 pages. 1982.

Vol. 963: R. Nottrot, Optimal Processes on Manifolds. VI, 124 pages. 1982.

Vol. 964: Ordinary and Partial Differential Equations. Proceedings, 1982. Edited by W.N. Everitt and B.D. Sleeman. XVIII, 726 pages. 1982.

Vol. 968: Numerical Integration of Differential Equations and Large Linear Systems. Proceedings, 1980. Edited by J. Hinze. VI, 412 pages. 1982.

Vol. 970: Twistor Geometry and Non-Linear Systems. Proceedings, 1980. Edited by H.-D. Doebner and T.D. Palev. V, 216 pages. 1982.

Vol. 972: Nonlinear Filtering and Stochastic Control. Proceedings, 1981. Edited by S.K. Mitter and A. Moro. VIII, 297 pages. 1983.

Vol. 978: J. Ławrynowicz, J. Krzyż, Quasiconformal Mappings in the Plane. VI, 177 pages. 1983.

Vol. 979: Mathematical Theories of Optimization. Proceedings, 1981. Edited by J.P. Cecconi and T. Zolezzi. V, 268 pages. 1983.

Vol. 982: Stability Problems for Stochastic Models. Proceedings, 1982. Edited by V.V. Kalashnikov and V.M. Zolotarev. XVII, 295 pages. 1983.

Vol. 989: A.B. Mingarelli, Volterra-Stieltjes Integral Equations and Generalized Ordinary Differential Expressions. XIV, 318 pages. 1983.

Vol. 994: J.-L. Journé, Calderón-Zygmund Operators, Pseudo-Differential Operators and the Cauchy Integral of Calderón. VI, 129 pages. 1983.

Vol. 999: C. Preston, Iterates of Maps on an Interval. VII, 205 pages. 1983.

Vol. 1000: H. Hopf, Differential Geometry in the Large, VII, 184 pages. 1983.

Vol. 1003: J. Schmets, Spaces of Vector-Valued Continuous Functions. VI, 117 pages. 1983.

Vol. 1005: Numerical Methods. Proceedings, 1982. Edited by V. Pereyra and A. Reinoza. V, 296 pages. 1983.

Vol. 1007: Geometric Dynamics. Proceedings, 1981. Edited by J. Palis Jr. IX, 827 pages. 1983.

Vol. 1015: Equations différentielles et systèmes de Pfaff dans le champ complexe – II. Seminar. Edited by R. Gérard et J.P. Ramis. V, 411 pages. 1983.

Vol. 1021: Probability Theory and Mathematical Statistics. Proceedings, 1982. Edited by K. Itô and J.V. Prokhorov. VIII, 747 pages. 1983.

Vol. 1031: Dynamics and Processes. Proceedings, 1981. Edited by Ph. Blanchard and L. Streit. IX, 213 pages. 1983.

Vol. 1032: Ordinary Differential Equations and Operators. Proceedings, 1982. Edited by W.N. Everitt and R.T. Lewis. XV, 521 pages. 1983.

Vol. 1035: The Mathematics and Physics of Disordered Media. Proceedings, 1983. Edited by B.D. Hughes and B.W. Ninham. VII, 432 pages. 1983.

Vol. 1037: Non-linear Partial Differential Operators and Quantization Procedures. Proceedings, 1981. Edited by S.I. Andersson and H.-D. Doebner. VII, 334 pages. 1983.

Vol. 1041: Lie Group Representations II. Proceedings 1982–1983. Edited by R. Herb, S. Kudla, R. Lipsman and J. Rosenberg. IX, 340 pages. 1984.

Vol. 1045: Differential Geometry. Proceedings, 1982. Edited by A.M. Naveira. VIII, 194 pages. 1984.

Vol. 1047: Fluid Dynamics. Seminar, 1982. Edited by H. Beirão da Veiga. VII, 193 pages. 1984.

Vol. 1048: Kinetic Theories and the Boltzmann Equation. Seminar, 1981. Edited by C. Cercignani. VII, 248 pages. 1984.

Vol. 1049: B. Iochum, Cônes autopolaires et algèbres de Jordan. VI, 247 pages. 1984.

Vol. 1054: V. Thomée, Galerkin Finite Element Methods for Parabolic Problems. VII, 237 pages. 1984.

Vol. 1055: Quantum Probability and Applications to the Quantum Theory of Irreversible Processes. Proceedings, 1982. Edited by L. Accardi, A. Frigerio and V. Gorini. VI, 411 pages. 1984.

Vol. 1057: Bifurcation Theory and Applications. Seminar, 1983. Edited by L. Salvadori. VII, 233 pages. 1984.

Vol. 1058: B. Aulbach, Continuous and Discrete Dynamics near Manifolds of Equilibria. IX, 142 pages. 1984.

Vol. 1059: Séminaire de Probabilités XVIII, 1982/83. Proceedings. Edité par J. Azéma et M. Yor. IV, 518 pages. 1984.

Vol. 1063: Orienting Polymers. Proceedings, 1983. Edited by J.L. Ericksen. VII, 166 pages. 1984.

Vol. 1065: A. Cuyt, Padé Approximants for Operators: Theory and Applications. IX, 138 pages. 1984.

Vol. 1066: Numerical Analysis. Proceedings, 1983. Edited by D.F. Griffiths. XI, 275 pages. 1984.

Vol. 1071: Padé Approximation and its Applications, Bad Honnef 1983. Prodeedings Edited by H. Werner and H.J. Bünger. VI, 264 pages. 1984.

Vol. 1072: F. Rothe, Global Solutions of Reaction-Diffusion Systems. V, 216 pages. 1984.

Vol. 1085: G.K. Immink, Asymptotics of Analytic Difference Equations. V, 134 pages. 1984.

Vol. 1086: Sensitivity of Functionals with Applications to Engineering Sciences. Proceedings, 1983. Edited by V. Komkov. V, 130 pages. 1984

Vol. 1100: V. Ivrii, The Precise Spectral Asymptotics for Elliptic Operators Acting in Fiberings over Manifolds with Boundary. V, 237 pages. 1984.